Telecommunications Protocols

OTHER McGRAW-HILL COMMUNICATIONS BOOKS OF INTEREST

Agosta, Russell *CDPD: Cellular Digital Packet Data Standards and Technology*
Azzam *Cable Modems*
Benner *Fibre Channel*
Best *Phase-Locked Loops, Third Edition*
Bush *Private Branch Exchange System and Applications*
Cooper *Computer and Communications Security*
Dayton *Integrating Digital Services*
Faynberg et al. *The Intelligent Network Standards Their Application to Services*
Fortier *Handbook of LAN Technology*
Gallagher *Mobile Telecommunications Networking with IS-41*
Goldberg *Digital Techniques in Frequency Synthesis*
Goralski *Introduction to ATM Networking*
Grinberg *Seamless Networks*
Hebrawi *OSI: Upper Layer Standards and Practices*
Held, Sarch *Data Communications*
Heldman *Competitive Telecommunications*
Heldman *Information Telecommunications*
Inglis *Video Engineering, Second Edition*
Kessler *ISDN, Second Edition*
Kessler, Train *Metropolitan Area Networks*
Lee *Mobile Cellular Telecommunications, Second Edition*
Lee *Mobile Communications Engineering, Second Edition*
Lindberg *Digital Broadband Networks and Services*
Logsdon *Mobile Communication Satellites*
Macario *Cellular Radio, Second Edition*
Roddy *Satellite Communications, Second Edition*
Rohde et al. *Communication Receivers, Second Edition*
Russell *Signaling System #7*
Simon et al. *Spread Spectrum Communications Handbook*
Smith, Gervelis *Cellular System Design and Optimization*
Tsakalakis *PCS Network Deployment*
Winch *Telecommunication Transmission Systems*

Telecommunications Protocols

Travis Russell

McGraw-Hill
New York San Francisco Washington, D.C. Auckland Bogotá
Caracas Lisbon London Madrid Mexico City Milan
Montreal New Delhi San Juan Singapore
Sydney Tokyo Toronto

Library of Congress Cataloging-in-Publication Data

Russell, Travis
Telecommunications protocols / Travis Russell.
p. cm.
Includes bibliographical references and index.
ISBN 0-07-057695-5
1. Data transmission systems. 2.. Computer network protocols.
I. Title..
Tk5105.R87 1997
621.382'12—dc21 97-17358
CIP

McGraw-Hill

A Division of The ***McGraw-Hill*** *Companies*

3 4 5 6 7 8 9 0 FGR/FGR 9 0 2 1 0 9 8 7

ISBN 0-07-057695-5

The sponsoring editor for this book was Stephen S. Chapman, the editing supervisor was Peggy Lamb, and the production supervisor was Suzanne W. B. Rapcavage. It was set in Vendome by Terry Leaden of McGraw-Hill's Professional Book Group composition unit.

Printed and bound by Quebecor/Fairfield.

This book is printed on recycled, acid-free paper containing a minimum of 50% recycled, de-inked fiber.

This book is dedicated to my wife, Deby, who gave so much of her time so that I might complete another dream of my own. Thanks for being so understanding and for pushing me along when I lost focus.

Contents

Introduction xvii

Acknowledgments xix

Chapter One The Fundamentals 1

- 1.1 Historic View of Telecommunications 2
  - 1.1.1 Data Communications History 2
  - 1.1.2 Telephony History 7
- 1.2 Standards Organizations 12
  - 1.2.1 The Standards Process 13
  - 1.2.2 Organizations Here and Abroad 14
    - 1.2.2.1 International Telecommunications Union 14
    - 1.2.2.2 International Organization for Standards (ISO) 15
    - 1.2.2.3 European Telecommunication Standards Institute (ETSI) 16
    - 1.2.2.4 American National Standards Institute 16
    - 1.2.2.5 Bell Communications Research (Bellcore) 16
    - 1.2.2.6 Exchange Carriers Standards Association (ECSA) 17
    - 1.2.2.7 Electronics Industries Association (EIA) 18
    - 1.2.2.8 Institute of Electrical and Electronics Engineers (IEEE) 18
    - 1.2.2.9 Federal Communications Commission 18
    - 1.2.2.10 Network Reliability Council (NRC) 18
    - 1.2.2.11 Federal Telecommunications Standards Committee (FTSC) 19
    - 1.2.2.12 National Bureau of Standards (NBS) 19
    - 1.2.2.13 Defense Communications Agency (DCA) 19
    - 1.2.2.14 Underwriters Laboratories (UL) 19
    - 1.2.2.15 Canadian Standards Association (CSA) 20
    - 1.2.2.16 ATM Forum 20
- 1.3. Digital Transmission Fundamentals 21
  - 1.3.1 From Electrical to Binary 21
  - 1.3.2 Alphabet Soup—ASCII and EBCDIC 22
  - 1.3.3 Digitizing Voice 25
- 1.4. The Basics of Telecommunications Protocols 28
  - 1.4.1 Protocol Services 28
    - 1.4.1.1 Protocol Tasks 30
      - 1.4.1.1.1 Segmentation and Reassembly 31

1.4.1.1.2 Encapsulation 33
1.4.1.1.3 Connection Control 33
1.4.1.1.4 Ordered Delivery 34
1.4.1.1.5 Flow Control 34
1.4.1.1.6 Error Detection/Correction 35
1.4.2 Layering and Its Advantages 35
1.4.3 The Open System Interconnection (OSI) Model 36
1.4.3.1 The OSI Layers 37
1.5. Networking Fundamentals 39
1.5.1 Evolution to Distributed Processing 39
1.5.2 Client/Server Environments 41
1.5.3 The Local Area Network 41
1.5.3.1 Services Provided 42
1.5.3.2 Routing Principles 42
1.5.4 Wide Area Networks (WANs)—The Outside Connection 43
1.5.4.1 Services Provided 43
1.5.4.2 Routing Principles 44
1.5.5 Switching Principles 45
1.5.5.1 Circuit Switching 45
1.5.5.2 Packet Switching 46
1.5.5.3 Cell Relay 47
1.5.6 Chapter Test 48

Chapter Two The Evolving Telephone Network 51

2.1. The Infrastructure 52
2.1.1 Predivestiture Bell System Networks 52
2.1.2 Postdivestiture Bell System Networks 55
2.1.2.1 New Switching Hierarchy 58
2.1.2.2 Local Access Transport Areas 58
2.1.3 About Divestiture and Its Reasoning—Winds of Change 60
2.1.4 New Telecommunications Law 60
2.2. The National Information Infrastructure 61
2.2.1 The Objective 62
2.2.2 The Promise of Equal Access to All 63
2.2.3 Cost and More Cost—The Reality 64
2.3. The North Carolina Information Highway 65
2.3.1 Model Citizen or Political Agenda 65
2.4 The Backbone 67
2.4.1 From Analog to Digital Trunking 67

2.4.1.1 Multiplexing 68
2.4.1.2 Time Division Multiplexing 69
2.4.2 The Digital Hierarchy—DS1 and DS3 71
2.4.2.1 T-1 Facilities 72
2.4.3 SONET—The New Fiber Backbone 75
2.5. The Private Network 77
2.5.1 Private Branch Exchanges 77
2.5.2 Features and Capabilities of Private Networks 78
2.5.3 Voice and Data Integration 81
2.5.4 Centrex Services 81
2.5.5 Computer Telephony Applications 82
2.5.5.1 TAPI 83
2.5.5.2 ASAI/SCAI 87
2.6. The Transport 88
2.6.1 The Evolution of ATM 88
2.7. The Subscriber Interface 89
2.7.1 Integrated Services—Pulling It All Together 90
2.8. Chapter Test 91

Chapter Three LANs to WANs 93

3.1. Evolution to Distributed Processing 94
3.1.1 An Overview of Mainframes and Their Applications 94
3.1.2 The Move to Personal Computers 96
3.2. LAN Technology—Connecting to the Desktop 97
3.2.1 Topologies and Basic Architecture 98
3.2.2 LAN Devices 101
3.2.2.1 Repeaters 102
3.2.2.2 Bridge 102
3.2.2.3 Routers 103
3.2.2.4 Other Network Devices 104
3.2.3 An Overview of Ethernet 105
3.2.3.1 Media Access Control 107
3.2.3.2 Logical Link Layer 110
3.2.3.3 Acknowledged Connectionless Service 112
3.2.4 An Overview of Token Ring 112
3.2.5 An Overview of FDDI 117
3.2.6 Client/Server 123
3.2.7 Network Operating Systems 124
3.3. Bridging the Gap with Wide Area Networks 124
3.3.1 Basic Architecture and Options Available 124

3.3.2 X.25 Packet Switching 125
3.3.3 Using T-1 for Connectivity 126
3.3.4 Switched 56 126
3.3.5 Frame Relay 127
3.3.6 ISDN 127
3.3.7 TCP/IP 128
3.4. Internet As A Model 129
3.4.1 Lessons to Be Learned from the Internet 129
3.4.2 Issues to Resolve—Corporate Policies and Legislature 130
3.4.3 Corporate Solutions—The Intranet 131
3.5. The Internet Infrastructure—Worldwide Networking 132
3.5.1 Who's In Control?—Supercomputer Centers 133
3.5.2 Direct or Indirect—Getting Connected 133
3.6. Internet Services 134
3.6.1 E-Mail—Global Delivery 135
3.6.2 Information Exchange—File Transfer 135
3.6.3 Cheap Remote Access—Terminal Emulation 136
3.6.4 Blessing or Curse?—Newsgroups 136
3.6.5 Commercialized Internet—World Wide Web 137
3.6.6 Fad or Reality—Voice on the Internet 138
3.7. Chapter Test 138

Chapter Four TCP/IP—Protocol of the Internet 141

4.1. Introduction 142
4.1.1 History of TCP/IP 142
4.1.2 Overview of Internets 143
4.1.2.1 Autonomous Systems 144
4.1.3 Description of TCP/IP 144
4.2. TCP/IP Standards 147
4.2.1 Standards Documentation 147
4.2.2 Standards Groups 148
4.3. Internet Protocol 149
4.3.1 IP Header 150
4.3.2 IP Addressing 155
4.3.2.1 Sockets and Ports 156
4.3.2.2 IP Addresses 156
4.3.2.3 Subnet Masking 159
4.3.3 Domain Name System 162
4.3.4 Routing in an Internet 164

4.3.4.1 Source Routing 165
4.3.4.2 Time Stamping 165
4.3.4.3 Circular Routing 166
4.3.4.3.1 Split Horizon 167
4.3.4.3.2 Poison Reverse 167
4.3.4.3.3 Triggered Updates 167
4.3.5 IP Routing Protocols 168
4.3.5.1 Address Resolution Protocol 170
4.3.5.2 Reverse Address Resolution Protocol 171
4.3.5.3 Routing Information Protocol 171
4.3.5.4 Open Shortest Path First 172
4.3.5.4.1 HELLO Protocol 173
4.3.6 IP Services 173
4.3.6.1 Fragmentation and Reassembly 173
4.3.7 Internet Control Message Protocol 174
4.4. Transport Control Protocol 175
4.4.1 TCP Header 176
4.4.1.1 Processing of Urgent Data 178
4.4.1.2 Processing of Push Data 179
4.4.2 TCP Ports and Sockets 180
4.4.3 TCP Services 180
4.4.3.1 TCP Error and Flow Control 181
4.4.3.2 TCP Management 182
4.5. User Datagram Protocol 183
4.5.1 UDP Header 184
4.6. Internet Application Protocols 184
4.6.1 TELNET 185
4.6.2 File Transfer Protocol 186
4.6.3 Trivial File Transfer Protocol 186
4.6.4 Simple Mail Transport Protocol 187
4.6.4.1 Post Office Protocol 188
4.6.5 Network News Transport Protocol 189
4.6.6 Hypertext Transport Protocol 190
4.6.7 SLIP and PPP 191
4.7. Network Management 192
4.7.1 Simple Network Management Protocol 192
4.7.1.1 Management Information Base 193
4.8. Chapter Test 194

Chapter Five Signaling System #7 197

5.1. From Signaling to Control 198
5.1.1 Signaling Methods—How They Evolved 199
5.1.2 Common Channel Signaling—The Advantages 201
5.1.3 After Signaling—Autonomous Network Control 203
5.2. Intelligent Networks 203
5.2.1 What is Intelligence? 205
5.2.2 Future Services 206
5.2.2.1 Intelligent Routing 207
5.2.2.2 Smart Custom Features 207
5.2.2.3 Database Access—Key to Intelligence 208
5.2.2.4 End-to-End Subscriber Services 209
5.2.3 Broadband Requirements 210
5.3. SS7 Architecture 211
5.3.1 Data Links 211
5.3.1.1 56/64-kbps Links 214
5.3.1.2 1.544-Mbps Links 214
5.3.1.3 ATM Links 215
5.3.2 Network Components 216
5.3.2.1 The Service Switching Point 216
5.3.2.2 The Signal Transfer Point 217
5.3.2.3 The Service Control Point 218
5.4. SS7 Protocols 219
5.4.1 Message Transfer Part 222
5.4.1.1 Network Management 224
5.4.2 Signaling Connection Control Part 228
5.4.3 Transaction Capabilities Application Part 230
5.4.4 Telephone User Part 235
5.4.5 ISDN User Part 236
5.5. Chapter Test 243

Chapter Six ISDN and Broadband ISDN 247

6.1. ISDN—An Overview of Its Capability 248
6.1.1 ISDN Standards 249
6.1.2 ISDN Features 250
6.1.3 Services and Applications 250
6.2. Subscriber Interface to SS7 254
6.2.1 End-to-End Signaling with DSS-1 255
6.2.2 Private Intelligent Networks 256
6.3. Early ISDN Issues 257
6.3.1 The Cart Before the Horse—Premature Offering 258

6.3.2 Interoperability—Where Did the Standards Go? 259
6.3.3 Configuration—The Consumer Nightmare 260
6.4. ISDN Network Architecture 261
6.4.1 Basic Rate Interface 261
6.4.2 Primary Rate Interface 262
6.4.3 Channel Usage 262
6.4.4 The Nodes and the Reference Points 263
6.4.4.1 ISDN Functions 264
6.4.4.2 ISDN Reference Points 264
6.4.5 Protocols of ISDN 265
6.4.5.1 Link Access Procedure for the D Channel 266
6.4.5.1.1 LAPD Sequencing 270
6.4.5.1.2 Management 271
6.4.5.1.3 Connection Establishment 271
6.4.5.1.4 LAPD Flow Control and Error Detection/Correction 272
6.4.5.2 B Channel Data Link Protocol 273
6.4.5.2.1 V.120 Connection Establishment 274
6.4.5.3 ISDN Layer Three 275
6.4.5.3.1 Q.931 Message Applications 277
6.4.5.3.2 Call Establishment Messages 278
6.4.5.3.3 Call Information Phase Messages 279
6.4.5.3.4 Call Clearing Messages 280
6.4.5.3.5 Miscellaneous Messages 281
6.4.5.3.6 Q.931 Message Parameters 281
6.5. Broadband ISDN—The Future 286
6.5.1 Overview of BISDN Advantages 287
6.5.1.1 BISDN Architecture 290
6.5.2 BISDN and ATM—What Do They Have to Offer One Another? 291
6.6. Frame Relay 292
6.7. Chapter Test 296

Chapter Seven The Cellular Network 301

7.1. From Radiotelephone to Cellular Telephones 302
7.1.1 Overview of Radiotelephone Networks 304
7.1.2 The Cellular Solution—Architecture and Distribution 305

7.2. Cellular Network Architecture and Protocols 306
7.2.1 The U.S. Network 309
7.2.2 The International Network 310
7.2.3 Cellular Operations 312
7.2.4 Time Division Multiple Access 320
7.2.5 Coded Division Multiple Access 322
7.2.6 Global System for Mobile Communications 324
7.2.7 CDPD - Packet Switching over Cellular 326
7.3. Personal Communications Services 326
7.3.1 New Network and New Services 327
7.3.2 GSM—To Be Or Not To Be 328
7.4. Specialized Wireless Solutions 328
7.4.1 One Number Service 329
7.4.2 Data Access 329
7.4.3 Alarm Services 329
7.4.4 Telemetering 330
7.5. Chapter Test 330

Chapter Eight The Fiber Backbone 331

8.1. From Copper to Fiber 332
8.1.1 Existing Digital Transmission Overview 333
8.1.2 SONET—the Solution 335
8.2. SONET Overview 337
8.2.1 SONET Network Nodes 337
8.2.2 The SONET Protocol 340
8.2.3 SONET Framing 345
8.2.3.1 Virtual Tributaries 346
8.2.3.2 Byte Interleaving 348
8.2.3.3 Automatic Protection Switching 349
8.3. Fiber in The Loop 350
8.3.1 Current Implementation Plans 350
8.4. Chapter Test 354

Chapter Nine ATM—Key to the Future 357

9.1. Integrating the Public Switched Telephone Network 358
9.1.1 The Reason for ATM 359
9.1.2 From the Network to the Desktop 361
9.1.3 From LAN to LAN 362
9.1.4 ATM Services and Applications 362

9.1.4.1 Voice Networks 363
9.1.4.2 High-Speed Data 364
9.1.4.3 High-Resolution Graphics 365
9.1.4.4 Video and Audio 365
9.1.4.5 Interactive Multimedia 367
9.1.4.6 ATM Services 367
9.1.4.7 ATM Bearer Services and Classes of Service 369
9.2. ATM Network Access 369
9.2.1 User-to-Network Interface 371
9.2.2 Network-to-Network Interface 371
9.3. ATM Overview 372
9.3.1 ATM Planes 373
9.3.1.1 OAM Messages 373
9.3.1.2 OAM Connectivity Verification 373
9.3.1.3 Interim Local Management Interface (ILMI) 375
9.3.2 ATM Layers 375
9.3.3 ATM Header and Payload 378
9.3.4 Routing in ATM—VCI/VPI 380
9.3.5 ATM Signaling 381
9.3.5.1 ATM Addressing 382
9.3.6 Adaptation Layer 384
9.4. Chapter Test 387

Appendix A Communications Evolution 389
A.1 History of Computing 389
A.2 History of Telephony 392

Bibliography 395

Index 399

Introduction

Some time ago, I was asked to teach a course on the fundamentals of telephony at a local college. As I began research for the course, I noticed that there were no text books available which covered all of the topics in the course. Most of the fundamentals books covered transmission theory very thoroughly, but did not discuss data networking or cellular. I found myself using several reference books to create my course material. When identifying a text book for the course, I was unable to find a book diverse enough to cover all of the topics which would be taught. So, I began compiling information and creating a course book to fill in the gaps. The course book was well received, and was the beginning of this book.

Some time later, my good friend John Faulkner came to me, and asked if I would write a book that would cover all of the topics which I had taught at the college. John is International Sales Manager for Tekelec and had many customers who would benefit from such a book. Thanks to John's persistence, this project got under way.

The intention of this book is to provide an overview of the technologies found in today's telecommunications networks. Many may argue that the technologies discussed in this book are old news, and do not warrant discussion here. However, it is these same technologies that will form the networks of tomorrow.

While these technologies may be very old, they are being used to build our next generation of telecommunications networks, and those in the telecommunications industry are scrambling to learn more about them. A good example is Signaling System #7 (SS7), which was developed in the late 1960s. SS7 has become one of the hottest topics in telephony circles today, and is the core of the Intelligent Network being built by major telephone companies worldwide.

This book will not go into great detail on any of the technologies, but will provide an overview of each one. The intent is to provide enough information that the reader will understand what the technology is, and understand the basic functions provided. If more detailed information is needed, the reader can then find a book which is dedicated to that topic. There are many books available which discuss one particular technology, such as TCP/IP or ISDN. These books will typically go into great detail, and are ideal for engineers and developers wishing to develop products based on these solutions.

This has been one of the toughest projects for me, because of the diversity of the topics. The industry is changing rapidly, and keeping current in all of the various areas of telecommunications is a challenge in itself. I hope you will find this book useful as a reference and as a text for learning new technologies. I know it has been an education for me as I researched for this book. My hope is that you will find it educational as well.

—TRAVIS RUSSELL

Acknowledgments

This book would have never been started, if not for my good friend John Faulkner. It was John's idea to write this book, when he found himself trying to explain the various protocols of our industry to his distributors and customers. Now you have a book to give them, John. Hope this helps close the sale!

Also many thanks to Allan Toomer, President and CEO of Tekelec, for all of his support.

Telecommunications Protocols

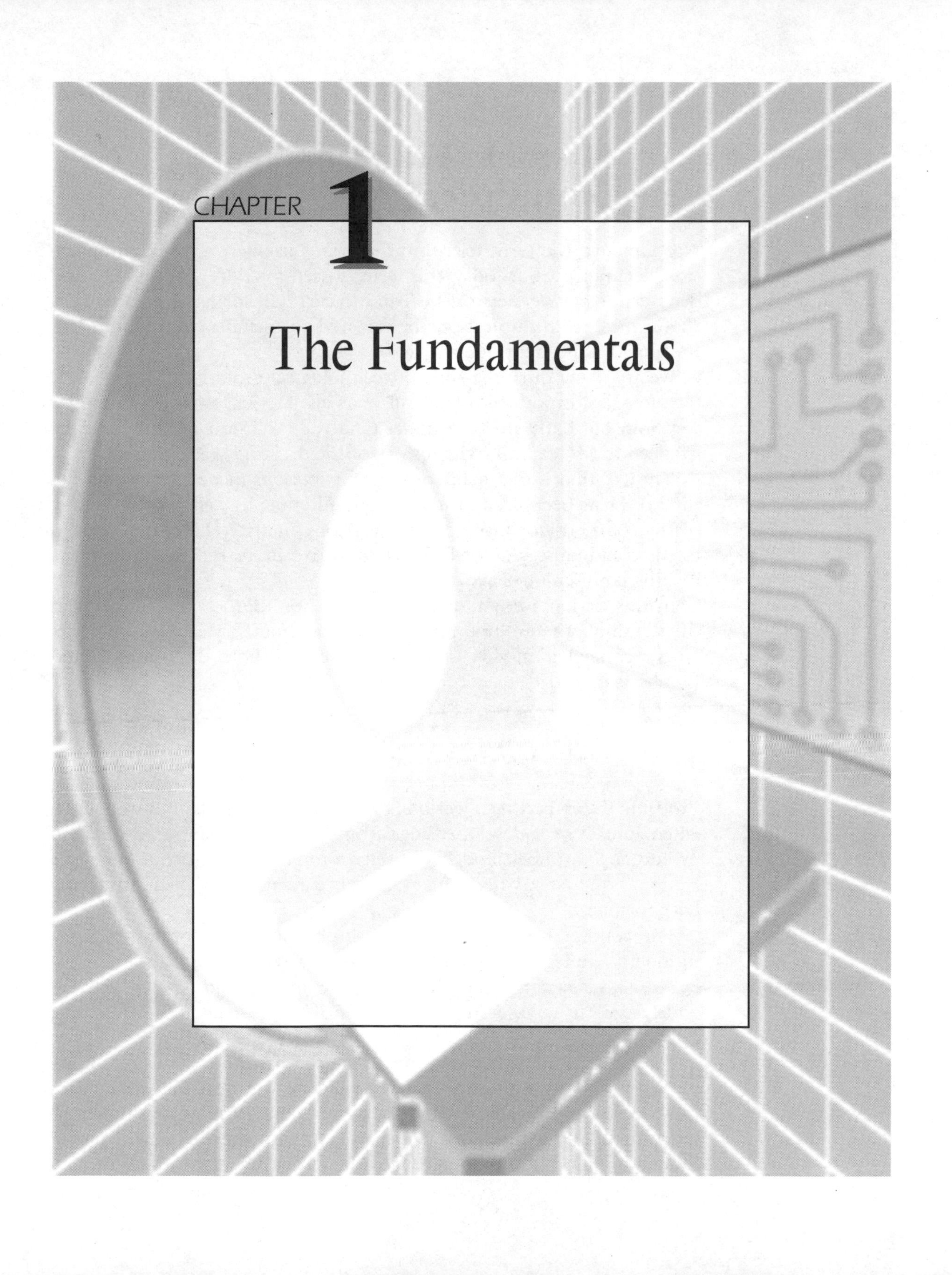

CHAPTER 1

The Fundamentals

1.1. Historic View of Telecommunications

Not long ago, the term telecommunications implied voice communications technology. But today, this term is used to address both voice and data technology. As networks expand to carry all forms of information, the term telecommunications has evolved to include voice, video, data, multimedia, and high-fidelity audio.

We are living during a time of technological evolution. Many of the technological innovations from 10 years ago are just now finding their way into our daily lives. To understand the evolution of this industry, we need to first examine the history of the data and voice industries.

You will find is that many of the contributions made to data communications have been used in voice communications as well. Likewise, many of the contributions made by the telephony industry have been applied to the data industry as well. Today, these two industries have combined, making it difficult to separate the two.

In this section, we will discuss first data communications history and then examine the evolution of the telephone industry. Appendix A provides a chronological view of all of the events in both the computer and telephone industries.

1.1.1. Data Communications History

Probably the earliest recollection of data contributions comes from 1614, when John Napier developed logarithms. Logarithms are used today to express large numbers and have been an important contribution to the computer industry, where large numbers can be tedious when working with formulas.

The concept was to make the multiplication and division of large numbers simpler. For example, $100 = 10^2$. The superscript 2 is the logarithm. This idea later led to the development of the slide rule, which was widely used in engineering practices until the advent of the electronic calculator.

Blaise Pascal, a mathematician, created the pascaline, a simple adding machine with a mechanism that was used for the next 300 years. Naturally, his adding machine was mechanical, consisting of a number of gears, dials, and a mechanical display. Turning the dials, which were located in different positions, turned the associated gears, which in turn would effect the output of the display. This same method was used in many other machines for some time to come.

One of the most infamous machines was Charles Babbage's Difference Engine, as well as his Analytical Engine. The Difference Engine was created in 1822 but was never built due to a lack of governmental support (as well as funding). The Analytical Engine fell from the same fate and was never built except for one small module. Only the blueprints were completed.

The problem was the amount of brass required to forge the various gears and wheels. The size of the Analytical Engine would have exceeded the modern day locomotive. This would have been the first "programmable" computer, but certainly not digital.

The Difference Engine was eventually built by the British inventor George Scheutz. Scheutz used Babbage's blueprints and funding from the British government to create the behemoth machine. The British government then used the machine to calculate insurance tables.

Some interest has been rejuvenated in Babbage's early machines, and enthusiasts have resurrected the Analytical Engine blueprints to see if a working machine could actually be built. The brass module built by Babbage currently resides in the Smithsonian Institute.

George Boole in 1847 developed what he called boolean algebra. Boolean algebra laid the foundation for the binary numbering system and modern day logic circuits. Much of what we learn today in digital electronics was the result of Boole's contribution. Boole believed that machines could be built that would use two states to represent almost any information. Those two states could be manipulated mathematically using boolean algebra, creating a rather simple device capable of handling information without knowing all of the various variables.

While most inventors were using wheels and dials to manipulate input into their machines, another inventor, Herman Hollerith, thought of a more portable approach. Using heavy paper cards, Hollerith devised a way to change the values of dials and counters according to the position of holes punched in the cards. His machine became nothing more than a tabulator, which would later prove its value to the U.S. government in tabulating the U.S. Census. By inserting holes into cards rather than writing on paper forms, the cards could be returned to the Census Bureau, which in turn would insert the cards into their tabulator and get instant results. Each hole position would carry a significant meaning and would correspond to a dial (or counter) on the machine.

The Hollerith card (or punch cards as they were called later) was widely used until the mid- to late 1970s for feeding programs into mainframe computers. The first usage of the Hollerith cards resulted in tripling the speed of the Census tabulation. Hollerith later formed his own company, which became known as International Business Machines, or IBM.

Mechanical computers were bulky, expensive, and hard to maintain. Work had already begun on developing a faster machine, using electricity rather than brute force. However, there were not a lot of electronic devices capable of providing what was needed by early computers. Electronics did not make any significant contributions until 1906 when Lee DeForest invented the vacuum tube.

The vacuum tube was used for a variety of applications. In computers, it was capable of switching current from one direction to another, much like transistors do today, based on the value of the current. They were used in amplifiers to increase power and in rectifiers to regulate voltage and current.

The vacuum tube was not an efficient electronics device and dissipated lots of heat. Because of the heat output, the tube did not have a long lifetime either. Early computers using vacuum tube technology could have as much as 22 hours (h) of maintenance per week, just maintaining the tubes. They were very fragile, made of glass and filled with a gaseous material, much like a light bulb. This made it difficult to use equipment in rugged environments.

Nevertheless, the vacuum tube provided a much needed function in early electronics and enjoyed a long life. Tubes were commonplace components of televisions and radios through the 1970s (remember tube testers at the local five and dime stores?).

In 1947 William Shockley, Walter Bratlain, and John Bardeen developed a small device they called the transistor. The transistor allows electricity to flow in in one direction and flow out in one of several other directions, depending on the type of transistor and the value of the voltage. This is the same basic function of the vacuum tube, but the transistor does not require the power that the vacuum tube does and is much cooler than a vacuum tube.

Constructed from a small wafer made from silicon, this tiny little device revolutionized electronics. Soon, transistors could be found in everything from pocket radios to kitchen appliances. This was the first step toward miniaturization of many devices, including the computer.

The first computer to become commercially available was the UNIVAC, made by Sperry Rand. Prior to the UNIVAC, computers were special-purpose devices, custom built to customer specifications. The UNIVAC allowed companies to purchase a computer system (mainframe) without going through the lengthy process of design and manufacturing. Sperry Rand could build computers according to a sales forecast and have them ready to ship when the customer ordered it.

These systems were not as we know them today. The UNIVAC was a mainframe computer and required special electrical and environmental controls. Companies had to hire programmers to customize software for their specific needs since software could not be purchased "off the shelf." Consequently, not many companies were able to afford such machines.

Meanwhile, development toward smaller electronics continued. In 1952, G. W. A. Dummer developed the integrated circuit. This tiny "chip" contained hundreds of microscopic transistors, providing a small package of many devices. Today, these ICs contain millions of circuits and are used in everything we touch. Televisions, radios, appliances, even automobiles are controlled by integrated circuits of all shapes and sizes.

Computers were still cumbersome, difficult to program, and certainly far from user friendly. Programmers were growing weary of writing the same instructions over and over. Many functions (such as accessing a disk drive) required instructions which had to be rewritten many times in a computer program.

To resolve this problem, Bob Patrick from General Motors and Owen Mock of North American Aviation created a program called the operating system. The purpose of the operating system is to provide a set of instructions used by computers on a routine basis. Accessing disk drives, printing on a printer, and managing files stored on mass storage devices are part of what operating systems do.

Programmers no longer have to write redundant instructions when creating a program; they only need to reference an instruction in the operating system, and the instruction is carried out by the operating system. This has been instrumental to the success of computer systems over the generations, and as software has evolved, operating systems have become more and more powerful.

Memory was still a problem in computer systems. The integrated circuit provided a means to consolidate memory circuits into smaller packages, but it was not until 1967 that Random Access Memory (RAM) was available as an integrated circuit. Developed by Fairchild Semiconductor, the RAM chip was instrumental in making computers even smaller.

Douglas Engelbart spent 10 years developing a device that many of us today take for granted. Engelbart recognized the need for a device that would allow users to control their computer without typing in commands. The mouse was born in 1968 and was first used as a standard interface by Apple Computers' Macintosh computer in 1989.

The Department of Defense (DOD) began development in the early 1960s of a military network capable of surviving any type of attack. The network had to be capable of routing data around failed nodes and intelligent enough to fix itself in the event that links should fail. This network, dubbed the ARPANET, later became what we know as the Internet. Using the Transmission Control Protocol/Internet Protocol (TCP/IP), the ARPANET was initially used for military data and was later opened to defense contractors and educational institutions.

The microprocessor was the last step in making computers small enough to fit on a desktop. Prior to the development of the microprocessor, computers relied on large circuit boards full of transistors and integrated circuits to provide the processing power it needed. With the microprocessor, all of this circuitry could be placed in one small little package, allowing computers to shrink even more.

The microprocessor has continued to grow in power and shrink in size. Many devices today have more processing power than the early UNIVAC and can be placed on top of your desk. Ted Hoff from Intel was responsible for the development of the microprocessor in 1971.

Also in 1971 the first floppy disk drive came from Alan Shugart and IBM. Shugart later went on to form his own company, Shugart Disk Drives. This was the first removable storage available for computers. The first floppy disk was an 8-inch (in) variety, capable of storing around 300K of data.

In 1973, the Winchester disk became the first "hard drive" available for computers. While not very large in capacity then, today a 3.5-in disk drive can store more than 1 gigabit of data.

The computer revolution had begun. Built smaller, with more processing power, the personal computer had become a formidable competitor to the mainframe. Companies started purchasing personal computers and attaching them to their mainframe computers as terminals. When the mainframe failed, the PC could continue operating on its own. However, these personal computers had no means of communicating with one another, without relying on the mainframe. In 1973, Xerox changed the computer industry with the Ethernet network. Robert Metcalfe developed the protocol, which is still in use today (although it has been modified to resolve several issues with the earlier version).

Personal computers were being found everywhere. Software could be purchased at the local computer store for almost any purpose. In 1975, Bill Gates and Paul Allen formed a company called Microsoft, today the world's largest software company (and Gates is one of the richest men in the world).

Another pair of entrepreneurs, Steve Jobs and Steve Wozniak, built a small personal computer called the Apple. While never being able to earn the lion's share of the computer market, Apple continues to struggle as a major competitor against archrival IBM even today. Based on its own proprietary operating system, Apple Computers has been forced to compete against another (perhaps more formidable) competitor, Microsoft.

The 5.25-in floppy disk was developed in 1978 by Radio Shack and Shugart Industries, providing an even smaller medium on which to store the huge amounts of data we would create over the next 10 years. The floppy disk continued to shrink in size while growing in capacity, and in 1984 the 3.5-in floppy disk was released with a whopping 1.4 megabyte of data storage.

This has been surpassed by the development of recordable CD-ROM in 1985, allowing millions of bytes of data to be stored on an optical disk. Today, CD-ROM has become the medium of choice for many software companies.

In 1993, the presidential election heated up with networks the hot topic in politics. The Information Highway was first dubbed by President Clinton and his staff as the next computer revolution. The Information Highway continues to evolve as the public continues to struggle with budget and applications for this generation's most talked about computer network.

1.1.2. Telephony History

Now that we have reviewed the evolution of the data communications industry, let us take a look at the telephone industry and how it has evolved. The evolution of the telephone industry has been very different from the computer industry. While a lot of the technology has been shared by both industries, the telephone industry has never made anyone rich.

In fact, many of the industry's inventors, including Alexander Graham Bell, have died as middle-class citizens, living a comfortable life but not rich by any means. Telephone companies have been started, merged, bought out, and split up. It is a wild industry and not one full of security.

In 1856 Western Union was founded. Western Union's primary interest was in telegraphy. The telegraph was the first means of communicating over distance. For many small towns, the telegraph was essential for communicating with other towns. Ironically, Western Union later became the Bell System's largest competitor.

In 1860, Philip Reis found that by applying electricity to a wire, sound could be transmitted through the wire. At one end of the wire he tied a cork wrapped with sausage skin (a crude microphone). He then wrapped the wire around a knitting needle and applied a battery between the cork and the knitting needle.

When the knitting needle was placed onto the strings of a violin, the strings vibrated each time the cork was thumped. If he had thumped the strings of the violin, he would have discovered that the same process would work in reverse.

A company called Western Electric was formed in 1872 and became the chief manufacturer for Western Union as well as other companies in the telegraph business. Western Electric later became part of the giant Bell System.

In 1876, Alexander Graham Bell filed his patent for an invention he called "Improvements to Telegraphy." The small box was not functional and was intended to transmit sound through a microphone. Just hours later, Elisha Gray filed his own patent for the telephone, which he was able to demonstrate as a working device. Unfortunately, Bell had beat him to the patent office, and Bell received credit for the invention of the telephone. Gray was the cofounder of Western Electric.

Bell's invention consisted of only a receiver and did not have a microphone. He later filed a patent in 1877 for a device with a combination receiver and microphone built into one unit. His first patent required two separate devices.

This was the year that Bell founded the Bell Telephone Company. This was the beginning of the Bell System as we know it today. Since Bell did not have the money necessary to start his company, he had to use money from his future father-in-law, Gardiner Hubbard, to start the company.

Bell was designated as the Chief Electrician of the Bell Company, while his assistant, Watson, was made superintendent and accountant. While his friend and future father-in-law received 1497 shares in the company, Bell only received 10 shares.

Western Union formed the American Speaking Telephone Company in 1877, becoming Bell's first big competitor. Competition in the telephone industry was fierce, with many small companies starting up all over the nation. American relied on the invention of Elisha Gray, as well as another inventor, Thomas Edison. Edison had invented a microphone that was much more efficient than Bell's.

Bell Telephone later sued American in 1878 for patent infringement. As a result, Bell Telephone acquired Western Union in 1910. A bitter legal battle ensued, with Western Union giving in and relinquishing all of its telephone patents, improvements, and a network of 56,000 telephones. As years went by, Bell Telephone pursued many small telephone companies, buying them out to increase their customer base.

In 1879, Bell left the board of his company after a dispute between the board and his now father-in-law Hubbard. William H. Forbes became president, and the Bell Telephone Company moved on without its founder.

In 1879, a doctor suggested that the telephone companies use numbers rather than names for connecting callers. As the popularity of the telephone grew, operators were having a difficult time keeping track of who was who. The doctor had been using a numbering system in his office for tracking patients. Hence, the birth of the telephone number.

In 1881, Bell Telephone Company purchased Western Electric. This became the manufacturing arm of the Bell System for years to follow, until the Bell System was finally divested in 1984. However, Western Electric was not the only telephone manufacturer. There were many companies manufacturing telephone equipment, including Stromberg and Carlson (later divided and sold to a number of companies).

The first long distance line was installed in 1884 between New York and Boston, and in 1915 the first transcontinental telephone call was made from New York to San Francisco.

The first telephone switch was invented in 1891 by a mortician. Angry because the town's operators were not connecting calls to his business (the operator's cousin also ran a funeral home), Almon B. Strowger vowed to rid the public of operators forever. He built the first step-by-step telephone switch in his garage. His invention did not get installed in a telephone office until 1921. There were many problems with the invention, and the device was sent to a company in Europe for further refinement. After the invention was improved, the Bell System began using the switch in its telephone offices.

The step-by-step was so reliable that many remain in service even today (but only in remote areas). Many private businesses installed step-by-step switches for private exchanges. In the Los Angeles area, many of the luxury hotels had their own switches installed. I had the pleasure of working on several of these switches during the 4 years I worked for the Bell System (a piece of one of those switches, called the line finder, hangs on my office wall).

In 1896 Strowger invented the rotary dial, which allowed callers to dial direct rather than rely on operators. However, operators were not made obsolete right away. Direct distance dialing was not allowed until 1951.

In 1900, the American Telephone and Telegraph (AT&T) corporation was founded. Its biggest competitor was Western Union. It acquired Western Union in 1910, pursuing not only the telephone business but the telegraph business as well. This did not last for long, however.

The Justice Department sued Bell Telephone in the first antitrust suit. The result was called the Kingsbury Agreement, by which Bell Telephone agreed to surrender Western Union and its telegraph services. It also forced Bell to allow independent telephone companies to connect to the Bell network, now spanning across the nation.

In 1922 Alexander Graham Bell died of diabetes. He was not a rich man by any means. He had lived a comfortable life in the upper middle class. Perhaps his greatest riches were in knowing that his was perhaps the greatest invention of all, the telephone. His assistant, Watson, resigned from Bell Telephone shortly after Bell left and pursued a career as an actor; he died in 1934.

The Bell System marched on. In 1925, the Bell Laboratories was formed. Bell Labs was responsible for research and development of telephone products. Their inventions reached far beyond the telephone. The transistor, the first sound picture, numerous audio inventions, and wireless radio systems were credited to Bell Labs.

In 1934, the federal government implemented a statute called the Communications Act of 1934. This statute governed voice communications and placed numerous regulations on the operations of telephone companies and how they maintained their communications networks. The Communications Act of 1934 was enforced until 1996, when it was finally superseded by the Telecommunications Act of 1996. The new telecommunications law would introduce sweeping changes to the industry, and removed barriers put into place by the Communications Act of 1934.

The Bell System moved into other services, laying the first coaxial cable in 1936. The cable was later used in 1940 for sending television signals from the Republican Convention. The Bell System laid many coaxial cables throughout their networks, sending voice, television, and radio signals from coast to coast.

In 1949, the Bell System found themselves before the federal government again. Another antitrust suit was brought against AT&T and the Bell System by the Justice Department. Once again they were under attack for monopolizing the industry.

The suit was settled in 1956, with the Bell System making patent-licensing concessions. As Bell Labs developed new technology, they were forced to license the technology to other competitors for further development. Many products such as UNIX and the transistor were marketed by other companies and included in their own products.

The agreement, called the Consent Decree, allowed Bell to keep its manufacturing division, Western Electric. Other concessions prevented the Bell System from entering into the computer industry. This was the second antitrust suit against the Bell System but not the last.

The first transoceanic cable was laid in 1956 from New York to Europe. This was a monumental project, providing transmission facilities across the Atlantic Ocean. This cable is still in operation today (although it has been replaced with fiber optics since its first installation). Fiber optics has also been laid across the Atlantic.

The entire industry changed in 1968 with the Carterfone Decision. Tom Carter, a ham radio operator and inventor, had developed a device allowing ham operators to connect their radios to the telephone network. The Bell System refused to allow him to connect the device to the network, preventing him from selling it to other radio operators.

Carter sued the Bell System, something unheard of considering the formidable size of the Bell System. Carter was able to find lawyers willing to volunteer their time, in the interest of beating the monolithic company. They were successful.

The Carterfone Decision opened the doors to other manufacturers who wanted a piece of the telephone business, and soon many companies were building telephones, modems, and other devices to be connected to the telephone network

MCI won a landmark lawsuit against the Bell System, opening the long distance market to competition. MCI was a microwave company that wanted to provide long distance services through their microwave network. Their lawsuit ended in victory and opened the doors for many other long distance companies to compete against AT&T. The industry began a revolutionary change.

In 1974, the Justice Department once again charged AT&T and the Bell System with monopolizing the telephone industry and pursued a lengthy suit. The result was the breakup of the Bell System in 1984. AT&T, Bell Labs, and Western Electric were divested from the operating companies, forcing the operating companies to be self-sustaining.

The company divided into seven regional operating companies: Pacific Telesis, NYNEX, Ameritech, Southwestern Bell, US West, Bell Atlantic, and Bellsouth Telecommunications. AT&T became an independent company and was not allowed to share assets, revenues, or even facilities with the Bell Operating Companies.

In 1995, AT&T announced it would break up into three new companies after struggling to compete in several industries. The three divisions would compete in three different industries: cellular, computers, and long distance services.

Probably the change that will have the most impact on the telephone industry is the decision in 1996 to change the Telecommunications Bill that had been in place for over 30 years. Sweeping telecommunications changes include allowing local telephone companies to compete with long distance companies in the same service areas.

A part of this change also affects how telephone companies charge other telephone companies for access into their networks. Much of the access charges allowed under the divestiture were alleviated with this new legislation. Telephone companies will be forced to find other sources for revenue to make up for the loss of access charges.

This is a quick overview of both the computer industry and the telephone industry. As you can see, while the computer industry highlights center around technology, the telephone industry has struggled with takeovers, lawsuits, and regulation. This continues to be the case even today.

1.2. Standards Organizations

Although standards are extremely important in the computer and the telephone industries, they do not happen overnight. Many technologies take as long as 10 years before they are completed, published as standards, and deployed in commercial products. It is very important for all who work in the telecommunications industry to understand how standards are created and who creates them.

In this section, we will discuss what standards are and how they are created and will look at the organizations responsible for creating and publishing standards. First, let us define what a standard is.

There are basically two types of standards. A de facto standard is one which is not really endorsed by a particular organization nor is it necessarily published. De facto standards are the result of monopolizing a market. For example, tissues are often referred to as Kleenex, even though not all tissues are made by Kleenex. The name has become a de facto standard whenever someone refers to tissues because of the market dominance enjoyed by Kleenex. IBM also shares market domination and has become a de facto standard in many areas.

A de jure standard is one agreed on by committee. De jure standards can take many years before completion because of the number of people involved and the very nature of the standards process. Some organizations have streamlined their standards process and are able to move much more quickly, while others have learned that anything by committee is a lengthy process.

Standards are developed as voluntary or regulatory standards. Voluntary standards are usually developed by members of the industry. There is no requirement that vendors adhere to voluntary standards, but there are many advantages in doing so. Voluntary standards ensure that vendors can interconnect with other vendors' equipment, providing opportunities to compete in otherwise proprietary networks.

Consumers are guaranteed that the equipment they purchase will perform in their networks according to the standards and can purchase equipment from multiple vendors to find the best solution for their money. Most of the standards in the telecommunications industry are voluntary standards.

Regulatory standards are those which must be adhered to by vendors. They are enforced by government agencies (like the Federal Communications Commission, or FCC) and are used to ensure that the industry works in the best interest of the public. This does not always work since many times a regulation works against the best interests of the public.

1.2.1. The Standards Process

There are many ways by which a technology can become a standard. De jure standards are the most difficult to develop. Typically, a standards organization will elect to pursue development on a particular technology based on a white paper submitted by a member or industry representative. The executive committee then assigns the task of establishing the standards details to the working committee which is responsible for the particular area in which the standard will be used.

For example, in the American National Standards Institute (ANSI), there are working committees that define furniture, plumbing, electrical, and telecommunications standards (to name a few). The working committee is assigned the task of developing the standard and begins forming various groups to address specific functional areas.

One group may be responsible for addressing the network management aspects of a protocol, while another group may be responsible for the development of the data link layer protocol. These individual groups receive contributions from the various members (of which there may be many) and then must vote on whether to accept or reject the various contributions.

The contributions hardly ever encompass an entire solution, but only a small area of the working group's interest. At regular meetings, they may have to vote on many different contributions, rejecting many and accepting few. Those that get accepted usually need reworking.

Most organizations require a unanimous vote before a contribution can be accepted. Some (such as the Asynchronous Transfer Mode, or ATM, Forum) have changed this rule, allowing a majority vote. This in theory should accelerate the standards process, although this has not necessarily been the case.

Once a working group has completed a particular part of a standard, it is presented to the General Counsel, which then gives final approval. The proposed standards are voted on once again and, if successful, are published as standards.

Some organizations (such as the International Telecommunication Union, or ITU) only publish standards at certain intervals. The ITU used to publish their standards every 4 years, lengthening the time that technology got to market. This has now been changed since the ITU recognized to the need to move more quickly. Their standards are now published after voting has been completed.

1.2.2. Organizations Here and Abroad

There are many different organizations responsible for the development and publishing of standards. Some standards organizations simply vote on standards developed by other organizations and endorse them for publication under their organization's charter. For example, ANSI publishes standards endorsed for use within the United States, even though they may not have developed the standard. Following are the standards organizations most involved with telecommunications.

1.2.2.1. International Telecommunications Union This was formerly known as the CCITT. They changed their name in 1992. The standards group responsible for telecommunications is called the Telecommunications Standardized Sectorization (ITU-TSS). Their members are usually government agencies from various countries represented in the United Nations (ITU is a UN treaty organization). The United States is represented by the Department of State, as well as by ANSI and the Institute of Electrical and Electronics Engineers (IEEE).

There are 15 working groups in the ITU-TSS. Each is responsible for a specific area, much the same as ANSI. Their contributions are then submitted to the General Counsel for election as a standard. Once approved, the standard is published.

This process used to take place every 4 years. The working committees would meet on regular intervals, and the General Counsel would vote on contributions every 4 years. Their publications were color coded, a different color representing each 4-year interval. The following list identifies the year of publication for the various standards (published as "books"):

- 1960—Red Book
- 1964—Blue Book
- 1968—White Book
- 1972—Orange/Green Book
- 1976—Orange Book
- 1980—Yellow Book
- 1984—Red Book
- 1986—Blue Book
- 1992—White Book

This process proved to be inefficient and delayed many technologies from reaching market in a timely manner. The ITU has since changed this policy and now meets at regular intervals, approving and publishing standards as they are completed. It is hoped that this new change will accelerate the standards process.

The ITU standards are used by countries who want to interconnect their networks with other countries. Each nation then has its own national standard for use within that country. For example, the United States uses ITU standards for connecting to the international telephone network, but within the United States, we use the ANSI standards.

1.2.2.2. International Organization for Standards (ISO) ISO is a voluntary organization, founded in 1946. Its members are standard organizations from all over the world. Its charter is to endorse standards for use in international networks and to promote interconnectivity between international boundaries.

The United States is represented by ANSI, which is responsible for endorsing and publishing standards for use here in the United States.

There are more than 5000 standards published by the ISO, including the Open System Interconnection (OSI) model, used by all in the industry. The Technical Committee 97 (TC97) is responsible for data communications standards. Perhaps the standard that has had the largest impact on the industry has been the ISO 9000 quality standard.

1.2.2.3. European Telecommunication Standards Institute (ETSI) Founded in 1988, ETSI establishes standards for use within European countries. They are a member of the ITU, and their members are from within, as well as outside of, Europe (such as telecommunications manufacturers). They were formed to take over the role of the European Conference of Postal and Telecommunications Administrations (CEPT).

Membership to CEPT was previously open only to administrations and public operators. This was later changed to include manufacturers. ETSI has adopted this policy to allow more influence from the public sector into their decisions.

1.2.2.4. American National Standards Institute The ANSI is responsible for determining which protocol and media standards will be endorsed for use in United States networks. They are a nonprofit organization consisting of members from the industry. Manufacturers and developers who are actively working in the industry submit both personal and company-represented contributions for publication as ANSI standards. The IEEE is one example of membership.

The T1 committee is responsible for establishing telecommunications standards. There are a number of T1 committees, each one responsible for various areas. The T1E1 committee is responsible for carrier-to-customer installation interfaces.

The T1S1 committee is responsible for telecommunications services, architecture of telecommunications networks, and signaling networks (such as SS7) used in telecommunications. This is one of the more active committees in the telecommunications industry.

Transmission standards are under the jurisdiction of the T1X1 committee. Transmission includes T1 facilities, data facilities, and SONET. Performance and quality standards are determined by the T1Q1 committee.

All of these committees must work with one another since many of these standards are related to one another. Some committee members may even belong to more than one committee.

1.2.2.5. Bell Communications Research (Bellcore) For those who work in the telephone industry, Bellcore standards are the most important aspect of their products. All network equipment purchased by the Regional Bell Operating Companies (RBOCs) and used within their networks are required by those companies to meet or exceed Bellcore standards.

Bellcore is a private organization, funded by the seven RBOCs. Their future is questionable because of the complexities of the RBOCs' relationships. In some ways they are family, while in the business sense they are competitors. This is especially true as telephone legislation changes, and the RBOCs begin offering services which cross their geographical boundaries into the sectors of their fellow RBOCs.

Bellcore is also a research firm, providing the research and development previously provided by Bell Laboratories (before divestiture in 1984). The standards they publish are typically the same as the ITU and ANSI standards, with many enhancements. Their primary focus is in the areas of network management and interoperability. In fact, many of the Bellcore standards are almost exactly the same as the ANSI equivalents. The differences are in enhancements made by Bellcore. This often means that Bellcore standard compliance is much more difficult than ANSI compliance.

Their standards are published in several stages. They are first published as a draft and distributed to vendors in the industry for comment. They are then republished as preliminary, which are then distributed again. Final documents are published as Generic Requirements (GRs). These were formerly known as Technical References (TRs).

Bellcore also publishes requirements for equipment interfaces as well as operating requirements for specific device types. Again, they typically concentrate on the areas of interoperability and network management.

1.2.2.6. Exchange Carriers Standards Association (ECSA) ECSA was formed in 1983 as an independent organization to define interface standards used by carriers to connect to regional telephone companies. After the divestiture of 1984, AT&T, MCI, and many other long distance carriers became independents not affiliated with the Bell System. This meant they wsere not represented by Bellcore and needed an organization of their own to define interconnect standards.

The industry recognized the need for an organization to standardize how these carriers would connect to the various networks. The interexchange carriers are the members of this organization. They also sponsor the T1 Committee, which is accredited by ANSI (which means they write standards which are later published as ANSI standards).

1.2.2.7. Electronics Industries Association (EIA) This organization is perhaps best known for its contribution to the computer world. The ever-popular RS-232 interface is an EIA standard. In recent years, the EIA has been actively publishing standards for use in the cellular market.

The Interim Standard-41 (IS-41), used in many cellular networks today, is an EIA standard, as are IS-94 and IS-95, both being deployed in new cellular networks. These technologies are explained in greater detail in Chap. 7.

1.2.2.8. Institute of Electrical and Electronics Engineers IEEE is the largest professional society in the industry. Its members are working professionals in the industry, primarily engineers and developers. There are over 275,000 members to date worldwide.

The 802 standards are probably the most familiar standards in the data communications industry. They include Ethernet and Token Ring standards, as well as the Fiber Distributed Data Interface (FDDI). The following list are some of the more well-known standards:

- 802.1—Higher-layer protocol interface standard
- 802.2—Logical link control (LLC)
- 802.3—CSMA/CD LAN (Ethernet)
- 802.4—Token Bus (ARCNET)
- 802.5—Token Ring
- 802.6—DQDB Metropolitan Area Network (SMDS)
- 802.7—Broadband
- 802.8—Optical fiber
- 802.9—Integrated voice and data Local Area Network (LAN) interfaces

1.2.2.9. Federal Communications Commission The FCC is a regulatory agency which was established in 1934 as a result of the Communications Act of 1934. Their charter is to regulate the radio and all wire communications within the United States. They play an active role in all of the telecommunications and cellular networks.

There are seven commissioners appointed by the president of the United States to reside over the FCC. Their decisions are far-reaching and affect everyone who uses the services of the telephone companies, radio broadcasters, television broadcasters, and even corporate networks.

1.2.2.10. Network Reliability Council (NRC) Chartered by the FCC, the NRC monitors all public telecommunications networks, reporting outages by the industry through special reports. The NRC also makes recommendations for protocol operations to guard against network outages.

The NRC was formed in 1992, after several network outages in New York City and Washington, D.C., blocked telephone calls for several hours. One vendor's software was found to have caused the outage, and due to the lack of any industry reports was not communicated to other companies. Several other outages were experienced by companies using the same software.

The FCC determined these outages could have been prevented, had there been a way to notify the industry of problems in the software. That is when the NRC was formed. Their original charter was for 5 years, but was extended in 1996 for another 7 years. All network equipment manufacturers who provide equipment for telephone networks are encouraged to participate in NRC sessions.

1.2.2.11. Federal Telecommunications Standards Committee (FTSC) The FTSC is also part of government procurement. Their task is to ensure the interoperability of lower-layer protocols (used at the network layers). They also verify that government network equipment meets the standards of the government.

1.2.2.12. National Bureau of Standards (NBS) When selling to the government, all technology must conform to this agency's standards. The protocols defined by this agency are typically upper-layer protocols, and they apply to all government procurement (with the exception of the military, which falls under the jurisdiction of the DOD).

1.2.2.13. Defense Communications Agency (DCA) The military has unique requirements and therefore uses a different agency (under the direction of the DOD) for establishing military standards. The DCA is responsible for defining all standards used within military networks.

1.2.2.14. Underwriters Laboratories (UL) This organization is really not a standards organization, but their role is an important one in the United States. All manufacturers of electrical equipment send their products to the UL for testing. Conformance is not mandatory, since this is not a regulatory agency. However, UL approval on any product assures the general public that the product has met the minimal requirements of the UL. This approval is beneficial to both manufacturer and consumer.

1.2.2.15. Canadian Standards Association (CSA) This is the Canadian equivalent to the Underwriters Laboratories. The primary difference is that all products sold for use in Canada must be CSA approved. Many vendors who manufacture products for use in the United States and in Canada must have both the UL approval and the CSA seals.

1.2.2.16. ATM Forum There are many forums like the ATM, but this forum receives the most publicity because of the hype surrounding ATM technology. The ATM Forum began with just four members, all of which were manufacturers of ATM products. Their interests were in establishing some implementation agreements by which they would all design their products, until the ITU could finish their work on defining the ATM standards.

It is important to understand what an implementation agreement is. It is not a standard but simply an agreement between a number of companies that they will develop their products to perform in a specific way. The intent is to establish interoperability with other vendors' products.

The ATM Forum consists of a seven-member Board of Directors and the following committees: The Market Awareness and Education Committee, the Technical Committee, and the Enterprise Network Roundtable Committee.

The Market Awareness and Education Committee is responsible for educating the user and developer's communities about ATM. They provide demonstrations of ATM networks through their Interoperability Demos Group, provide educational seminars through the Education Group, and support end users through their End-User Focus Group. They also promote ATM and the ATM Forum through the Marketing Communications Group.

The Technical Committee is divided into the various working groups, which define the various implementation agreements. These working groups focus on a specific area in the network. Following is a list of the working groups in the Technical Committee:

- Traffic Management Group
- Physical Layer Group
- Data Exchange Interface Group
- Signaling Group
- B-ICI Group
- Testing Group
- Service Aspects and Applications (SAA) Group

- Private Network Node Interface (NNI) Group
- Network Management Group
- LAN Emulation Services (LES) Group

The Enterprise Network Roundtable Committee consists of the following working groups:

- Steering Committee
- Requirements Focus Group
- Education Group
- Membership Group

As mentioned before, there are many other forums within the telecommunications industry, including the Integrated Services Digital Network (ISDN) and the Frame Relay Forums. Their roles are somewhat different from that of the ATM Forum. The ATM Forum is mentioned here because of the role they are playing in ATM development.

1.3. Digital Transmission Fundamentals

Before talking about specific technologies, we need to first understand the fundamentals of digital transmission. When we speak about protocols, we talk about parameters, fields, and various values within these fields as if they were plain text. However, what is actually transmitted over wire, through airwaves, and through computer circuitry is nothing more than electrical current (or optical, depending on the medium).

In this section, we will define how text, speech, and video are transmitted from an analog source over digital facilities. Transmission is the lowest form of communications within a network, which is another reason we are covering this first. We will not discuss digital transmission in great detail. There are plenty of books that cover the engineering aspects of digital transmission

1.3.1. From Electrical to Binary

All transmissions within a network must be converted from binary code to electrical or optical signals. All information, including the information appended by protocols, must be transmitted at this lowest form of transmission. In some cases, additional information is appended at this layer to facilitate timing and error control (the concept of layering is introduced later in this book).

First, let us examine how text and voice are transmitted over transmission facilities; then we will examine some of the additional controls used by some transmission equipment.

1.3.2. Alphabet Soup—ASCII and EBCDIC

Characters must be converted into binary numbers so that they can be converted into electrical current or optical transmission. There are two standards used for representing just plain text, without any formatting.

It is important to understand that these standards are not used for the text which is generated by modern-day word processors or desktop publishing systems. When text is word processed, additional information must be provided by the source so that the receiver knows how the text is to look (italics, bold, underlined, specific fonts, etc.).

This formatting is not represented in these two standards, but in proprietary formats handled at the upper layers of protocols. Applications receive envelopes of data which include binary information regarding the formatting of the text.

The two standards which deal with plain text are the American Standard Code for Information Interchange (ASCII) and Extended Binary Coded Decimal Information Code (EBCDIC). ASCII was developed by ANSI, and EBCDIC was introduced by IBM and is used predominantly by their terminal equipment.

ASCII code is much like Morse code. There are 7-bit codes for each character (both upper- and lowercase), supporting 128 characters. Some of the codes represented do not appear on screens but are used as control characters. For example, "EOT" is the code for end of text, which is the same as end of transmission. Today, modern protocols provide the necessary control information, and the ASCII characters are now encapsulated within the protocol envelope.

The purpose of the code was for use in terminals, which have no processing capability. A terminal receives a serial bit stream of characters in ASCII code and displays those characters as they are received. Today, ASCII code is still used, but as mentioned above, the characters are now encapsulated within the protocol used to transmit the data. Table 1.1 illustrates the entire ASCII code set. While numbers are represented in

TABLE 1.1
ASCII Code Set

				6,5,4	000	001	010	011	100	101	110	111
3	2	1	0	Hex	0	1	2	3	4	5	6	7
0	0	0	0	0	NUL	DLE	SP	0		P		p
0	0	0	1	1	SOH	SBA	!	1	A	Q	a	q
0	0	1	0	2	STX	EUA	"	2	B	R	b	r
0	0	1	1	3	ETX	IC	#	3	C	S	c	s
0	1	0	0	4	EOT	RA	$	4	D	T	d	t
0	1	0	1	5	ENQ	NAK	%	5	E	U	e	u
0	1	1	0	6		SYN	&	6	F	V	f	v
0	1	1	1	7		ETB	`	7	G	W	g	w
1	0	0	0	8			(	8	H	X	h	x
1	0	0	1	9	PT	EM	0	9	I	Y	i	y
1	0	1	0	A	NL	SUB	•	:	J	Z	j	z
1	0	1	1	B		ESC	+	;	K	[	k	
1	1	0	0	C	FF	DUP	,	<	L	\	l	
1	1	0	1	D		SF	-	=	M	]	m	
1	1	1	0	E		FM	.	>	N	^	n	
1	1	1	1	F		ITB	/	?	O	_	o	

TABLE 1.2

EBCDIC Code Set

<table>
<tr><th></th><th>7,6</th><th colspan="4">00</th><th colspan="4">01</th><th colspan="4">10</th><th colspan="4">11</th></tr>
<tr><th>Bits</th><th>5,4</th><th>00</th><th>01</th><th>10</th><th>11</th><th>00</th><th>01</th><th>10</th><th>11</th><th>00</th><th>01</th><th>10</th><th>11</th><th>00</th><th>01</th><th>10</th><th>11</th></tr>
<tr><td>3,2,1,0</td><td></td><td>0</td><td>1</td><td>2</td><td>3</td><td>4</td><td>5</td><td>6</td><td>7</td><td>8</td><td>9</td><td>A</td><td>B</td><td>C</td><td>D</td><td>E</td><td>F</td></tr>
<tr><td>0000</td><td>0</td><td>NUL</td><td>DLE</td><td></td><td></td><td>SP</td><td>&</td><td>-</td><td></td><td></td><td></td><td></td><td></td><td></td><td></td><td></td><td>0</td></tr>
<tr><td>0001</td><td>1</td><td>SOH</td><td>SBA</td><td></td><td></td><td></td><td></td><td>/</td><td></td><td>a</td><td>j</td><td></td><td></td><td>A</td><td>J</td><td></td><td>1</td></tr>
<tr><td>0010</td><td>2</td><td>STX</td><td>EUA</td><td></td><td>SYN</td><td></td><td></td><td></td><td></td><td>b</td><td>k</td><td>s</td><td></td><td>B</td><td>K</td><td>S</td><td>2</td></tr>
<tr><td>0100</td><td>4</td><td></td><td></td><td></td><td></td><td></td><td></td><td></td><td></td><td>d</td><td>m</td><td>u</td><td></td><td>D</td><td>M</td><td>U</td><td>4</td></tr>
<tr><td>0101</td><td>5</td><td>PT</td><td>NL</td><td></td><td></td><td></td><td></td><td></td><td></td><td>e</td><td>n</td><td>v</td><td></td><td>E</td><td>N</td><td>V</td><td>5</td></tr>
<tr><td>0011</td><td>3</td><td>ETX</td><td>IC</td><td></td><td></td><td></td><td></td><td></td><td></td><td>c</td><td>l</td><td>t</td><td></td><td>C</td><td>L</td><td>T</td><td>3</td></tr>
<tr><td>0110</td><td>6</td><td></td><td></td><td>ETB</td><td></td><td></td><td></td><td></td><td></td><td>f</td><td>o</td><td>w</td><td></td><td>F</td><td>O</td><td>W</td><td>6</td></tr>
<tr><td>0111</td><td>7</td><td></td><td></td><td>ESC</td><td>EOT</td><td></td><td></td><td></td><td></td><td>g</td><td>p</td><td>x</td><td></td><td>G</td><td>P</td><td>X</td><td>7</td></tr>
<tr><td>1000</td><td>8</td><td></td><td></td><td></td><td></td><td></td><td></td><td></td><td></td><td>h</td><td>q</td><td>y</td><td></td><td>H</td><td>Q</td><td>Y</td><td>8</td></tr>
<tr><td>1001</td><td>9</td><td></td><td>EM</td><td></td><td></td><td></td><td></td><td></td><td></td><td>i</td><td>r</td><td>z</td><td></td><td>I</td><td>R</td><td>Z</td><td>9</td></tr>
<tr><td>1010</td><td>A</td><td></td><td></td><td></td><td></td><td></td><td>!</td><td>|</td><td>:</td><td></td><td></td><td></td><td></td><td></td><td></td><td></td><td></td></tr>
<tr><td>1011</td><td>B</td><td></td><td></td><td></td><td></td><td>.</td><td>$</td><td>,</td><td>#</td><td></td><td></td><td></td><td></td><td></td><td></td><td></td><td></td></tr>
<tr><td>1100</td><td>C</td><td>FF</td><td>DUP</td><td></td><td>RA</td><td><</td><td>•</td><td>%</td><td>@</td><td></td><td></td><td></td><td></td><td></td><td></td><td></td><td></td></tr>
<tr><td>1101</td><td>D</td><td></td><td>SF</td><td>ENQ</td><td>NAK</td><td>(</td><td>)</td><td>_</td><td>‘</td><td></td><td></td><td></td><td></td><td></td><td></td><td></td><td></td></tr>
<tr><td>1110</td><td>E</td><td></td><td>FM</td><td></td><td></td><td>+</td><td>;</td><td>></td><td>=</td><td></td><td></td><td></td><td></td><td></td><td></td><td></td><td></td></tr>
<tr><td>1111</td><td>F</td><td></td><td>ITB</td><td></td><td>SUB</td><td>|</td><td></td><td>?</td><td>“</td><td></td><td></td><td></td><td></td><td></td><td></td><td></td><td></td></tr>
</table>

TABLE 1.3

BCD Code Set

0000	0
0001	1
0010	2
0011	3
0100	4
0101	5
0110	6
0111	7
1000	8
1001	9

the ASCII code set, there is another method of representing numbers which we will explain a little later. For now know that numbers can be represented in ASCII or another code, called Binary Coded Decimal (BCD).

The EBCDIC code set is almost identical to the ASCII code set, with the exception of an extra bit. EBCDIC supports more graphical characters than does ASCII. However, in today's networking environment, EBCDIC is rarely used. Modern desktop publishing applications have provided a new means on conveying how information is to be displayed. Table 1.2 shows the entire EBCDIC code set.

There is another code set, BCD, used today to represent numbers. Originally, it was a 6-bit code set used to represent alphanumeric characters. Today, it is a 4-bit set used to represent numbers only. You will find it used within many protocols, where digits must be represented. Its advantage is in the number of bits to represent a number. The code set only covers digits 0 through 9, but these can be combined to support every number conceivable, with fewer bits than what would be required with ASCII or EBCDIC.

BCD is used in ISDN, where telephone numbers must be represented, and in Signaling System #7, a protocol used by telephone companies to convey control information between telephone switches. Table 1.3 shows the entire BCD code set.

One last code set which is used in many telecommunications protocols is the International Alphabet Number 5 (IA5) code set. This is an ITU standard and is very close to the ASCII code set.

1.3.3. Digitizing Voice

Voice transmission in today's telephone networks is almost all digital (at least between telephone company offices). Even many Private Branch Exchanges (PBXs) support all-digital transmission. There are many advantages to digitizing voice for transmission within a network.

Analog voice is susceptible to noise (conveyed as static and pops). As the transmission is passed through the network, it passes through a series of amplifiers. Analog amplifiers do not have the ability to determine what is actually voice and what is noise, so they amplify everything.

In digital transmission, the voice is not amplified; it is regenerated. This is possible because of the nature of digital signals. The original waveform may deviate from the source as it passes through the network

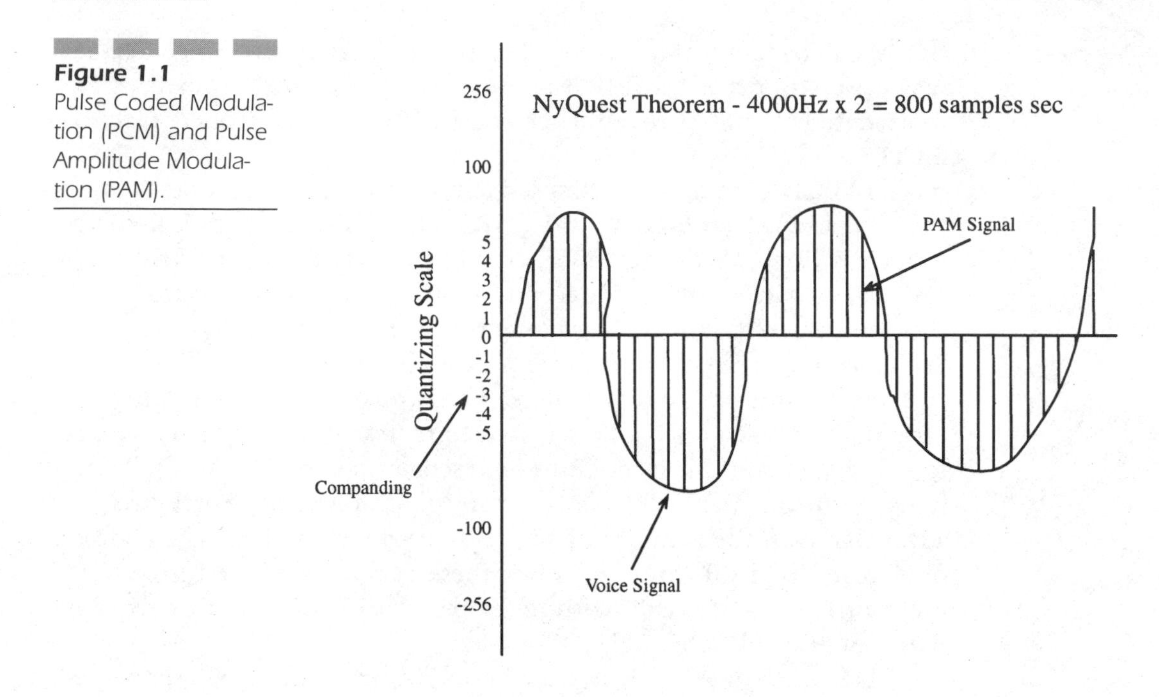

Figure 1.1 Pulse Coded Modulation (PCM) and Pulse Amplitude Modulation (PAM).

and may become distorted when noise is inducted, but the original signal can be regenerated by repeaters by looking at the received waveform and regenerating it based on its highs and lows. This is explained in more detail in the section on digital transmission.

Digitizing voice is not a complicated problem. The voice is sampled at regular time intervals. Each time the voice is sampled, the amplitude (height) of the signal (waveform) is compared to a scale.

The scale consists of numbers, beginning at zero and ending at 256. There is both a positive side of the scale (0 to +256) and a negative side of the scale (0 to −256). Figure 1.1 shows the scale and how an analog waveform is sampled. Each time a sample is taken, a pattern is created. This is indicated by the shaded bars in the figure. The bars represent what is called the Pulse Amplitude Modulation (PAM) signal. This signal indicates the amplitude of the signal at the time the sample was taken.

The trick is to take samples often enough so that the PAM signal is close to the original analog sine wave. The more often samples are taken (frequency) the better the regenerated signal. For a voice signal, this is 8000 samples per second. We will look at how that was determined later.

The next step after deriving the PAM signal is to create the digital equivalent of the PAM. The numbers on the scale are converted into 7-bit binary numbers, with the most significant bit (the first bit transmitted) identifying whether the number is a positive or negative value on the scale.

The outcome of the digital conversion is called the Pulse Coded Modulation (PCM) signal. It is a binary stream of 8-bit words, each byte equaling one PAM sample. There are some issues with PCM, which we will discuss next.

If samples are not taken often enough, the original waveform will not be digitized accurately. This is referred to as aliasing. To prevent aliasing, samples are taken at more frequent intervals. The NyQuest Theorem was written by a scientist (NyQuest) who determined that the normal voice frequency is from 300 to 3200 hertz (H_z; cycles per second). To prevent interference from adjacent facilities, 4 kHz is allocated for analog voice signals.

This being the case, NyQuest figured that the sampling rate should be twice that of the frequency, which in the case of voice would be 8000 samples per second. While this theory helps prevent aliasing, electrical properties can still introduce noise which can affect the PAM signal. Low-pass filters built into the electronic circuitry of the devices which provide analog-to-digital conversion also prevent aliasing by smoothing the PAM signal and making it more "readable."

The scale which is used to create the PAM signal is called the quantizing scale. This scale is limited in the steps it can represent, which limits it to the amplitude it can represent. If all of the steps are equally divided on the quantizing scale, the waveform again may be misrepresented.

Since the amplitude in most voice conversations is in the midrange of the quantizing scale, it makes sense to put more steps in the midrange and fewer steps in the upper and lower ranges. This provides better representation for "everyday" conversation, while still accommodating the infrequent signals at higher or lower amplitudes. This is referred to as companding. The trade-off is distortion of sorts when the amplitude goes beyond these ranges, but the distortion is acceptable for voice communications. If it were used for high-fidelity audio, this would not be acceptable. There are two forms of companding, μ-law and A-law. μ-law is the method used here in the United States, while A-law is the method used in international networks. The two are very similar but not compatible.

There are other types of PCM which obtain the same results but offer compression. Adaptive Differential PCM uses a 4- rather than an 8-bit word. Each 4 bits represent a change in amplitude. This allows two devices to be connected to one port, sending pairs of 4-bit words (one pair for each device).

Another form of compression is called Digital Speech Interpolation (DSI). This compression technique is found in many digital recording devices used in telecommunications networks, including voice mail and voice recognition systems. DSI deletes pauses where speech is not present, saving bandwidth and eventually disk space at the host device. Now that we understand how voice is converted to digital, we can examine the purpose of a protocol. Data does not require conversion, because it is already in digital form. There are other forms of analog signals which must be converted to digital, such as audio and video.

There are many techniques used to convert these signals to digital form, similar to the techniques used in PCM. For now, know that these signals are digitized and converted to a serial stream of binary bits.

A protocol is used to maintain order in the serial streams of bits, and to manage any connections established between two communicating entities. This next section looks at protocol functions in greater detail.

1.4. The Basics of Telecommunications Protocols

A protocol is best defined as a set of rules. In digital communications, protocols determine where certain information will be found in the binary bit stream. Addresses, control information, user data, and various other fields must be clearly defined and consistent within each transmission. This is the job of the protocol.

Not all protocols are alike. Many protocols provide similar features, but all are very different both in format and in implementation. Here, we will outline the common features of a protocol and then look at a model used in developing protocols even today.

1.4.1. Protocol Services

Protocol functions are divided into layers. This layered approach allows for better segregation of protocol functions as well as software modularity. Layers are important in communications networks because they allow software upgrades to be deployed without affecting every node in the network. Only those devices which utilize functions within the specific layer need to be upgraded. In a communications network, the first three layers are the most critical. All layers above the first three have no effect on the network itself. They are resident in end node software and are transparent to network devices (such as routers and switches).

The first layer is commonly referred to as the physical layer. This is where the user data and the information appended by the protocol are converted to either optical or electrical form and transmitted over the network. In some cases, the physical layer may append information as well (as is the case with T-1). This appended information is referred to as overhead and is usually minimal.

The second layer is known as the data link layer. The function of this layer is to provide node-to-node communications. The protocols which operate at this layer are not concerned with the contents of the user data or the data which resides at higher layers. In fact, protocols at this layer have no visibility to a data unit's final destination. The data link's only concern is the transmission of data between any two devices within a network.

The network access layer is responsible for the transfer of data between the host computer and the network. It is the function of this layer to ensure reliable transfer of data from the source, through the network, to the final destination. Addressing for both the source and the destination can be found at this layer.

There are two other layers which interact with the previously discussed layers. They reside at the layers above the network access layer and provide services between the source and the destination. They are not concerned with the transmission of the data unit through the network; they work at a higher level.

The transport layer ensures that the data transmitted is received in the same order. This is not a problem if the source and destination are connected directly to one another. In packet-switched networks, this is not the case. Related data units can take multiple paths, arriving at the destination out of order. The transport layer has the task of placing the received data units back into the order in which they were sent.

The process layer supports various applications within a host. For example, to send a file from one computer to another requires a protocol to manage the file transfer. To connect to another computer remotely and operate programs resident on the remote computer requires a protocol capable of managing terminal communications between the two devices. Process layer protocols are typically unique to the specific application they serve.

Addressing in protocols takes many forms. In all protocols, both the physical source and destination devices must be identified by an address (commonly called the machine address). The machine address can be hard coded in an interface card or administered through software. When a computer is connected to a LAN, the network interface card (NIC) provides the machine address for each device.

In addition to the machine address, a logical address used by the various network devices is often used. For example, in TCP/IP networks, machine addresses cannot be supported. Instead, a logical address is assigned which is later converted to a machine address by a network server. To identify the millions of computers connected to the Internet would require an address field too large to be efficient.

In reality, this concept works quite well. Network devices do not need to know machine addresses. They only need to know network addresses (which are often part of the logical address). If only the network address is needed, device databases can be consolidated, minimizing the memory resources required.

The logical address typically identifies a user rather than an actual machine, which allows for more flexibility as well. Users can move from machine to machine, identified by their logical address rather than the address of their machine. A user could have a computer at home, at their office, and a portable computer. The user's logical address could be assigned to all three computers, providing the user access to e-mail and other network services regardless of which computer is used.

Once a data unit reaches its destination, the host must determine which application within software the data unit should be sent. This is identified by a process address. There are a number of different methods used to identify processes. Operating systems use "sockets," which are nothing more than logical ports used to connect to applications.

If we reexamine addressing, we can see that the machine is assigned an address (used by the second layer), the user is assigned an address (used by the network layer), and the process is assigned an address (used by the operating system to determine how to route the data unit internally). This further illustrates how layering works within various protocols, allowing functions to be addressed using an hierarchical approach.

1.4.1.1. Protocol Tasks Protocols perform a number of specific tasks. Not all protocols provide the same tasks, and if they do, they rarely provide them in the same fashion. Protocols are developed to meet a specific need within a network and are often designed with the network topology in mind. For this reason, protocols can be very different from one another.

1.4.1.1.1. Segmentation and Reassembly Unfortunately, user data cannot usually be sent in one large block. To do so would mean the network would be burdened with one transmission, blocking other transmissions until the block was completely sent. Protocols are limited by design to specific segment sizes. Remember that protocols are layered, which means the protocol at each layer may have different size limitations for data blocks.

Segmentation allows a protocol to take the user data and divide it into smaller blocks of data before transmitting it over the network. There are many considerations when determining the size of the data block. Different protocols have different requirements and may support different applications.

The transmission medium can also play a factor when determining the maximum size of a protocol packet. Some media may be limited to the amount of data which can be sent reliably in one block. Others may not have any limitation at all. Ironically, some fiber optic facilities use protocols which only support small data blocks. This is usually because of the types of applications being supported by the network.

For example, the ATM developers struggled with the issue of packet size. Data users wanted to be able to send very large data blocks, using the argument that the more data sent in one packet, the more throughput gained. This argument does not work when video or voice is being sent through the same network.

A factor called *latency* must be considered when designing networks to support real-time applications such as voice and video. If large blocks of data were to be supported, the receiver would have to wait until the entire block of data had been received before processing the data. Once all of the data was received, it could begin processing it, but if there were a large amount of data to process, there could be noticeable delay.

In the case of voice, this would be apparent by pauses in the voice transmission. The receiver would have to wait for the next large block of data to be received before it could begin processing the next transmission. This introduces more delay as the receiver waits to process the large data blocks. Video shares the same issues.

If an error were to occur, and data was retransmitted, large data units would take a long time. This is another reason why protocols favor smaller data units. They do not require large receive buffers, and retransmission does not burden the network.

To counter this problem, the developers voted on small blocks of data. These can be routed quickly, they do not require large buffers to store them before processing, and they can be processed quickly once received by the destination host. Small data blocks work well for real-time applications.

This means the protocols used must be capable of breaking up large amounts of data into smaller chunks which can be transmitted over the network and reassembled at the destination. In some types of networks (such as packet networks), data is not guaranteed to be delivered in the same order it was sent. This is because data units can travel many different paths, depending on the status of the various network elements.

The protocol must be capable of determining the original order in which the data units were sent and whether or not segments are missing from the original transmission and must be able to put the data units back together before giving them to the application.

This process may take place at different levels. For example, an application may pass a large amount of data down to the next layer, where a protocol is interfaced. This protocol may divide the data units into smaller data units while appending protocol control information (overhead). The protocol may then pass each new data unit to the next layer, which could repeat the process by further dividing the data unit into yet smaller data units, appending its own overhead, and passing it to the next layer. This could take place at four layers before the data is finally sent.

Each layer works independently, the upper layers being transparent. This means the lower layers know nothing about the contents of the data units at the upper layers. They simply pass the data along and never process the information appended by upper-layer protocols.

This can add to the complexity of a network when network boundaries must be crossed. In the case of the Internet, TCP/IP provides control of segmentation at various layers and allows data to be segmented at various points in the network and reassembled by the destination host without error. This can be very complex when one considers the number of segments in the Internet and the number of times data can be segmented at the various nodes throughout the network.

Reassembly is accomplished through a variety of techniques. A protocol will usually assign a number or identity to each data segment, which, when received at the destination, identifies when the data was sent (by order). Another method is to identify where in the original block of data the data segment belongs. This is done by providing an *offset,* which identifies to which byte the segment belongs. For example, if a data segment starts at the thirteenth byte of data, the offset would be byte 13. The receiving host knows when reassembling data to place this segment after the twelfth byte.

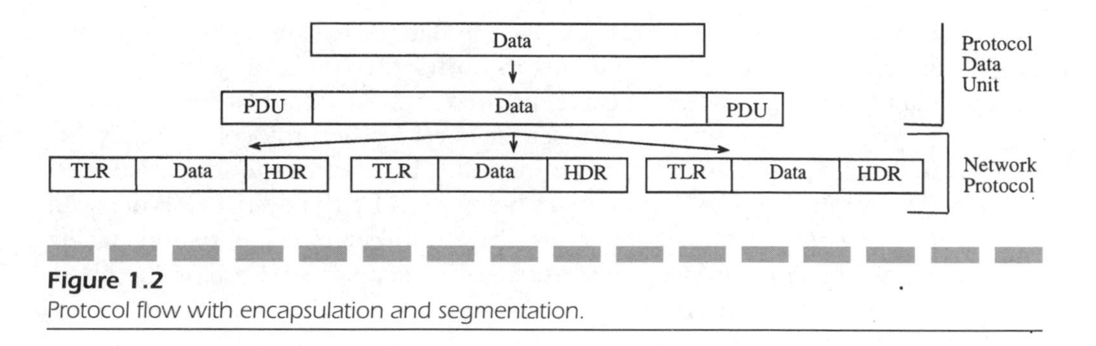

Figure 1.2
Protocol flow with encapsulation and segmentation.

1.4.1.1.2. Encapsulation All protocols perform some form of encapsulation. This is the placing of data into an "envelope" of sorts, surrounded by protocol control information. An example of control information may be source and destination addresses, as well as error-checking data. This forms a *packet,* frame, or protocol data unit, depending on where the data is being processed within the protocol stack.

As the data is passed from one protocol layer to another, additional protocol information can be appended around the existing data, causing the data unit to grow as it passes through each layer. Figure 1.2 illustrates how user data is passed to a file transfer protocol, which adds control information and then passes the data unit to the next layer. The protocol at the next layer appends information necessary for the host to maintain error control and addressing so that the receiving host can route the data to the appropriate application. This information is then passed to the next layer, which encapsulates the original data as well as all of the appended information into its own envelope.

This next layer may add addressing for routing through a network. In addition, error control information is needed as well as sequence numbering (for reliable data transfer). This information is then passed to yet another layer, and so on. As one can see, as data is passed through the various layers and finally makes its way onto the network, the size of the data unit grows.

1.4.1.1.3. Connection Control Connection control is found in connection-oriented protocols only. It is the responsibility of the protocol to first establish a connection with the destination. This is not a physical connection, but a logical one. This is accomplished by sending a variety of protocol messages to the destination host and waiting for acknowledgments.

Once these messages have been acknowledged, user data can begin flowing through the network. As the data is received, the destination sends periodic acknowledgments to notify the originating host that the data has been received. This is explained in more detail later.

When data transfer is complete, it is up to the protocol to notify the destination of connection termination. This means the logical connection is released, and the resources being used by the destination host can now be used for another *session*. In circuit-switched networks, where physical connections are established, this means the release of the physical link between devices.

1.4.1.1.4. Ordered Delivery This is accomplished by numbering each data unit as it is passed to the network. The receiving host then keeps track of each of the *sequence numbers* as they are received. When sending acknowledgments, the receiving host identifies the sequence numbers it has received.

Most protocols do not require an acknowledgment every time a data unit is received. Instead, the receiving host waits until after several data units have been received. It then sends one acknowledgment for all data units received. This means that both the receiving host and the originating host must maintain buffers.

A transmission buffer keeps all data units which have been transmitted until they are acknowledged. Once they have been acknowledged, they can be dropped from the buffer. If a retransmission is requested, the originating host retransmits everything in its transmission buffer.

A receive buffer is used to store all data units until they can be processed. As resources become too busy to handle the received data units, congestion occurs, and buffer overflow causes errors. This buffer is also used to collect associated data units, which are those that have been segmented and must be reassembled by the receiving host.

1.4.1.1.5. Flow Control Flow control is important at all layers of the protocol stack. Think of a printer. As the memory (receive buffer) becomes full, the printer must be able to notify the sending host that congestion has occurred and that it should wait before sending more data. The same is true within a network. Protocols must be able to control the flow of data through the network to prevent errors.

Some protocols use the sequence numbers for flow control. Others add an additional parameter which indicates how many data units can be received before an acknowledgment is required. In all cases, special messages can be used to stop the flow of data units in the event the receiving host is no longer capable of processing data units.

1.4.1.1.6. Error Detection and Correction Error detection and correction is another process provided by protocols. Sequence numbers are added to the header of a packet and are used to ensure ordered delivery. This allows the receiver to determine whether the data units have all been received and if they have been received in the same order they were sent.

Also part of the header contains information used to check the integrity of the data received. There are several methods used for checking integrity, but they all use the same basics. An algorithm is run before the data is actually transmitted, and the results are placed in the header. When the data has been sent, the same algorithm is run again, and the results checked against the value placed in the header.

In both cases, when an error is detected by the protocol, it cannot be fixed. Instead, the protocol discards the data unit and returns an error message to the sender. Depending on the protocol, this message is treated as a request for retransmission. In some cases, the actual message is called a retransmission request, while in others the message is simply an error message.

1.4.2. Layering and Its Advantages

We have already discussed some of the benefits of layered protocols. Each layer can be independent, transparent to the layers above and below. The only requirement is that the interface used to communicate between the layers must be compatible with the layers.

The operating system provides the interface between layers, which helps provide some standardization. In the upper layers, a proprietary interface is typically used to support the specific functions of the protocol. The lower layers can be connection-oriented, while the upper layers are connectionless (and vice versa). By allowing the network layer to be connectionless, data units can be transmitted over any available route, helping prevent congestion over any one link.

In a connectionless protocol, data units are sent with all of the information necessary to process the data unit when it is received. There is no guarantee that the data unit will actually be received by the destination since these protocols do not use any sequence numbers and do not provide ordered delivery.

In a connection-oriented protocol, a request must be sent first. The request is used to ensure that the destination has the necessary resources to process the data unit once it has been received. The protocol will also support sequencing, to ensure ordered delivery. Connection-oriented protocols typically establish a session with the application at the destination, and they maintain that session until the data transmission is complete. This may require transmission of many data units, depending on the size of the user data.

It should be noted here that a session is a logical connection, not a physical one. This is controlled by software, with interaction by the operating system. A host may support multiple sessions simultaneously, depending on the operating system and the platform (hardware).

In a connectionless protocol, as mentioned before, no session is requested, and sequenced delivery is not guaranteed. This makes connectionless protocols less reliable than connection-oriented ones. When using highly reliable transmission facilities (such as fiber optics), this is not usually an issue.

In many network protocols, the lower-layer protocol is connectionless, while the upper-layer protocols are connection-oriented. This shifts the responsibility of higher processing to the hosts rather than to the network devices, lowering the cost of the network devices (such as routers) and providing faster and more efficient transmission.

1.4.3. The Open System Interconnection (OSI) Model

The OSI model was first released in 1984 and serves as a model for all new protocols. The OSI model divided the functions of network communications into seven layers. Each layer communicates with the same layers in the various devices along the network (peer-to-peer communications). Some devices only operate at specific layers (such as a network router, which supports layers 1 through 3). These devices have no visibility to the data contained in the upper layers, and they treat the data as user data.

Different protocols can be used at each layer, providing a service to the layers directly above. For example, data can be transmitted over an X.25 network, which provides the functions of layers 2 and 3. A protocol such as TCAP (an SS7 protocol) can use this network to send information used by cellular or telephone switches.

It is also possible to encapsulate a protocol into another protocol that provides the same services at the same layer. For example, X.25 can be encapsulated into ATM, even though both provide the same services (understand that ATM may support some "features" not supported by X.25, and vice versa). In this discussion, services are protocol services such as ordered delivery and error detection and correction. This is often the case when two networks are geographically separated, and a different type of network is being used to interconnect the two. When this is the case, the "bridging" network is transparent to the two interconnected networks.

1.4.3.1. The OSI Layers Layer 1 is referred to as the physical layer. This layer converts the data unit, including the protocol headers, into an electrical or optical signal for transmission over the network. It does not add a significant amount of overhead to a transmission (except in the case of the Synchronous Optical Network, or SONET). Examples of layer 1 include T-1, Switched 56, and SONET.

Layer 2 is referred to as the data link layer. It provides error detection and correction (to a limited degree) and controls node-to-node communications. Consider this analogy. When traveling across the United States, you contact a travel agent to arrange the trip. The travel agent must coordinate the trip from the origination point to the final destination.

The airline coordinates the trip from one network to another network (including transitions between networks). It is not concerned with the ultimate destination, only the entry points (origination airport to destination airport and any intermediate airports). Likewise, the pilots on the various flights used do not know the ultimate destinations of their passengers.

The pilots in the above analogy are only concerned about one leg of the trip. They route the passengers from one airport to the next one (node to node). This is the same as layer 2. Layer 2 protocols control data units transmitted from one node to the next but not beyond that point.

Layer 3 is referred to as the network layer. Using the above analogy again, the airline is responsible for transporting the passenger from the origination to the final destination airport. However, the airport is not responsible for rental cars or hotel accommodations (at least mine is not). This is analogous to the network layer, which moves data units from one network to another, routing data through intermediate networks if necessary.

Layer 4 is referred to as the transport layer, which is analogous to the travel agent. The travel agent takes care of the details once travelers have reached their final destinations, arranging for rental cars and hotel accommodations. Layer 4 controls data from the origination host to the destination host.

Layer 4 is resident in host software and does not interact with network devices. The originating host software communicates with the destination host by sending protocol messages and negotiating sessions. The data unit contains control information sent by the software that is providing the layer 4 functions. An example of a layer 4 protocol is TCP.

Layer 5 is referred to as the session layer. It is the responsibility of this layer to maintain a dialog with the destination host application in a connection-oriented protocol. In the days of mainframes, the session layer was what started the session (remember logging in?) and then releasing the session when the dialog was through. A dialog is the communications between two entities. This layer is only needed when connection-oriented services are provided.

Layer 6 is referred to as the presentation layer. There is not a lot of use for this layer today. Remember that OSI was developed during the mainframe era, where terminals communicated with mainframes. Terminals do not have any processing capability, so a protocol was needed to identify how text should appear on the terminal screen (graphics could not be supported on terminals, at least not as we know them today). The session layer was responsible for providing code and character set translation (such as ASCII or EBCDIC) and is now used for compression and encryption.

Layer 7 is referred to as the application layer. This layer provides management functions to support distributed applications. This is also the interface to user applications where network communications are necessary. For example, an e-mail application will need to communicate with layer 7 to prepare an e-mail for transmission over the network.

In summary, the first three layers are used for network communications. The functionality of these first three layers can be found in the various network devices, such as routers and gateways. Layer 4 provides reliable connections regardless of what the lower three layers provide.

An example of how layer 4 provides reliable connections is in the case where the lower three layers are providing connectionless services while layer 4 is providing connection-oriented services. Remember also that layer 4 resides in the hosts, not network devices.

The upper layers are involved with the exchange of data between users, determining how the data should be displayed and managing the dialog between two hosts. These upper layers are also host-resident.

It should be noted that in a LAN, only the first two layers are needed. LAN protocols do not provide the services of the upper layers, although another protocol can be used over a LAN to support these services. We will discuss LAN protocols in more detail later.

1.5. Networking Fundamentals

All telecommunications today rely on communications between computers and computer peripherals. These systems must be able to communicate with one another and share data regardless of location and system type. The challenge we face as developers of new technology is standardization.

The second challenge faced by developers is interoperability. It would seem that if one were to use a standard network interface, unlike devices could communicate and interact with one another. However, there is much more to network communications than just passing data from one computer to another. Network management procedures must be compatible, and communications links must work the same way on both ends of a connection.

To understand these challenges, and to understand what telecommunications is all about, you need to understand what types of networks exist today and how they evolved. In this next section, we will talk about the different types of networks, how they evolved, and what lies ahead.

1.5.1. Evolution to Distributed Processing

In the early days of computing, the processing power was placed in a central processing unit, the mainframe. There are many advantages to centralized processing. Maintenance is easier, as is standardization of applications. Cost is usually higher, however, because of the investment required to purchase and maintain a mainframe.

Users communicated with the mainframe using "dumb" terminals. They were called dumb because they were not capable of running applications themselves. They were very inexpensive (around $250 will buy a good terminal even today), and they were reliable. However, terminals could only communicate with mainframes at low speeds. The terminals could not be connected directly to the mainframe because mainframes only support a limited number of ports. Controllers were connected to the mainframe and acted as a "traffic cop," managing the time-sharing of the processors in the mainframe. All terminals and printers were connected to the mainframe through a controller.

Communications to the controller was often limited to 9600 baud. This would be unacceptable by today's standards in many applications, especially when LANs are capable of communicating to servers at speeds up to 10 megabits per second (Mbps). The more users logged onto the same controller, the slower the communications to the mainframe.

If the controller were to fail, all devices attached to that controller would lose communications to the mainframe. This is critical in many operations where reliable connections are paramount. When all of the applications needed by users are resident on a centralized processor, and that processor cannot be reached, there is no productivity.

Despite disadvantages of mainframes, there are many advantages. Software is standardized. Everyone accesses the same program, making version control very simple. When a program is updated on a mainframe, it is updated for all of the users.

Files are stored in one central location, available for all users on the network. This is important where databases are used. Large databases require large storage devices, which have not been affordable for desktop users until recently.

When personal computers first became available in the 1970s, they were slow and limited in capability, and applications were not attractive. The PC market was targeted at the technically inclined because of the expertise needed to use them.

As PCs became more powerful, applications became more user friendly. Microprocessor and memory prices fell as the demand for PCs increased, which began a revolution in the computing industry. Soon, PCs started making their way to the corporate desktop.

The first hurdle of the PC industry was to create a technology that would allow PCs to talk with one another. They could be connected to the mainframe of course, but in such an arrangement, PCs could only be an expensive terminal. Users wanted more than that; they wanted the ability to run applications of their choice right on their desktop. Development soon began on ways to tie computers together.

The birth of the LAN created a whole new industry. Finally, PCs could be linked together over a common network, and users could share files with one another. It was then that users began to recognize the disadvantages of distributed processing. Applications had to be purchased in multiples. No longer could users all share the same application. They now had to purchase individual copies.

PCs were still expensive in comparison to mainframe terminals. The cost per employee continues to rise today as we demand more processing power from our computers. The average cost per employee today is around $3000 for a simple computer. As the tasks increase in complexity, so do the processing requirements and the cost.

Files could be shared, but there was no longer a centralized processor for storing databases. There was still a need for one central processing unit to store and retrieve databases. This lead to a new computer network environment known as client/server.

1.5.2. Client/Server Environments

In the client/server network, the PCs on the network use client software which allows them to access specific programs on the server and run them on their desktops. For example, to access a database on the server, users load client software on their PCs; the software provides the functions necessary to communicate with the server, query the database, and display the information in a predefined format.

On the server, software is used to receive queries from the client, search the database for the queried data, and return the data to the user over the network. The server software also provides version control to maintain the database. This is only one example of many possible uses of client/server networks.

The advent of the client/server network sounds like a retreat back to legacy mainframes. However, client/server networks provide something that mainframes could not. If the server fails, users can still continue to work, as long as they are not reliant on files located on the server. In many client/server networks, files stored on the server are downloaded by the users, updated, and then uploaded back onto the server.

Several companies have introduced "network computers." These scaled back PCs have plenty of memory and processing power, but no disk drives. They run applications and store files on servers. This new breed of computer is an attractive alternative to many companies who must provide expensive computers to all their employees.

1.5.3. The Local Area Network

The LAN evolved out of user demand. Personal computer users wanted the ability to exchange files from one computer to the next, much like they did on mainframes, without the problems they experienced with mainframes.

Early descriptions of a LAN used geography to define its boundaries.

This has grown more difficult as LANs have expanded beyond the walls of an office building to include multiple buildings. The best description of a LAN is a network which connects multiple computers together without using a router. What this means is that all computers are on the same network and do not have to use a layer 3 protocol to route them to another network.

Only one address is needed to route data over a LAN, the machine address (although there is also another address within the LAN protocols to identify the application to receive the data). Routing over a LAN is very simple, as we will find in our discussion below.

1.5.3.1. Services Provided A LAN provides connection to other computers on the same network. Files can be moved over the LAN from one computer to the next at very high data rates. A data rate of 10 Mbps is not uncommon, and in modern networks the speeds have grown to 100 Mbps and higher.

Only layer 2 is needed to move data over a LAN. LAN protocols provide the end-to-end delivery of data over the network, ensuring reliable delivery through error detection and correction and in some cases sequencing.

Network operating systems provide additional services not provided by typical LAN protocols. These operating systems allow a central network administrator to control the various elements of the network, while monitoring LAN performance. If servers are involved, the LAN administrator can configure the network to allow access to the servers by a limited number of computers.

An example of LAN protocols includes Token Ring, Appletalk, and of course, Ethernet. An example of Network operating systems includes Novell and Microsoft LAN. We will discuss LAN protocols in more detail in Chap. 3.

1.5.3.2. Routing Principles Routing over a LAN is simple. Data is encapsulated in a protocol header, which includes the address of the source and the destination. The data is then broadcast over the network. All nodes on the network read the header of the data as it passes over the network.

If a node recognizes the address as its own, it copies the data into memory. The data is left on the network to be absorbed or deleted by the originator or the network, depending on the network topology used (Token Ring or Ethernet). This is explained in fuller detail in Chap. 3.

There are no sophisticated routing algorithms needed for LANs. All nodes are part of the same network. To ensure that each node receives a unique address, manufacturers program their NICs with unique machine addresses. Network administrators need not worry about configuring the machine address on each node.

1.5.4. Wide Area Networks (WANs)—The Outside Connection

To connect LANs together, one needs WAN technology. WANs are used to interconnect LANs using a layer 3 protocol. This also means a router and/or gateway is required. These devices are responsible for routing data between networks.

The geography of a WAN may be within a city, or it may span the globe. There once was a description for Metropolitan Area Networks (MANs), which would cover a metropolitan area, and a WAN, which would interconnect MANs. In today's industry, there is little usage of the term MAN since WANs now span the globe, providing the same types of services.

Layer 4 protocols are also required in most cases to ensure reliable transfer of data. Most WAN technologies today use as little processing as possible, making the network faster and more efficient. The philosophy in today's network is to perform as much processing as possible in the upper rather than the lower layers of the protocol stack. This is possible in part because of better transmission facilities (fewer errors require less error detection and correction).

The best example of this is Frame Relay. This technology provides very little error detection and correction. Either data is acceptable or it is not. There is no retransmission request in Frame Relay. If the data is in error, the network discards it. An upper-layer protocol will take care of retransmission.

A comparison of Frame Relay with X.25 shows a stark difference in the amount of processing required. This is discussed more in Chap. 3 in the discussion about Frame Relay.

1.5.4.1. Services Provided A WAN provides interconnection from one local network to another, sometimes traversing the globe. The types of services supported depends on the protocol used. For example, a Frame Relay network does not provide a lot of error detection and correction. This would not be a good choice for data requiring reliable transfer over copper lines but is very suitable for data sent over fiber optics.

Some other protocols (such as X.25) may be more suitable for data which requires reliable transfer and connection-oriented services. The choice of which technology to use in a WAN will vary depending on the types of data transfer services needed. A WAN may also consist of several different solutions over various parts of the network.

Perhaps the best way to think of a WAN is as a network of networks, interconnected by any number of technologies. A good example of a diverse WAN is the Internet, which spans the globe, connecting many different networks together into one seamless WAN. The Internet uses a suite of protocols called TCP/IP.

The TCP/IP protocol suite consists of over 100 different protocols, each providing a different service to the network and the user. Many of the protocols are used for routing, network management, and file transfer. The user has very little interaction with these protocols, since one of the goals in TCP/IP was to make the protocol suite transparent to the user.

1.5.4.2. Routing Principles In a WAN, routing is a bit different than in local networks. Because there may be thousands of users connected to a WAN, it is more desirable to know only the network address for a particular user rather than the machine address. The devices which route data through WANs are routers. They are called layer 3 devices because they support the services of OSI layer 3.

The router does not care about the machine address, which is identified by the layer 2 protocol (such as Ethernet). When a data unit arrives at a router, the layer 2 protocol is stripped from the data unit, leaving only the upper-layer information. The router will then process the layer 3 portion of the data unit, encapsulate the entire data unit into another layer 2 packet, and send the data unit over one of its ports (determined by the destination network address).

A router may support more than one layer 2 protocol. For example, it is possible to have a router which can be connected to an Ethernet network, a Frame Relay network, and an ISDN network. The router can also support different types of layer 3 protocols, depending on the complexity of the product.

The address contained in layer 3 is more accurately defined as a user address. Some portion of the address must contain the user's unique address, which can be translated into a machine address by local servers. The rest of the address is considered the network address and is used by routers to route the data unit through the WAN.

Routers maintain route lists in memory which contain all of the network addresses known to the router and which ports should be used to reach these networks. Depending on the type of protocol and network being connected to, these routing procedures can be simple or complex.

For example, in a Frame Relay network, addressing is kept simple. Most Frame Relay networks are point to point and do not require any sophisticated routing algorithms. A TCP/IP network in comparison requires a sophisticated routing algorithm and a method of automatically updating all route lists autonomously.

The TCP/IP protocol suite provides several protocols used by routers for advertising their route lists. This allows routers to learn of network addresses without someone typing them in on a regular basis. It also allows routers to maintain their route lists in real time. Not all network protocols support this function.

It is important to understand the difference between routing in a LAN and routing in a WAN. As we discussed above, WANs route data units based on a network address. The user address is included as part of the network address but is not used by the routers in WANs for routing. The user address is used by the destination to determine how to route the data unit over the destination local network.

In the LAN, a machine address is used for routing. Unlike WANs, LANs do not use a routing algorithm to deliver data units; they broadcast the data unit over the entire network. Each node on the network has the opportunity to "read" the header of every data unit transmitted over the network to determine if the machine address matches its own.

1.5.5. Switching Principles

In this section, we will begin to discuss how various types of "switching" work. This applies to both voice and data. There are three types of switching used in networks today: circuit switching, used in voice networks; packet switching, used in data networks; and cell relay, now being deployed for both voice and data transmission.

1.5.5.1. Circuit Switching The telephone network is a circuit-switched network (although there is also a packet-switched network used for signaling data). In a circuit-switched network, a dedicated circuit must first be connected. Once the circuit has been "nailed up," transmission can begin. When the transmission is complete, the circuit is released for the next transmission.

Let us look at a simple telephone call. When you remove the receiver from the telephone and dial a telephone number, the telephone company switch searches its database to determine which circuit should be used to deliver the telephone call. If it is a long distance call, the switch knows it must connect to another telephone company office, where a switch called a *tandem* is located. The tandem switch will then use a circuit that connects it to another office, the toll office switch.

This process continues until there are circuits connected from the originator to the destination. These circuits cannot be used for any other telephone call; they are dedicated to this one call until the call is complete. Once the call is complete, the circuits can then be released and used for another call.

Circuit switching is not an efficient method for routing any kind of data, whether it is digital voice or user data. The circuit is wasted much of the time because no transmission is using the bandwidth of the circuit 100 percent of the time. Any time there are idle periods on the circuit, the circuit is being wasted. It would be much more efficient to have a transmission facility capable of transmitting many different "conversations" over the same circuit at the same time.

This was achieved (sort of) through multiplexing. We have not discussed multiplexing yet, but for the sake of this conversation, we will look at one advantage of multiplexing. A circuit can be divided into channels, with each channel used for a transmission. Digital telephone circuits are multiplexed and are capable of transmitting several different conversations at the same time on the same circuit.

There is one catch; each channel then becomes dedicated to the conversation until the caller disconnects. Then the channel can be released for another transmission. So in the case where there are 24 channels (the common denominator in today's digital facilities), there can be 24 different conversations going on at once over the same facility.

This is better than wasting the circuit for one transmission, but it could still be better. Imagine having no channels. Transmissions are sent over the same circuit as needed, but there are no limits to the number of conversations that can be sent over the same facility at the same time. We will discuss this possibility a little later.

1.5.5.2. Packet Switching In packet switching, there are no dedicated circuits. Each circuit in a packet-switching network carries many different transmissions at the same time. The only rule is that every data unit sent through a packet-switching network must have enough information in the header that the nodes in the network can determine how to route the data unit. This tends to add overhead to the data unit, but the trade-off is well invested.

Another advantage of packet switching is the ability to route data units over any route, rather than a fixed route. For example, if I have a lot of data to send, the data will have to be divided into many different data units. These data units do not have to follow the same route in a packet-switching network.

The trick is being able to place the data units into the proper order when they are received. If data units are routed over different paths, it is highly likely that the first data unit may be received after subsequent data units, which means the order of transmission is now mixed up. The protocols used in packet-switching networks have the ability to reassemble the data units into their proper order.

Packet switching is favored over circuit switching for many different reasons. It is more reliable than circuit switching because if a particular circuit in the network should fail, the routers in the network simply route data units over different circuits, taking a different route altogether. In a circuit-switched network, this is not possible. If a circuit fails in the middle of a transmission, the entire connection must be released and a new one established, which means the conversation must start over again (think of being disconnected from a telephone call; the whole process of connecting must be repeated).

Packet switching is not new. The industry recognized the need for a more efficient way of transmitting data over long-haul networks and deployed the first X.25 networks in the 1960s. These packet-switching networks were used by many corporations for years, and many still use them today. Many corporations are looking toward the Internet and a packet-switching network using TCP/IP as their WAN solution.

1.5.5.3. Cell Relay Cell relay is a newer approach to data networks. Standards are now being completed for ATM, which is a cell relay technology. The concept is simple. Rather than use channels as we discussed earlier, use a transmission facility capable of supporting millions of transmissions at one time. In reality, data is still sent in serial fashion, but at such high data rates that it appears to be simultaneous.

ATM has not been deployed as a large network yet. Telephone companies are offering ATM services in select areas, as they carefully evaluate its performance. Soon, ATM will replace all of the circuits in the telephone company's networks, supporting everything from data to voice and video. All of this transmission will take place over high-speed fiber optics (already in place in most metropolitan areas).

The *cell* is the data unit. In order to support voice, the data units must be small so that they can be processed quickly and sent through the network with minimal delay. This is not true with data, which favors large data units. We will discuss the principles of data unit sizes and transmission delays further in Chap. 9. For now, understand that voice requires small data units, and data favors large data units.

In a cell relay network, the facility is used when needed. Whenever there is information to be transmitted, the switch simply sends the data units. There is no need to negotiate for a connection (as is the case in circuit switching), there is no need for a channel to be allocated (there are no channels in ATM), and as long as there is enough bandwidth to support it, there can be unlimited transmissions over the same facility.

The move to cell relay has been a long one. Development first began on this concept in 1969 at Bell Labs and has continued ever since. The development of industry standards is taking even longer. It will not be until the end of this decade that widespread deployment of ATM will be near completion.

1.5.6. Chapter Test

1. Alexander Graham Bell receives most of the credit for inventing the telephone, but there were two other major contributors. Who were they?
2. The transistor was invented by whom?
3. What is the national standards body for the United States?
4. What organization represents the United States in the International Telecommunications Union (ITU)?
5. A de jure standard is:
   a. A type of soup
   b. A standard caused by monopoly
   c. A standard created by a committee
6. The ATM Forum is responsible for creating international standards for ATM.
   a. True
   b. False
7. How many bits does ASCII use to represent a letter?
8. How many bits does BCD use to represent a single digit number?
9. What is the number 7 in BCD?
10. What is the function of layer 2 in the OSI model?

11. A router is considered a layer______device.
12. Connectionless protocols provide sequencing of messages and guaranteed delivery.
    a. True
    b. False
13. Connection-oriented protocols require a dedicated circuit for data transfers.
    a. True
    b. False
14. A LAN is any network which:
    a. Is contained within one building
    b. Connects all computers in the network over the same transmission facility
    c. Uses a layer 2 protocol, with no routers
    d. Spans the globe, providing interconnection between many different networks.
15. A WAN is any network which:
    a. Is contained within one building
    b. Connects all computers in the network over the same transmission facility
    c. Uses a layer 2 protocol, with no routers
    d. Spans the globe, providing interconnection between many different networks
16. Which of the following best describes the method of switching where packets are routed based on the addressing in their headers and may take any route depending on network status?
    a. Circuit switching
    b. Packet switching
    c. Cell relay
17. Which of the following best describes the method of switching which requires a dedicated circuit to be established end-to-end before transmission can begin?
    a. Circuit switching
    b. Packet switching
    c. Cell relay

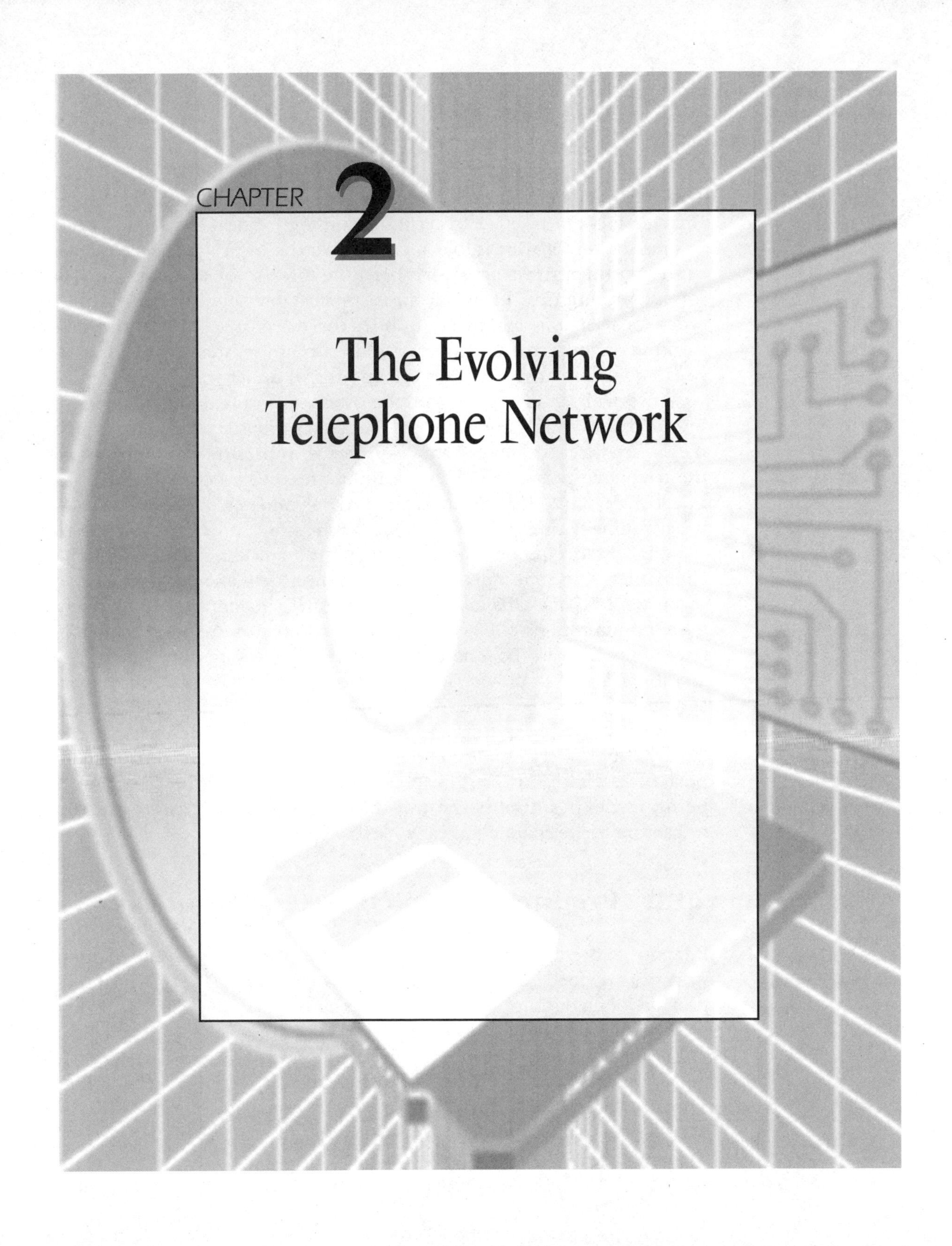

CHAPTER 2

The Evolving Telephone Network

2.1. The Infrastructure

The telephone network as we know it is going through a lot of change. Change does not take place overnight, and a great deal has already been in preparation for what is to come. Probably the most significant change in the existing infrastructure has been the move to fiber optics. The reason for making this change involves a much more sophisticated plan. In order for telephone companies to offer the many new services they are working on today, the underlying infrastructure must change. This plan began deployment several years ago and is still under way today.

As we will see in this chapter, the divestiture caused another change in our nation's telephone network. The segregation of telephone companies as separate businesses had a significant impact on the topology of the telephone network. Suddenly, calling areas were being divided into Local Access and Transport Areas (LATAs), and telephone companies were finding their existing service areas were now under different rules.

Long distance service was no longer the market for local telephone companies, who were now forced on finding new avenues for creating revenues. In 1996, this changed again with the signing of the new telecommunications bill. Local telephone companies are once again able to seek revenues in the long distance market, and long distance carriers are now allowed to provide local telephone service in any of their market areas.

All of this change affects the telephone network. As new services are deployed by the telephone companies, the network must change to support these services. In this chapter, we will look at all of the changes being made and discuss the impact on the network, as well as the impact on the telephone industry itself.

2.1.1. Predivestiture Bell System Networks

Previous to the divestiture of the Bell System in 1984, the telephone network was divided into two separate business units. AT&T Long Lines provided long distance services to all of the Bell System Operating Companies, as well as the independents.

Local service was provided by 20 telephone companies, once operating as independents and bought out by the Bell System early in its life. These 20 companies and AT&T formed what was known as the Bell System. Together, they were able to provide complete telephone services to the nation.

In addition to the Bell System, there were still a large number of independents who provided telephone service in those areas which did not interest the Bell System. When the Bell System was busy buying out other independents, it elected to leave out some more remote areas, concentrating on the areas which were dense in population and provided more revenue (see Table 2.1).

TABLE 2.1

Predivestiture Bell System Operating Companies

Operating Companies	Operating States
Bell Telephone Company of Pennsylvania	Pennsylvania
Diamond State Telephone Company	Delaware
The Chesapeake and Potomac Telephone Company	D.C., Maryland, Virginia, West Virginia
Cincinnati Bell	Ohio, Kentucky
Illinois Bell Telephone Company	Illinois, Indiana
Indiana Bell Telephone Company	Indiana
Michigan Bell Telephone Company	Michigan
Mountain States Telephone and Telegraph Company	Arizona, Colorado, Idaho, Montana, New Mexico, Texas, Utah, Wyoming
New England Telephone and Telegraph Company	Maine, Massachusetts, New Hampshire, Rhode Island, Vermont
New Jersey Telephone Company	New Jersey
New York Telephone Company	New York
Northwestern Bell Telephone Company	Iowa, Minnesota, Nebraska, North Dakota, South Dakota
Ohio Bell Telephone Company	Ohio
Pacific Telephone and Telegraph Company	California
Bell Telephone Company of Nevada	Nevada
South Central Bell Telephone and Telegraph	Florida, Georgia, North Carolina, South Carolina
Southern New England Telephone Company	Connecticut
Southwestern Bell Telephone Company	Missouri, Oklahoma, Texas, Arkansas, Kansas, Illinois
Wisconsin Telephone Company	Wisconsin

In addition to the operating companies, the Bell System also consisted of other business units. Western Electric became the manufacturing unit for all of the Bell System. There was a time when the only types of telephones which could be connected to the Bell System network were those created by Western Electric. This later changed, of course, after the Carterfone Decision.

Bell Laboratories provided research and development for all of the Bell System. There have been a great number of inventions from the Labs. The transistor was one of the greatest accomplishments from the Labs, starting the evolution of the microcomputer industry. Of course, many more great inventions have come out of the Labs, all of which were available to the industry for a nominal licensing fee.

Some independents have grown quite large because the areas that were once remote have become large metro areas as the population spread from the inner cities to sprawling communities. General Telephone (GTE) has been providing service in the Ventura area of California for many years. In the last 20 years, there has been a lot of growth in this area, and today they have some of the highest-revenue-generating markets in California, including Orange County.

The Bell System network was a bit more sophisticated than most, offering many levels of redundancy and diversity. Figure 2.1 shows the hierarchical approach taken by the Bell System in their network design. The concept was simple: provide multiple paths for all transmissions in the event a particular path became congested or unavailable. This hierarchy was developed to economize on trunking. This was accomplished by connecting end offices to tandem switches, which act as hubs to the

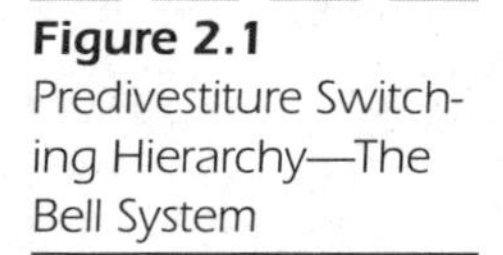

Figure 2.1 Predivestiture Switching Hierarchy—The Bell System

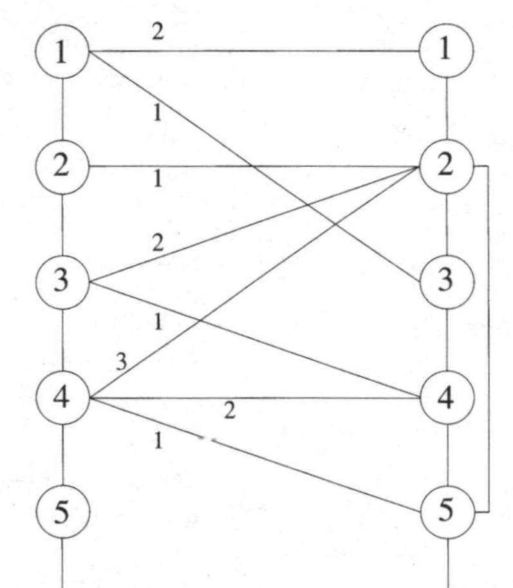

local network. The tandems then provide high-capacity trunks to other tandems, making better usage of facilities.

Each office type was given a number. Class 5 offices are end offices, which provide local telephone service to subscribers. They may be collocated with switches of a higher classification, such as a class 4. The class 5 office connects through a class 4 office, which is a tandem switch (also referred to as the toll center). The idea is to have all of the class 5 offices within a geographical area connect to a class 4 office in the same area. The class 4 office was also the point at which operators assisted in outbound calls (connecting to other networks).

The class 1, 2, and 3 offices were referred to as control switching points. Their job was to provide diversity to the long distance (toll) network. Each of these offices had to have at least one office of higher or lower ranking "homed" to it, which means there was a path for calls to get to this office.

Alternate routing was also provided by this network approach. As shown in Figure 2.1, if the class 5 office could not connect to another class 5 office through its local class 4 office (tandem), the call would be routed through a class 3 office, which could then connect back to the class 4 office or to a class 2 office in the other network.

This plan was ditched after divestiture in favor of a more robust plan. The hierarchy was flattened to eliminate intermediate switches.

2.1.2. Postdivestiture Bell System Networks

After the divestiture of the Bell System, many changes began to take place. AT&T became AT&T Long Distance and was no longer a part of the Bell System. This in itself created some problems because much of AT&T's equipment was collocated within Bell System central offices. With the divestiture, the employees from the local telephone company were no longer allowed to work within the areas of AT&T equipment.

In fact, yellow lines were painted in central offices to indicate where the AT&T property started. Local telephone company employees were not allowed to cross over the yellow lines, and AT&T personnel were not allowed to cross over onto the local telephone companies side. If this sounds ridiculous, it is not near as silly as trying to navigate your way through a central office without ending up in AT&T territory. Today, most of AT&T's equipment is within its own areas, and the lines have disappeared in many central offices. I mention this here so you can understand the difficulties faced by the Bell Operating Companies in implementing the many changes that divestiture brought about.

Western Electric, which manufactured all of the equipment for the local telephone companies, became a part of AT&T. The divestiture prevented local operating companies from manufacturing their own equipment and forced them to look for outside vendors to manufacture their equipment. The intent was to open up competition in the industry and force the Bell companies to use other vendors instead of monopolizing the market. Bell Laboratories also went to AT&T.

In 1996, AT&T underwent another restructuring. This time, AT&T divided itself into three different companies. All three companies operate independently of one another. The new companies are AT&T, Lucent Technologies, and AT&T Global Information Solutions.

AT&T consists of the Communications Services Group, providing long distance services. AT&T Universal Card Service continues to operate under the AT&T name as well, providing credit card services to long distance subscribers. AT&T Solutions provides consulting and systems integration, while AT&T Wireless provides cellular communications.

Under the Lucent Technologies name, there are a number of entities. AT&T Network Systems develops telephone switching equipment and transmission equipment for the telecommunications industry. AT&T Global Business Communications Systems provides business switching systems (Private Branch Exchanges, PBXs) around the world. AT&T Consumer Products provides consumer-level communications equipment such as telephones and answering machines. AT&T Paradyne continues developing modems, while AT&T Microelectronics provides electronics components and photonic components.

AT&T Global Information Solutions is an independent consulting company for transaction intensive systems. This includes retail, financial, and communications markets.

As a result, the operating companies were forced to create a new organization for research and development. All of the operating companies contributed heavily to this new organization, called Bell Communications Research (Bellcore). There are a lot of questions today regarding the value that Bellcore adds to the local operating companies. Many of the companies feel uncomfortable sharing their trade secrets with other operating companies since they are all now competing against one another.

Bellcore recently went up for sale, in hopes of making it a stand-alone company. The future for Bellcore is not clear, although they have begun making many organizational changes in hopes of surviving on their own. One of the changes made at Bellcore is their new consulting services, providing all of the industry with network consulting and reliable testing for manufacturers of telecommunications equipment. Certification by Bellcore is an important feature of any telecommunications equipment, and Bellcore is capitalizing on this by providing testing services to all manufacturers.

The local telephone companies were divided into seven regional operating companies. These regional operating companies are:

Pacific Telesis
US West
Southwestern Bell Telephone
Ameritech
Bellsouth Telecommunications
Bell Atlantic
NYNEX

There are a number of operating companies which are owned by these Regional Bell Operating Companies (RBOCs). However, in recent years the RBOCs have begun changing their identity, as well as the identity of their member operating companies. It was once thought that the use of the Bell logo was important to success, but now the RBOCs have begun dropping the Bell logo in favor of assuming their own identity, and the operating companies have also changed their identity to that of the RBOC.

To further compound the problem of identity, some of the RBOC's have begun mergers (recently, Southwestern Bell merged with Pacific Telesis), making it difficult to tell who is in charge, and broadening the subscriber base of both companies. If these mergers continue, there will be one huge telephone company again, the same as before the divestiture of the Bell System. The ironic part is that the same companies will have once again joined forces to create one big monopoly.

The long distance market has become a mass of providers, many of whom lease facilities from other long distance companies and resell the services. The signing of the telecommunications bill in 1996 has brought about many different changes in both the local and the long distance telephone markets.

With local telephone companies now able to provide long distance services and the long distance companies now able to provide local services, there will be a lot of confusion over who owns what network. In most cases (for the exception of AT&T, who has had a nationwide network all along), long distance companies will have to buy access from local telephone companies, cable television companies, and even electric utilities to make their local networks possible.

The local telephone companies will have to lease facilities from long distance providers for the same reason. This of course creates a whole new revenue opportunity for both the RBOCs and the long distance providers.

2.1.2.1. New Switching Hierarchy Figure 2.1 showed the Bell System network predivestiture. One of the changes made in many of the RBOC networks was the hierarchical change to their switching network. The hierarchy was made flatter, with the addition of a Point of Presence (POP), to facilitate equal access to the local network by all long distance carriers (interexchange carriers, or IXCs).

As shown in Figure 2.2, the end office still connects to a tandem switch (Access Tandem), which acts as a hub. The Access Tandem (AT) is used to connect to other end offices within the same LATA. This replaced the class 4 office and connects all of the end offices within a LATA. However, the end office also has the ability to bypass the tandem and connect directly with the POP for calls handled by the IXC. This eliminates unnecessary traffic through the tandem.

If a connection must be made outside of the LATA, the call is routed through the POP to the IXC network. Choice of which IXC to use depends on the caller's preference, which is stored in a database in the SS7 network and accessed by local switches. The Tandem Office (TO) centralizes switching functions within a network. Notice that the end offices have a connection to a TO as well as to the AT.

2.1.2.2. Local Access Transport Areas After divestiture, what was once calling areas defined by the various telephone companies became redefined as LATAs. The division between LATAs was defined by the courts in an effort to provide all telephone companies with equal and fair territory.

The decision of where to place LATA boundaries was based on the demographics of the area as well as on revenue potential. All LATAs are supposedly equal in revenue potential, creating a fair market area for each of the operating companies and independents. Investments already made by the telephone companies were also taken into consideration since many telephone companies had already made significant invest-

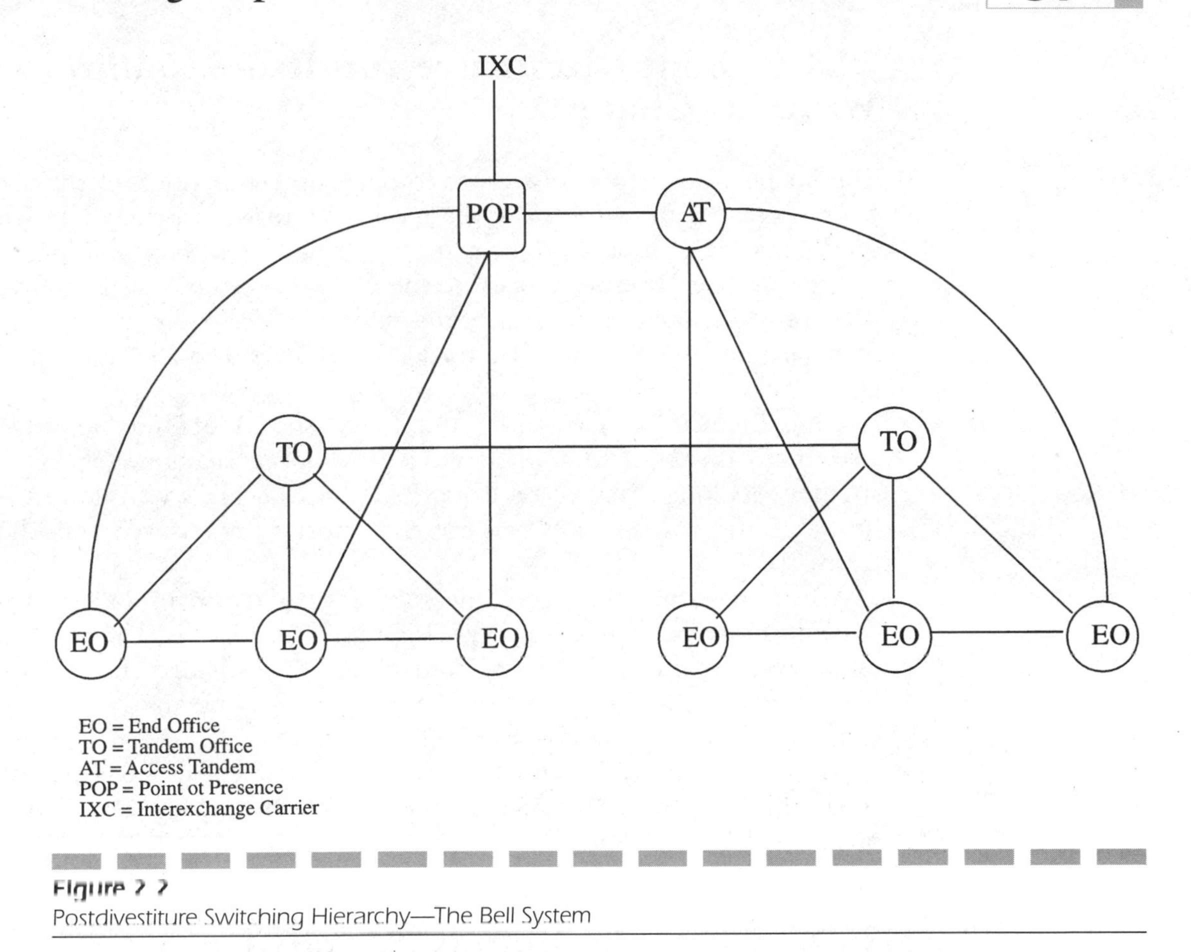

Figure 2-2
Postdivestiture Switching Hierarchy—The Bell System

ments in their networks and could lose those investments to another company.

There are some 164 LATAs defined today. Until the signing of the telecommunications bill in 1996, a local exchange carrier (LEC) could not carry calls outside of its LATA, even if it provided service in more than one LATA. Instead, the call had to be routed through an IXC.

After the telecommunications act of 1996 was passed, the LECs were allowed to provide services outside of their LATA, and the IXCs were allowed to provide services within a LATA (local telephone services). This bill opens up a lot of competition in local and long distance telephone services and provides an opportunity for cable television companies and power utilities to start using their networks for telephone services.

2.1.3. About Divestiture and Its Reasoning—Winds of Change

The purpose of the divestiture was to open up competition in the telephone industry by eliminating the monopoly that the Bell System had on the market. Competition for manufacturing of telephone equipment opened up after divestiture because the Bell System was forced to divest itself of its various entities and begin buying services from other equipment vendors (Western Electric no longer belonged to Bell Operating Companies).

Long distance competition became fierce since AT&T was no longer able to rely on the revenues it received from local telephone services to augment its long distance network. Likewise, the LECs were forced to become profitable quickly because they no longer received subsidies from AT&T.

All of the companies faced massive reorganization in an effort to become more profitable quicker. This reorganization is still evolving today, as the telephone industry continues to face changes. It is highly likely that the IXCs will begin making deals with one another as they compete against the LECs for local telephone service.

In dividing the LECs into LATAs, the courts created a fair playing field. This was carefully designed to ensure each LEC was given equal opportunity to compete.

2.1.4. New Telecommunications Law

In 1996, new legislature changed the way telephone companies and cable companies do business. Challenging earlier telecommunications legislation, this new bill passed the House of Representatives and the Senate with much debate. Its passing allowed long distance telephone companies to provide local telephone service to any subscriber in any area.

This bill also allowed local telephone companies to enter into the long distance business, previously prohibited by the divestiture. In addition to allowing the Bell Operating Companies to provide local and long distance service, the new legislation also allowed cable television companies to begin offering telephone service.

The intention of this bill was to open the industry to competition and to allow telephone service providers to offer more competitive services. It has opened the door to many exciting innovations. We will begin seeing the impact of this bill over the next few years.

The changes which will affect telephone companies the most are changes to the way telephone companies charge one another for access to each other's networks. Many of these access charges have been alleviated, forcing telephone companies to provide access to their databases and network components at no charge.

Another mandate issued by the FCC as part of this new legislation is something called Local Number Portability (LNP), which means subscribers can shop around for the cheapest local telephone service, without changing their telephone number. The mandate goes further to say that subscribers must be able to move from one geographic location to another and keep their telephone number.

This represents big changes in the way calls are routed through the network and how telephone numbers are allocated. Presently, a block of telephone numbers is assigned to each telephone company, and these numbers are further divided and assigned to specific central office switches.

Under the new changes, this practice will no longer work. Telephone numbers will have to the allocated by a central authority (still undecided) on an as-needed basis. Area codes will no longer indicate which city or state a caller is calling from, since the number may have been assigned in one state and then ported to another location when the subscriber moved.

The support of LNP will represent enormous costs to telephone and to cellular companies. Their equipment must be upgraded, and new databases will have to be deployed to support this feature. Many telephone companies are protesting the implementation of these changes, especially the time allotted for the mandates to be implemented. LNP must be implemented in the top 100 market areas by the year 1997.

2.2. The National Information Infrastructure

During the presidency of Bill Clinton, a new term was cast. The Information Highway became the new "buzz word" that would captivate our industry for the next decade. Never before has so much interest been shown in data networking by politicians around the nation. It has become a race to see who could build the best and most powerful network, *best* being the network which could provide the most services to the largest community of users and *powerful* being the network which could deliver the most information to all neighborhoods, regardless of financial status or education. The White House began a campaign to create a nationwide network which would reach into every city, every school, and every neighborhood. This network is known as the National Information Infrastructure (NII).

It is somewhat ironic that this network has been under way for many years. The telephone companies have been slowly deploying fiber optics and digital transmission facilities throughout their networks in an effort to provide new services to their subscribers. What the politicians have provided is new emphasis on the purpose and objectives of this network, and they have escalated the completion schedule.

This campaign also served to promote the Internet, a network already in place but somewhat obscure from the public eye. What was once reserved for research and science is now a part of everyday American life. It is important to understand that the Internet is not the NII, and while the Internet will continue to play a major role in the Information Highway, it is not the solution to the problem. The Internet and the technology it uses (Transmission Control Protocol/Internet Protocol, or TCP/IP) cannot support real-time applications such as voice and video in suitable fashion.

One of the most important roles that Congress has played in this campaign has been the signing of the telecommunications bill, which challenges the separation of cable television, telephone, and computer networks and provides for the opportunity for all of these to be joined together into one cohesive network.

2.2.1. The Objective

The objective is fairly comprehensive. Americans should be able to live where they want to without consideration of job opportunities. This would be accomplished through telecommuting over a nationwide network capable of supporting high-speed data, voice, and video.

Schools would become available to everyone, regardless of where they lived. By attending via on-line "classes," students would no longer be forced into situations such as segregation or attendance in a low-income area at a school deprived of more advanced programs.

Social and health services will be made available over the network, accessible by anyone owning a computer and a modem. This means that hardware and software will have to be affordable to everyone. This is the challenge to computer vendors, who are eager to participate in this program.

The government's role is to promote investment by the private sector using such vehicles as taxation and regulatory policies. At the same time, it will ensure that all Americans have equal access to this network at affordable prices. To accomplish this, the government will assist companies developing new technologies to be used in this new network by providing government grants and research programs dedicated to the Information Highway.

The government has already begun providing access to government records, government agencies, and Congress. The White House has its own Home Page on the World Wide Web (WWW), congress has provided a list of e-mail addresses for representatives and senators, and the federal government is now mandating that all companies which do business with the U.S. government be able to access the procurement agencies of the government through the Internet.

President Clinton has already formed the Information Infrastructure Task Force (IITF) to oversee the activities of the government in this project. The IITF is advised by the Advisory Council on the NII, a private sector committee. This committee consists of 25 members named by the Secretary of Commerce in December of 1993.

To ensure the reliability of the network, the Federal Communications Commission (FCC) has extended the charter of the Network Reliability Council (NRC), which oversees the nation's telephone networks and reports all network outages to Congress and the industry. The NRC was established after a software bug in switching software caused several major network outages in New York, Washington D.C., and California in 1992.

2.2.2. The Promise of Equal Access to All

The "National Information Infrastructure: Agenda for Action," (accessed over the Internet) says it best. "The NII can transform the lives of the American people—ameliorating the constraints of geography, disability, and economic status—giving all Americans a fair opportunity to go as far as their talents and ambitions will take them." We are already seeing much of this take place. Universities and colleges are now offering courses over the Internet, and anyone can obtain a degree through correspondence courses (distance learning). All one needs is a computer, modem, and Internet account.

According to the NII, this will be achieved using a variety of equipment. Cameras, scanners, telephones, fax machines, computers, compact discs, video and audio tape, cable, wire, satellites, fiber optics, and virtually every piece of equipment used to convey information in some form or fashion will become a part of the NII.

This means that the information provided can take any form, which of course is the biggest challenge to network providers. Their networks must be capable of transporting all forms of information to every household. Pacific Telesis is the first telephone company to take this challenge seriously by converting all of the homes in Southern California over to their new fiber optics network. In addition, every home will be rewired with coaxial cable to prepare for the oncoming NII.

Probably the most significant portion of this project is extending access to the government over the Internet. Already many agencies have begun providing important information over the Internet (such as the Library of Congress), and for the first time we can send e-mail to the White House.

2.2.3. Cost and More Cost—The Reality

Sounds like utopia? It will be if the government can get its way. However, what is often overlooked is the cost of such a project. While the NII has set agendas and milestones to be met, the government is finding that projects of this magnitude do not happen overnight. The development of ATM standards has been somewhat accelerated, thanks to the efforts of the ATM Forum, but we are still not close to completing these standards and deploying a nationwide ATM network based on them.

Our schools are finding that the biggest roadblock is finances. Even if the private sector provides deep discounts on equipment, the schools must be wired and the staff trained on computer technology. This expertise does not exist in our schools today, and unless the districts begin implementing programs soon that are targeted at their teachers and administrative staff, the NII will not fulfill its major objective: access by all schools.

Only two states have aggressively pursued this project. A nationwide network for all of its schools has been started in Texas. This network was deployed by a single employee without school funding and uses the Internet for interconnection. In essence, it was a "pet project" by one individual without support from the state.

In North Carolina, a more aggressive program was started, dubbed the North Carolina Information Highway (NCIH). This project has received a lot of press in the North Carolina area, but little has been achieved.

2.3. The North Carolina Information Highway

Governor Jim Hunt took President Clinton's challenge to heart and obtained funding from the state legislature for a statewide network connecting key universities, colleges, and schools, as well as hospitals and research organizations. At first cut, some 100 sites were named for connection, representing a diverse cross-section of the state's population.

Some 36 vendors have been involved in putting this network together. The state's telephone companies, GTE, Southern Bell, and Sprint, are the owners of this network and have purchased switching equipment from local vendors (Nortel and Fujitsu have locations in North Carolina) based on ATM technology.

The governor formed a committee to oversee the deployment of this network and promote the concept to the private sector. The committee is responsible for helping state agencies deploy ATM technology and assists in training at the various sites which are connected to this network.

The state legislature provided funding which would allow the schools to receive a Synchronous Optical Network (SONET) fiber optics connection (OC-3) during the first phase of the project, to be upgraded at a later phase. The NCIH also worked with telephone companies to set fees for all participants, with all fees being the same regardless of location. This means no long distance charges when using the network. All schools pay the same monthly flat rate.

2.3.1. Model Citizen or Political Agenda

Once the project began, the model began to erode. What was at first a list of some 100 schools and other agencies became whittled down to some 35 sites. The original cost estimates did not account for the installation of fiber optics and coaxial cables in the schools, which in many parts of the state are over 60 years old.

Many parents questioned why the state was funding a computer network for a school which was condemned and in need of massive repairs. What happened? The same thing that happens to all great ideas: lack of detail and focus on the dream of what was to be.

Had the state looked a bit closer at their facilities, and paid closer attention to the realities of the current economic status of the state, it would have recognized the enormous cost to deploy such a network. Many corporations have been guilty of the same malady. Failure to look at the real costs involved in deploying sophisticated networks at the most fundamental levels, such as cabling, will kill any network project.

The NCIH project has not failed, but there is much still to be accomplished. If anything, this project should serve as a model for other states wishing to do the same. Education of all participants is critical to its success (I visited several state libraries connected to the network, only to find it is not used because the staff does not know what to use it for).

Cost should be calculated at the lowest levels. Schools and other facilities should be evaluated first to determine the feasibility of deploying a sophisticated network and all of its equipment. A crumbling ceiling during a rainstorm caused by a leaking roof can quickly cost thousands of dollars in network equipment.

I mention all of this here to make a point. Oftentimes when evaluating cost of a new network, cost is based on equipment purchases and access fees. The installation and training costs are often overlooked, or expectations are too high and inflated. Not enough money is allocated to installation and training, resulting in poor performance and lack of expertise to properly operate and maintain the network. If anything, these areas should be the bulk of any networking budget.

Whatever the outcome of the NCIH project, there is no debate that this important network serves as a model of how government can play a role in the development of the Information Highway. Such support will prove to be paramount to the success of any network with such global ambitions.

With the signing of a new telecommunications bill in the United States, the government has shown support to the industry by opening up new opportunities for telephone and cable companies. What lies ahead for these companies is the further development of new technologies to serve their customers and provide services one could never imagine possible in earlier decades.

2.4. The Backbone

Before telephone companies can begin providing services touting limitless bandwidth, the backbone of this nation's telephone network must be redesigned. This is a formidable task in itself, given the amount of copper already in place throughout the major cities of the United States. Perhaps our foreign neighbors have an advantage here, in that they have little already invested in their infrastructure and can begin building a state of the art network without worrying about existing infrastructure.

The drive for a new infrastructure actually began back in the 1970s, when the existing cable plant proved inadequate for many major metropolitan cities such as New York and Los Angeles. There, the copper cables already in place were near capacity, and the communities were growing at an alarming rate.

The only solution was to either invest millions of dollars in placing new copper cables or to replace existing cables with new high-capacity fiber optic cables. The fiber optic cables took much less space and carried a hundred times the capacity of the copper cabling. The fiber optics solution also supports long-term plans for an infrastructure capable of carrying voice, data, and video over the same facilities.

In this section we will talk about the migration from analog copper facilities to digital copper facilities and then to fiber optic facilities capable of handling far more information than ever before possible. We will also discuss the primary drivers behind this evolution and look at the advantages of each of the technologies used to achieve higher bandwidth in less infrastructure.

2.4.1. From Analog to Digital Trunking

The first step to facility consolidation is to permit more than one transmission over the same cable pair. Previously, when only analog transmission was available, each copper pair was dedicated to a single voice transmission or maybe a single data connection. If several transmissions can be assigned to the same cable pair, less cable is required to serve the same number of customers. This is known as pair gain.

In order to transmit more than one conversation on a single copper pair, the voice transmission must first be converted to digital. We have already discussed the process used to convert voice transmission to digital form. The Pulse Code Modulation (PCM) transmission is inserted into one of several channels transmitted over the cable pair and is dedicated to this channel for the duration of the call.

Once we have converted voice to digital form, any medium can be used to transport the transmission to its destination. Fiber optics has proven to be the best alternative for a number of reasons. The cabling itself is flexible and is not subject to external interference caused by electrical equipment. Copper on the other hand is very susceptible to external interference.

2.4.1.1. Multiplexing To send multiple transmissions over the same copper pair, the transmissions must be divided. This is referred to as multiplexing. There are several forms of multiplexing: Frequency Division Multiplexing (FDM) and Time Division Multiplexing (TDM). Both of these techniques achieve the same result: multiple transmission over the same facility.

FDM divides the bandwidth of an analog facility into blocks of frequencies. If you remember our discussion about digitizing voice, you will remember that voice requires 4 kHz of bandwidth. In FDM, the transmissions are divided into 4-kHz frequency blocks, starting at 0 to 4 kHz. The next block begins at 4.1 kHz and ends at 8 kHz, and so on. Table 2.2 shows how FDM is divided into blocks, as well as the total number of transmissions supported by FDM.

Transmissions are sent simultaneously, but the multiplexers change

TABLE 2.2

Frequency Division Multiplexing Hierarchy

Mux. Level	No. of Voice Circuits	Formation	Frequency Band (kHz)
Voice channel	1		0—4
Group	12	12 voice circuits	60—108
Supergroup	60	5 groups	312—552
Mastergroup	600	10 supergroups	564—3084
Mastergroup mux.	1200—3600	various	312/564—17,548
Jumbogroup	3,600	6 mastergroups	564—17,548
Jumbogroup mux.	10,800	3 jumbogroups	3,000—60,000

the actual frequency of the voice transmission according to this hierarchy. This means that a demultiplexer at the terminating end must convert the transmission back into its original frequency. A 12-channel multiplexer will output 48 kHz of bandwidth, divided into twelve 4-kHz blocks of transmission. Each block is considered a channel, supporting one transmission.

FDM is more expensive than TDM, which is a digital technique. There are other problems with FDM. Because it is analog, it is susceptible to external interference. Noise (in the form of static and "pops") is simply amplified and cannot be effectively filtered out of the transmission. There are very few FDM facilities used today.

2.4.1.2. Time Division Multiplexing The most common form of multiplexing is TDM. This is a digital form of multiplexing and is used in virtually every digital network today. The transmission facility is divided into time slots. Each time slot is used to carry one transmission.

TDM was developed for digital transmission facilities such as T-1. TDM uses the same concepts as FDM, but instead of using blocks of frequencies, it uses time slots in 64-kbps increments. Digital transmission is linear in fashion, meaning that bits are sent in serial order.

Each transmission is assigned a time cycle, or time slot. When this time slot occurs, data from the assigned device is transmitted. Think of the gates used in horse racing. All of the horses are lined up in the gate. In TDM, only one gate opens at a time, allowing data in the queue to be transmitted.

Once the gate opens and the queued data is sent, the gate is closed and the queue is filled with new data. When the time slot comes around again, the gate is opened and the data in the queue is transmitted.

Again, looking back at our discussion of PCM and voice digitization, we remember that digital voice (PCM) is sent in 8-bit segments. If we take that 8-bit PCM "word" and place it into a digital time slot, we can support up to 24 individual time slots, each carrying one 8-bit PCM word from a different transmission.

The time slots are repeated; that is, channel (or time slot) 1 is repeated every 24 channels. A telephone call sent over a digital facility is assigned to a dedicated channel for the duration of the call, which means the conversation will be converted to digital form and transmitted in the same channel until the call is terminated.

A 24-channel block is called a frame. If we look at Table 2.3 in the section below, we can see that one frame is called a DS1 in the United

States. Time slots, or channels as they are also referred to, are fixed in length and can be used to send voice or data.

There are two methods of TDM: asynchronous and synchronous. In synchronous TDM, the time slots are assigned to a device. If the device has nothing to send, the time slot passes with no data. It is considered "idle." This is not considered efficient use of bandwidth because no device is transmitting 100 percent of the time, and many time slots will pass with nothing in them.

In asynchronous TDM, time slots are dynamically assigned. If a device has something to transmit, it is assigned the next available time slot. If there is nothing to transmit, the time slot goes by with nothing sent (unless another device has something to transmit, in which case the time slot can be assigned to that device).

Asynchronous Time Division Multiplexing (ATDM) is the concept used for newer technologies such as Asynchronous Transfer Mode (ATM). The transmission facility uses bandwidth efficiently, with every time slot being used. Only when no device has anything to send are time slots empty (idle).

While ATM is based on ATDM, ATM is not channelized. In other words, there are no time slots in ATM. Data is inserted in a fixed-size cell, and the cell is then transmitted. Parameters within the cell (header) provide the necessary information for routing and processing the data.

Another form of ATDM is statistical ATDM. Statistical ATDM requires some intelligence in the multiplexing equipment. When a device has data to transmit, it is determined how much data must be sent and how many time slots will be needed for that transmission. The multiplexer then reserves these time slots for the transmission.

The result is several consecutive time slots assigned to the device, rather than the device being assigned one time slot at a time and waiting for another idle time slot to become available. Devices with less data are allowed to transmit in between the time slots assigned to the "chatty" device, preventing devices from becoming congested with data waiting to be sent.

In ATM, another parameter has been added to this technique. Referred to as Quality of Service (QoS), this parameter indicates whether or not data in the queue waiting to be transmitted is delay-sensitive data. Voice and video are considered delay-sensitive since delaying their transmission will affect the quality of the received transmission (in other words, these are real-time applications, and delays will cause pauses in the playback of the transmissions). QoS works like a priority, indicating that a device may have delay-sensitive data and should be sent ahead of any other data waiting in queues.

Now that we understand various multiplexing techniques used in telecommunications, let us look at the various digital transmission facilities used in telephone and data networks.

2.4.2. The Digital Hierarchy—DS1 and DS3

The first step toward consolidation in the United States was the development of a digital hierarchy which could support voice transmission with very little modification to the existing infrastructure. Telephone companies already had invested in repeaters and amplifiers which were installed outside of the central offices. These devices were not capable of supporting digital transmission without modification.

Even with modification, they could only support 24 digital time slots, which was less than our European counterparts, who deployed devices capable of supporting 32 time slots per cable pair. This is the principal reason for the disparity in digital networks between the United States and our foreign neighbors.

The hierarchy used in the United States, which is shown in Table 2.3, starts with the Digital Signal 0 (DS0). This is the lowest common denominator in the digital hierarchy. Each DS0 is used to carry a transmission, usually voice or data. As shown in the table, 24 DS0s make a DS1. A DS0 supports 64 kbps. In a DS1, which uses 24 DS0s, 1.544 Mbps is supported.

A DS1 will have specific bits which identify where the first and the last DS0s are located. The entire DS1 is referred to as a frame. Protocols

TABLE 2.3

North American Digital Signal Hierarchy

Digital Signal Designation	Bandwidth	Channels (DS0s)	Carrier Designation	Medium (typical)
DS0	64 kbps	1 time slot	None	Copper, fiber
DS1	1.544 Mbps	24 channels	T-1	Copper, fiber
DS1C	3.152 Mbps	48 channels	T-1c	Copper, fiber
DS2	6.312 Mbps	96 channels	T-2	Micro, fiber
DS3	44.736 Mbps	672 channels	T-3	Micro, fiber
DS4	274.176 Mbps	4032 channels	T-4	Micro, fiber

such as T-1 provide the framing bits as well as some other signaling information required to delineate the 24 separate transmissions.

A device such as a channel bank or digital cross connect allows the DS1 to be terminated in a central office or subscriber premise, and it separates the various DS0s from the frame. Once separated, they can then be connected to various pieces of equipment, such as a telephone switch or router. Some devices have the capability of connecting directly to a DS1 without the use of a channel bank or cross connect by providing the same functionality within the device.

A DS3 is made of 28 DS1s, which must be demultiplexed before connecting to a device. Again, a channel bank or cross connect is needed to perform this function. All of these devices use TDM, taking the digital voice or data and placing it into time slots. When multiplexed, the individual DS0 time slots are placed into a frame and transmitted to the distant end, where the DS1 is demultiplexed and the individual DS0s are extracted from the frame and routed to the network equipment.

The digital hierarchy was designed to consolidate facilities between central offices. Instead of using a pair of wires for only one transmission, multiplexing allows telephone companies to send many transmissions over the same pair of wires. However, the digital hierarchy does not support the types of services under development today. Video, for example, requires far more bandwidth than the digital hierarchy can support. For this reason, a new technology was developed to replace the digital trunking already in place throughout the United States. This new technology uses fiber optics and a special protocol called SONET. We will discuss SONET in the next section.

It should be mentioned here that Europe also uses digital trunking. However, the digital hierarchy used throughout Europe and the rest of the world does not match that used within the United States. Theirs consists of 32 channels, providing more bandwidth within one frame. The European hierarchy also uses 64-kbps building blocks (see Table 2.4).

We will not go into the formulas which demonstrate how this bandwidth is contrived; this is covered in other books which delve into the engineering aspects of these technologies. The important thing to understand here is the purpose of the digital hierarchy and what it provides to the network.

2.4.2.1. T-1 Facilities T-1 facilities use the DS1 carrier as a transport. The T-1 protocol adds necessary overhead to the DS1 for framing and loopback testing, among other things. When extended to a subscriber premise, T-1 is used to provide an interface between the subscriber and

TABLE 2.4

CEPT Digital Signal Hierarchy

Digital Signal Designation	Bandwidth	Channels (DS0s)	Carrier Designation	Medium (typical)
Signal Level 0	64 kbps	1 time slot	None	Copper, fiber
Signal Level 1	2.048 Mbps	30 time slots	E-1	Copper, fiber
Signal Level 2	8.448 Mbps	120 time slots	E-2	Copper, fiber
Signal Level 3	34.368 Mbps	480 time slots	E-3	Micro, fiber
Signal Level 4	139.264 Mbps	1920 time slots	E-4	Micro, fiber
Signal Level 5	565.148 Mbps	4032 time slots	E-5	Micro, fiber

the telephone company. Other protocols can be used over the T-1 because the T-1 does not attempt to process any information other than the specific bits used for framing and signaling at the physical layer.

T-1 circuits use what is referred to as a *D*-type channel bank. The channel bank provides the multiplexing/demultiplexing functions required for digital carrier systems. The T-1 uses specific bits within the digital transmission for signaling and synchronization of the frames.

Synchronization is critical in any digital carrier system. If a device is connected at the far end of a circuit, and it is not synchronized with the near end channel bank, the far end device will not know when a T-1 frame begins or ends. It will misinterpret the boundaries of the frames, resulting in transmission errors. Synchronization is achieved through the use of highly accurate clocks. These clocks are then synchronized to the same reference, using satellite to access the reference signal. Clocking is critical in any digital trunking network.

A typical T-1 circuit consists of at least four wires, two for transmission and two for receiving. Remember from our discussion that voice when digitized is sent in PCM format. A PCM "word" consists of 8 bits for each sample of the voice. This 8-bit word is then assigned to a channel, or time slot, within the T-1.

During the life of a telephone call, the voice transmission for the call will always be sent over the same channel. The channel is then dedicated to the voice call. The same is true for data, when connection-oriented services are used or when providing a logical connection between two devices.

Figure 2.3 shows how bits from the transmission are "stolen" and used for framing and signaling. There is no degradation of transmission from

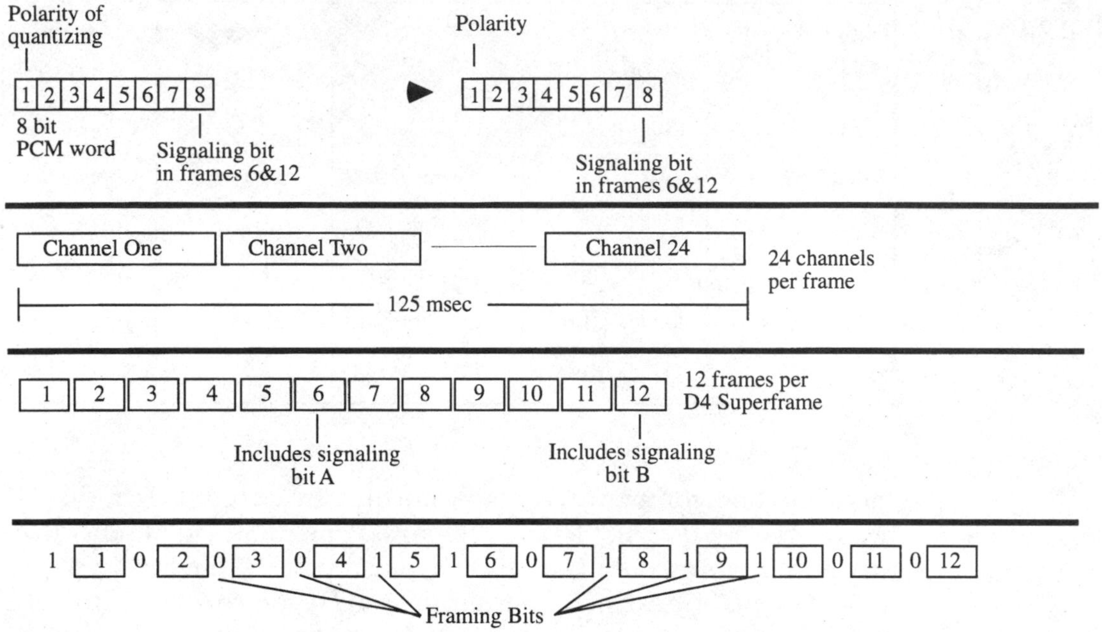

Polarity of quantizing

1	2	3	4	5	6	7	8

8 bit PCM word

Signaling bit in frames 6&12

Polarity

1	2	3	4	5	6	7	8

Signaling bit in frames 6&12

Channel One | Channel Two — Channel 24

24 channels per frame

125 msec

1	2	3	4	5	6	7	8	9	10	11	12

12 frames per D4 Superframe

Includes signaling bit A

Includes signaling bit B

1 [1] 0 [2] 0 [3] 0 [4] 1 [5] 1 [6] 0 [7] 1 [8] 1 [9] 1 [10] 0 [11] 0 [12]

Framing Bits

Figure 2.3
D4 Framing

this bit robbing, at least none discernible by the human ear. This is because only 1 bit out of an 8-bit word is taken, representing only one sample of the analog voice. The eighth bit from the sixth and twelfth channels are used for signaling in the D2, D3, and D4 channel banks.

Twelve frames of DS1 are referred to as a superframe. The type of channel bank used at both ends of the circuit determines where the framing bits are located within a frame. It is important for a T-1 to use the same type of channel bank at both ends of the circuit.

The Bell System has defined specifications for channel banks, specifying which bits are used for signaling and which are used for framing. For example, in the first generation of channel banks, D1, bit 8 of every channel was used for signaling (on hook, off hook, etc.). However, in later generations of channel banks (D2, D3, and D4) every sixth and twelfth frame of a superframe (defined below) contain the signaling bits. The eighth bit in these channels is used for signaling.

Signaling bits in T-1 are used to establish and terminate a connection (supervision and signaling). It should be noted here that signaling in this form has been replaced by a more sophisticated signaling system within the telephone network. T-1 signaling is used between subscribers and telephone switches, but interoffice signaling uses Signaling System #7, or SS7 (defined in Chap. 5). There are 192 bits in a frame, plus 1 extra bit used for synchronization, for a total of 193 bits in a frame.

The CEPT version of T-1 (E-1) does not use bit robbing. This is why CEPT channels support 64 kbps. All signaling is provided over a separate channel (or time slot as it is called in the CEPT hierarchy).

An Extended Superframe (ESF) is a combination of 24 frames, each frame being 24 DS0 channels. In other words, 24 DS1s become an extended superframe. Framing bits are inserted in frames 18 and 24. There is no difference in the way the DS0s are used; it is simply a matter of being able to combine many more DS0s into a collective group.

This should not be confused with the other levels of the digital hierarchy. ESF is simply an additional multiplexing function for the transmission of DS1s. If additional bandwidth is required, a DS3 or higher is used to support applications (such as video, which uses DS4 and DS5).

2.4.3. SONET—The New Fiber Backbone

The telephone industry has been busy converting its existing outside copper plant to newer fiber optics. There are many advantages to using fiber rather than copper. Higher bandwidth is the foremost advantage; quieter communications is another. There are a number of other economical advantages to using fiber optics.

In the existing copper infrastructure, amplifiers are required at regular intervals. These amplifiers are placed in cabinets in various locations (outside large subdivisions, along highways). The amplifiers are analog devices, unless recently upgraded to support newer digital services (such as Integrated Services Digital Network, or ISDN). The problem many telephone companies encounter when offering new services is in upgrading existing equipment (including these amplifiers and also multiplexers) to support the newer digital services.

With fiber optics, there are no amplifiers, only repeaters. The repeaters can be placed at greater distances from the central office and are easier to upgrade when necessary. If a fiber optic cable is cut, service can often be switched to another backup fiber. This of course depends on the network configuration. Most fiber optic operators are deploying SONET rings, which provide a self-healing network.

Another advantage of SONET is its standard multiplexing format, capable of carrying DS1 transmissions, or even DS3. SONET itself is a standard for fiber optics transmissions, making interconnection with other equipment easier. SONET also uses a flexible architecture, allowing for future growth.

One of the unique advantages of SONET is its capability to access a single transmission frame without demultiplexing the entire transmission. This is not always possible with the digital signal hierarchy.

For example, if a company is using a DS3 transmission facility, and they need to extract a single channel (DS0) from the DS3, the DS3 must first be demultiplexed down to DS1 levels. This is because special framing bits are inserted into the DS3 transmission, and they must first be extracted before the DS0 can be interpreted. The same is true if superframes are used. Special framing bits are inserted into the transmission stream, which must first be extracted before the DS0 can be accessed.

As seen in Fig. 2.4, SONET uses a more flexible architecture, allowing more diversity when interfacing to various networks. The illustration shows that any number of various signals can be input into a SONET

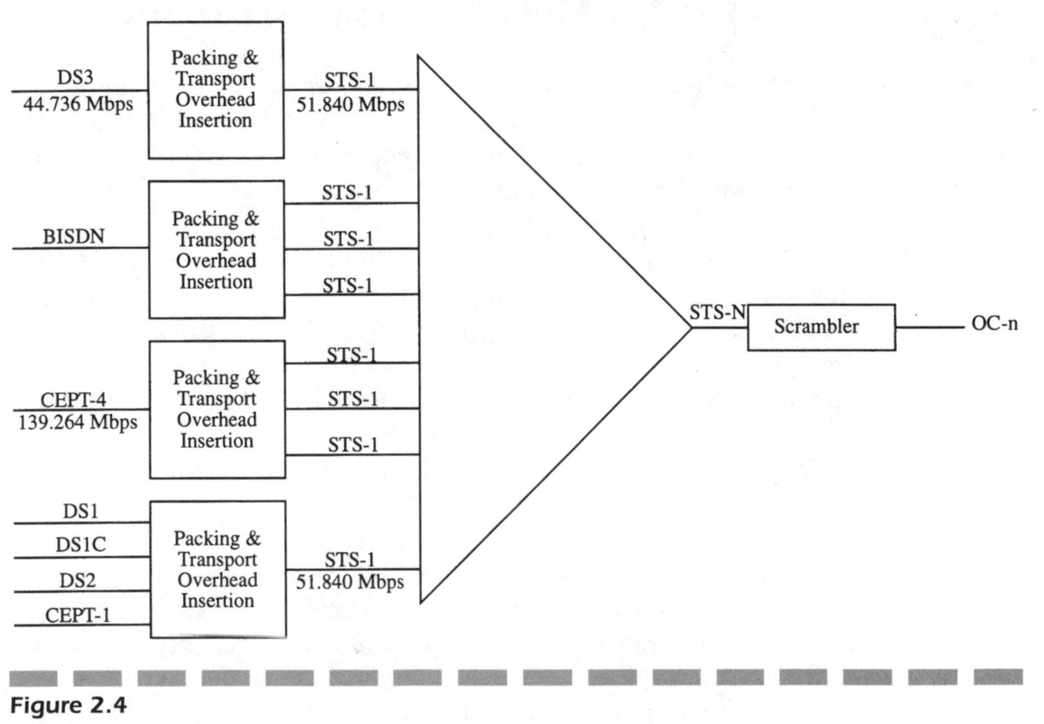

Figure 2.4
SONET Multiplexing

device and can be extracted as easily. This means that a telephone company can easily incorporate SONET as a backbone network to support their many smaller networks.

SONET can also be used in smaller networks, such as in a campus environment. With the standardization of ATM, this becomes critical. ATM applications are bandwidth-intensive, requiring fiber optics to support them. For networks too small to warrant the cost of SONET, Fiber Distributed Data Interface (FDDI) can be used in its place.

2.5. The Private Network

A private network is any network which uses privately owned equipment and does not provide the public access to the network. For example, a large corporation will usually have their own private telephone network. Callers from the outside can call into the network, but they must dial a full telephone number. Callers from within the network can dial abbreviated numbers, referred to as extension numbers. All users within the private network are connected to the same switch or to a series of switches interconnected through special facilities.

The telephone switches used within private networks operate as central offices. They provide all of the necessary functions, as well as many features not available through the central office. The purpose of using private switches is to consolidate telephone trunks. Rather than have a separate circuit for every telephone in the network, a group of circuits is connected to the switch, and users calling outside of the network are connected to any available circuit on an as-needed basis.

For example, a company with 300 telephones would need 300 telephone lines from the central office. If they own their own private switch, they could need only 10 telephone circuits, depending on the amount of traffic coming in and going out of the switch (determined through traffic studies). There are many different types of telephone switches available today for private networks.

2.5.1. Private Branch Exchanges

A PBX is like a central office switch. In addition to providing the basic features required to make and receive telephone calls, a PBX may provide many additional features. These features include things like call forwarding, voice mail, and automatic routing features. Originally, PBXs were analog switches, using mechanical relays. These were soon replaced with electromechanical switches.

As digital electronics progressed, digital switches became available, providing many new and unique features. These digital switches can also take advantage of digital circuits from the central office without the use of channel banks. Another added feature of digital PBX is the ability to switch data through the PBX rather than through a separate data network. Unfortunately, the data rates through a PBX are still very limited in comparison to a data network, but for things such as modem pooling and printer sharing, the digital PBX offers a very viable solution.

Early PBX consisted of a manual switchboard, where all calls were answered by a central operator, who then connected the caller to the proper extension. As switching systems became more automated, this functionality was taken over by the automatic PBX. With that came a number of features, which are outlined in the next section.

2.5.2. Features and Capabilities of Private Networks

A private network comes with many features. When a PBX is used (especially a digital PBX), there are many features available that make PBXs attractive. Besides features such as call forwarding, there are many integrated features served through adjuncts such as voice mail and Automatic Call Distributors (ACDs).

One of the most basic PBX features is Automatic Call Routing. This feature is called many different things by many vendors. There are two types of call routing, incoming and outgoing call routing. Incoming calls are typically routed to specific extensions, using an internal routing table. They can also be forwarded, when busy or when there is no answer, to other extensions or to adjunct systems, such as voice mail.

Outgoing calls can be routed based on digits dialed, time of day, or relative cost of the call. When a hub is used in the private network (central PBX connecting to all other PBXs in the same corporate network), long distance calls may be routed over leased lines to the hub PBX, where a connection is established over a local telephone trunk. The pricing for the call is then based on the cost from the hub to the destination number (often a local call).

This kind of routing can save a company thousands of dollars in long distance charges and is a common practice in large corporate voice networks. Complex routing tables are used to determine the routing for all outgoing calls.

Voice mail is usually provided through an external computer. An analog extension is connected to the computer (or several analog extensions), and the extensions in the system are call-forwarded to the analog extensions. Typically, these extensions are part of a hunt group.

In a hunt group, an extension is assigned as the "pilot" extension, and the other extensions are assigned as members of the hunt group. When the pilot extension is dialed, the call is routed to one of the first available hunt group members. In the case of voice mail, these extensions are capable of connecting to the voice mail computer and receiving Dual-Tone Multifrequency (DTMF) tones. When a connection is made to the voice mail system, the PBX will then send DTMF tones to identify the extension from which the call was forwarded.

Each extension in the PBX is assigned a mailbox number (usually with the same number as the extension) in the voice mail system. When a call is forwarded to the voice mail system, the PBX sends DTMF tones to identify the mailbox number to be connected. Software then controls a series of recordings which provide instructions on how to leave a message on the voice mail. The messages are stored in digital form on the hard disk of the voice mail system.

Voice mail systems are capable of alerting the owner of an extension in a number of ways. If the phone has a message lamp, the voice mail system can send a signal which lights the message lamp. In the case of an analog system, the voice mail must send voltage to the phone over the analog line, which in turn causes an analog lamp to flash.

In the case of digital phones, the voice mail system sends a digital message to the PBX, identifying the extension which has a message, and the PBX then causes the light on the digital phone to flash (or some other form of indication). Voice mail systems can also dial other numbers programmed into memory. These numbers can be home-phone or pager numbers.

Another feature related to voice mail is voice announcers. A voice announcer is an analog or digital recording system which is connected to the PBX through an analog extension. When the extension is dialed, the caller hears a series of announcements. These extensions are usually assigned to a hunt group, and the caller has no idea what the extension number is. One use for voice announcers is to alert callers to business hours. For example, after close of business, an operator may forward the attendant console to a voice announcer, which will then answer all incoming calls and give the business hours or maybe an alternative number for emergencies or after business hour calls.

An automated attendant allows incoming calls to be forwarded to a series of recordings, which allow callers to dial specific numbers for call routing. These are very popular in environments such as customer service and catalog sales. The caller is forwarded to the pilot number of a hunt group, which then connects the caller to a recording. The recording may provide a number of choices and a single digit number to dial for each option.

When the caller dials the single digit number, a routing table in the PBX identifies an extension to connect the caller to. The extension is usually another recording or maybe the pilot number to another hunt group. This feature allows incoming calls to bypass an operator, and if callers know which extension numbers within the system they are trying to reach, they can either dial the extension numbers themselves or be routed through the maze of recordings.

An ACD is used to route incoming calls to a large group of "agents." Each agent is a member of a hunt group and receives calls based on the amount of time he or she has been idle (off of the telephone). These systems are often adjuncts to the PBX, although many PBXs offer ACD features as integrated solutions.

The difference between an ACD and a normal hunt group is in the computing side of the system. There are a number of reports available with an ACD system that are designed to help managers of call centers to manage their resources. Since call centers deal with high volumes of calls (either incoming or outgoing), managing the resources of the call center is critical. Resources are either agents available to receive calls or available trunks.

The reports used by the manager of a call center identify productivity (or lack of it) within the group. This means that agents must "log on" to the system when they are first available and log off of the system when they are not available to receive calls. Some ACDs even provide several categories, such as "paperwork" or "break" to identify why the agent is not available to receive calls. These options then appear on the reports received by the manager.

The agent's telephone has a number of buttons unique to ACD functions. There is a log on button, which is used by the agent when reporting to work. Once the agent is logged on, calls can be routed to the agent's phone. The time the agent logged on is recorded in the system and recorded on the various reports made available to the call center manager.

Other buttons are used to remove an agent from the hunt group for various reasons, such as "break" or "paperwork." This allows the call center manager to closely monitor the productivity of all the agents in the system.

ACDs use an adjunct computer to store call activity data and produce real time reports for the call center manager. The screen on the computer terminal can be configured to display real time traffic statistics, as well as other call activity data, enabling managers to monitor their resources closely. Detailed reports can also be generated on the ACD terminal.

The reports can be quite sophisticated, depending on the system. Managers can compare agents with other agents or even an agent against a group of agents. Some systems even provide productivity reports in graph format.

2.5.3. Voice and Data Integration

Many digital PBXs offer integrated voice and data features. This allows both voice and data to be transmitted through the PBX. One would think this would eliminate the need for Local Area Networks (LANs) to interconnect corporate computers, but in reality, these PBXs cannot offer the same bandwidth and data rates as LANs.

Data devices such as modems, printers, and computers can be connected to the PBX through digital telephones or other data adapters. Both the voice and data are then sent over the same pair of wires, allowing companies to consolidate their wiring plans.

Printers can be assigned an extension in the system and can be reached by dialing their extension numbers on a special data line. A user's PC (connected to the digital telephone) can then direct print jobs to the printer without accessing the LAN. Likewise, modems can be attached to extensions assigned in a hunt group and can be used by internal callers when needed. This is referred to as modem pooling. Using modem pools eliminates the need for individual modems on every desktop.

While all this sounds attractive, PBXs can only support data rates up to 19 kbps, far below the data rate of 100 Mbps supported by LANs. Client/server applications are not supported in the PBX environment either because they require a faster data rate than the PBX can provide.

2.5.4. Centrex Services

Some companies have elected to use Centrex service rather than buying their own PBX. Centrex is offered by many local telephone companies and provides cost benefits to companies that are not willing to invest a lot of capital in their voice networks.

Centrex is like having a PBX, except the equipment is located in the central office. It still requires several lines coming into the corporate office but not as many as would be required for supporting individual telephone lines on every desk. Multiplexing equipment is used to support multiple transmissions over the same facilities, consolidating the cabling requirements.

Some say that the maintenance costs are much lower as well. Since the equipment is owned by the telephone company, repair costs are the responsibility of the telephone company. On the other hand, changes that are typically routine for PBX owners require service orders when using Centrex service. PBX owners usually have trained personnel to perform these changes, allowing them to make routine changes to the PBX configuration without unnecessary delay and service order costs. Changes to a Centrex service may take several days (or even weeks) depending on the complexity of the changes. A cost is associated with each change, driving the administration cost of Centrex far above that for a PBX.

2.5.5. Computer Telephony Applications

A new industry has been cultivated that uses computers in telephone networks. Users of private telephone systems have tried for many years to marry the computer with the telephone system in their company. Until the last 5 years, the industry has been unable to meet the needs of the user in this area.

Several manufacturers of PBX and computer equipment finally got together and forged agreements to develop interfaces that would allow desktop computers to control applications in PBX equipment. Some of the first applications were things like voice mail systems and calling center applications. Today, one can even buy plug-in computer boards that allow your computer to act as a minitelephone system.

What was missing was a standard interface to the PBX that any computer vendor could connect to, allowing desktop applications access to the powerful features of the telephone system. Access was limited to a dumb terminal and proprietary programming interfaces that required PBX owners to get trained on their system.

The first pioneers of computer telephony developed an interface that would allow computer equipment to control various PBX features, using

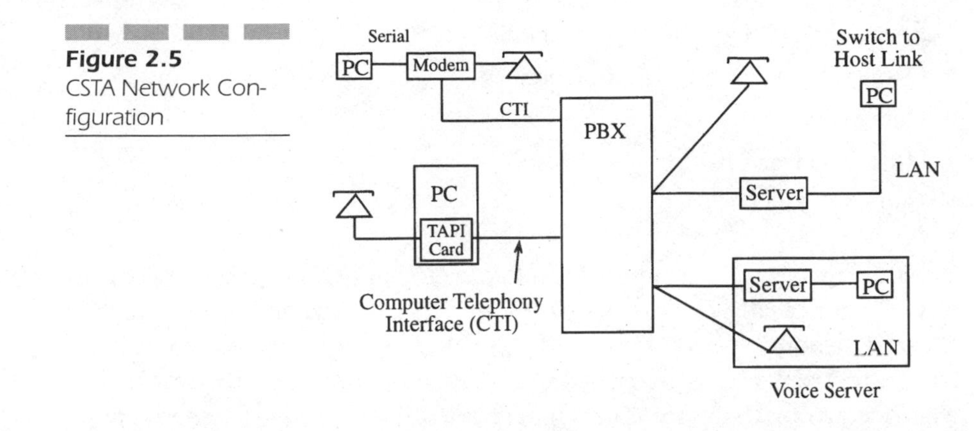

Figure 2.5 CSTA Network Configuration

a Graphical User Interface (GUI) instead of proprietary programming language to access the PBX and change its configuration.

Today, you can change the routing in your PBX, change voice mail configurations, and even receive sophisticated reports for your calling center, all through a desktop computer. While there is now a standard, not everyone has endorsed this standard. In fact, the European vendors were the first to pursue and provide a standard interface to their PBX systems for computer applications.

2.5.5.1. TAPI The Telephony Application Programming Interface (TAPI) was developed by the European Computer Manufacturers Association (ECMA) starting in 1988. Understanding the complexities of creating such a standard, and having difficulty focusing on any one solution, the ECMA formed a technical group (TG11) that is responsible for defining the standard.

The ECMA drafted TR52, which is not a standard but rather a report citing the various solutions found most favorable by TG11. This approach was taken because they found many different approaches to solving the same problems, and it was becoming difficult to agree on any one solution. To expedite the standards process, this report provided important insight to each solution.

The work group's first task was to outline the specific applications to be targeted so that everyone would have a clear understanding of what they were trying to accomplish. Efforts up to this point had led to many different proprietary solutions for the same applications.

The applications targeted by TG11 were:

- Call centers (inbound and outbound)
- Customer support environments

- Emergency call applications (such as 911)
- Data collection and distribution
- Data access
- Hotel applications
- Switched data

Call centers are commonly found at airline and hotel reservation centers. They typically consist of many operators, or "agents," who have specialized telephone sets. Every call handled by an agent is tracked by computer reports, allowing the call center manager to determine whether or not there are enough agents on duty to handle the flow of traffic.

These reports can also alert the call center manager to traffic patterns, which are then used to determine when additional 800 numbers need to be assigned to the group. Even trunk utilization can be determined by the reports offered by many systems.

Productivity reports allow call center managers to monitor the group's activities and compare agents with other agents, groups with other groups, or even agents with entire groups. All of these reports, and in many cases the actual configuration of the call center (number of trunks, number of agents, etc.), are accessible through a desktop computer, which in turn communicates with the PBX through the TAPI interface.

For outbound calling centers, a database selects telephone numbers randomly (usually based on some demographic or geographic parameter selected by the calling center management). Before the call is dialed by the computer, the next available agent is determined. The computer server then activates a screen on the agent's computer, showing the name, address, and telephone number that the computer is about to call. The agent can allow the call to proceed or interrupt the call and go to the next selection.

In a customer service environment, the desktop computer can be linked to network management software. As a computer user calls the help desk to get assistance with a computer application, or possibly to report network connection trouble, the help desk agent can be looking at the user's network connection.

The interface in this type of environment can provide information regarding the caller's extension number, which when linked to a database, can automatically trigger the network management software to identify the network address of the caller's computer. From there, the help desk can quickly troubleshoot the network connection. This saves the agent the time of looking up the caller's network specifics before troubleshooting.

The computer used by the help desk can even link calls to contact management applications so that a history of the telephone call can be kept by the customer service representative. If a database for customers is available, this can also be activated with the caller's data on the screen, making it simple for the customer service representative to handle the call.

We are all familiar with emergency call applications such as the 911 system used in most every U.S. city today. Another application for this can be found in some condominium complexes and hotel systems. When someone dials 911 from a PBX system, the 911 system cannot identify the extension number of the caller, and the only information provided is the main billing number of the business owning the telephone system. This can be a perplexing problem in a large hotel, making it almost impossible to identify where the distress call was made from.

At least one manufacturer addressed this through the use of a computer. The computer is attached to a TAPI interface. Whenever someone dials 911 from a PBX extension, the PBX sends the extension number of the caller to the computer. The computer searches a customer-defined database for the extension and finds the room number of the caller. Attached to the computer is a digital display, which is used to display the actual room number of the caller who dialed 911. This works equally as well in a condominium complex where a PBX is used to resell dial tone to the condo owners. The display unit is placed in a conspicuous area where emergency workers are most likely to see it when they arrive.

Data collection and distribution was one of the earliest applications for computers in telephone systems. For example, every PBX sends out calling data in the form of detailed reports identifying every telephone call made. These reports provide the extension number that placed the call, the duration of the call, the time the call was placed, and the number that was dialed.

This information is difficult to use in its raw format and would require someone to collate and organize each report since there is no sorting in the raw report. This is where the computer comes in. The computer is attached to the PBX and receives the detailed report at certain intervals or in real time. The computer then sorts the data and provides a call activity report which can be sorted by extension number, department number, or even user name.

These reports can even have costs assigned to various types of calls for cross-charging departments or charging guests (as is the case in hotel systems). This is but one example of data collection and distribution in a telephone system.

In calling centers, agents need access to databases that provide them with information about the caller (or in the case of outbound call centers, information about the called party). These databases can provide important information, including the customer's address, billing information, and even purchasing profile.

Some of these call centers have even automated the task of placing orders by asking the caller to enter a customer number or his or her own telephone number for routing to an automated voice response system. Once into the voice response system, the customer uses a dialpad to enter in ordering information, such as product number. The database provides the rest of the information needed to process the order. The billing information and shipping address are already known.

Hotels have many applications which require computers. We have already mentioned the call accounting system, used to track and record telephone calls made by guests. These call accounting systems automatically mark up the calls and send the information to a room billing system to be added to the room bill automatically.

Hotels also use automated directories. These directories allow operators to connect calls to guests without having to look up the guest's name on a sheet. Every guest is entered into the computer upon check-in. The computer creates a directory, which the operator can then search by entering in the guest's name. Once found, the call can be transferred to the room by simply pressing Enter or a similar key on the computer keyboard.

Switched data was not addressed by the ECMA because there was not a lot of interest in this area. There were a few switch vendors who did provide this functionality in their PBXs, however, allowing computers and peripherals to be attached to the PBX rather than a separate LAN. The problems with using a PBX to switch data are capacity and speed. PBXs can only transfer data at 19,200 kbps. This makes it difficult to compete with LANs running at 10 Mbps and up.

Once the applications were defined, a model was established. The model had to define the boundaries in which the standard would operate. This was important for the developers to understand, since otherwise they would spend far too much time trying to understand the operating parameters of their standard. By defining the domains in which the various aspects of the standard would apply, the focus of the various committees could remain within their area of domain.

Three domains were identified: the computing, the switching, and the application domains. The various committees worked within specific domains and identified the interfaces between the various domains.

Within each domain, objects were identified. These objects would require some form of identification so that the software could maintain the status of that object and be able to correlate the object with a specific call. For example, in the switching domain, a connection identifier acts as a reference for endpoints outside of the switching domain to specify a specific call. A connection is considered a transaction between an endpoint (a device) and a call. The state of the connection is reported to the computing domain so that the computing domain knows what is happening with any one connection.

A device can be a telephone, a button on a telephone, an operator, an ACD hunt group pilot number, a trunk, or any other physical (or logical) entity that is involved in a call. These are all treated as objects within their own domains.

A call represents communications between one or more devices. In any typical situation, a call could be associated with several connections, all which report their connection states (busy, alerting, connecting). In addition to being associated with connections, a call would also be associated with devices. The TAPI standard defines how this is reported from one domain to the other and how the references are managed.

In a nutshell, TAPI is a standard that defines a set of control messages understood by the telephone system and managed by the computer attached to it. Applications which interface to these control interfaces can then be written, allowing programmers to write applications with interfaces to the PBX functions.

2.5.5.2. ASAI/SCAI These protocols were developed for ISDN switches to allow them to send certain parameters passed from the ISDN interface down to databases and adjunct processors. For example, when a call is connected to a PBX over an ISDN interface, the calling party number is passed from the PBX to a database over the ASAI or SCAI interface.

Other parameters relating to the origin of the call can be passed over this interface as well. The most common application of this interface is for call centers, where the calling party information is passed over a data communications network to the desk of a customer service representative or reservation agent.

ASAI is an AT&T specification, while SCAI is an IBM specification. Both are similar in operation, providing slightly different parameters and protocol structures.

2.6. The Transport

In the information age, data is critical. Unfortunately, we live in a time when there is too much information. The telephone network infrastructure has not been capable of transporting all of this data into our homes. It has only been in the last 5 years or so that the telephone networks have started evolving into major pipelines capable of supporting the services and data transmissions that we now face.

SONET has been the key technology in telephone company networks, allowing them to consolidate all of their various transport networks into one cohesive facility capable of transporting data of all types from one point to another. The digital hierarchy as we have known it is now obsolete, and the optical hierarchy is quickly replacing it.

Previously, data transmission was routed through separate data networks, bypassing the voice switching equipment and forming a network of its own. Other special circuits were routed through separate networks, where the only commonality between them and the voice network was the interoffice trunking used to move them from one office to another. Sometimes, even the interoffice trunking has separate voice and data facilities.

This is rapidly changing as SONET is deployed through the networks. The intent is to use one technology throughout the network to handle all transmissions, whether they are voice, data, or video. SONET is capable of providing this capability, but it is only a pipeline. Another technology is needed which can route the various transmission over SONET and provide critical information regarding the type of transmission, its destination, and the quality of transmission required (delay, no delay, etc.).

2.6.1. The Evolution of ATM

ATM began development in 1969, when engineers at Bell Laboratories began experimenting with alternative switching techniques for routing data of all types through a common switching point. This would alleviate the need for multiple types of equipment in the central office and would provide a much faster, more efficient method by which data could be transmitted.

In the current telephone network, facilities are run from the central office to the subscriber over analog or digital lines. In most metropolitan areas, digital transmission is used up to the subscriber. This requires a multiplexer in the central office, as well as repeaters in the field. DS1s and DS3s are not used for facilities to the residential subscriber, so a different method of multiplexing is required.

These multiplexers (such as D4 and SLIC96) can be found in the field, close to the subscriber. However, using this method of distribution has proven costly to telephone companies due to maintenance and upgrades to support newer technologies. One example of this cost is ISDN, which cannot be used in many areas because the multiplexers used outside the central office will not support it. These must be replaced with compatible multiplexers before ISDN services can be offered. This has resulted in delays in providing ISDN in many rural areas.

ATM can provide a more efficient and cost-effective approach to distribution. By using one common facility throughout the network, telephone companies can rid themselves of the various pieces of equipment they own and maintain today. Using nothing but ATM switches and routers provides a standardized approach to transmission facilities.

ATM is capable of supporting any type of transmission, whether it is voice or video, making it an attractive choice for those companies that are looking to broaden their service offerings. Many telephone companies have already begun looking at cable television services and video on demand. Cable television companies, which already have a fiber optics network, have also been looking at ATM to deliver voice transmissions to their cable customers, competing directly with the local telephone companies.

2.7. The Subscriber Interface

There are a number of alternative types of access to the public network that are now being explored by telephone and cable television companies. While the existing infrastructure is considered an inefficient way to support the many new services being offered, there does not appear to be a lot of consensus as to which solution is the best one to replace it. There are as many solutions as there are Bell Operating Companies (BOCs), with everyone appearing to go a different direction.

One thing is clear; all of the solutions involve fiber optics somewhere in the "local loop." The local loop is that stretch of cable which extends from the local exchange to the residential customer. This stretch of cable represents the bulk of infrastructure cost and is where the telephone companies lose most of their money. It is small wonder that telephone companies are reluctant to replace the existing infrastructure with a costly solution such as fiber optics; recouping that expense in a competitive world is very difficult.

The difference between the various fiber solutions lies in the distance of the fiber. Some alternatives require the fiber to be extended all the way to the subscriber's home. Others extend the fiber only to the point of distribution to a collection of homes (100 is often used as a model). In Chap. 8, we discuss these various Fiber-in-the-Loop (FITL) solutions in more detail.

2.7.1. Integrated Services—Pulling It All Together

ISDN originally described the entire telephone network. The intent was to combine all services into one network, using technologies such as ATM, SONET, and ISDN. The purpose of ATM and SONET was to provide a transmission facility between central offices, capable of carrying any type of signal. SS7, a proprietary network used in telephone company infrastructures, was designed to control the various switching functions within the network and to provide access to databases used by telephone switches to route traffic.

ISDN was later changed to describe the interface between the subscriber and the telephone company. Today, ISDN is an interface between the telephone company's and residential and commercial subscribers and is capable of supporting data, voice, and in some cases video. However, ISDN is not capable of carrying cable television (or broadcast) video service.

ISDN is being modified to support higher bandwidths along with ATM and is being referred to as Broadband ISDN (BISDN). BISDN will be capable of supporting video and high-speed data and will be available for both commercial applications and residential services.

By now, it should be clear that all of these technologies cannot stand alone. SONET is required to pipe the traffic from one location to another. The speed of fiber optics allows for far more digital traffic than conventional facilities do. But SONET cannot deliver digital transmissions end to end and needs another technology to support the transmission from one end through the network to the other end. That is the job of ATM.

ATM provides the services necessary to deliver digital traffic from the originating node to the destination node, but it does not carry information necessary to process the digital traffic once it reaches its destination. This information includes dialed digits, originating number, and possibly even service requests. This is the job of BISDN, which will provide the information necessary for a node to process the digital traffic once it is received. However, BISDN must use the services of ATM to reach the destination node. Within the network (interoffice) there is also the need to deliver important switching information. This information is currently handled through a separate network, SS7. The SS7 network consists of three entities, the Service Switching Point (SSP), the Signal Transfer Point (STP), and the Service Control Point (SCP).

The SSP is actually a software function within the voice switch itself. The STP function will change with ATM, providing access to important databases in the telephone network which support new services offered by the telephone companies. For more detailed information about this network, see Chap. 5.

The whole evolution of the Public Switched Telephone Network (PSTN) has moved from analog facilities and some digital facilities (especially interoffice trunks) to all-digital transmission facilities using fiber optic cables and high-speed protocols such as ATM to deliver the traffic from point to point. The subscriber does not share the same bandwidth requirements as the telephone company, so it is highly unlikely that many subscribers will have direct ATM connections. Instead, subscribers will have interfaces supporting ISDN, BISDN, Frame Relay, and TCP/IP. These will be offered over digital facilities (such as DS1) and transferred to the telephone company's ATM network. This may sound like crystal ball stuff, but remember that the ATM network began as an interoffice trunking solution developed by Bell Laboratories and has since been embraced by the entire industry.

2.8. Chapter Test

1. Independent telephone companies such as General Telephone did not exist until after the divestiture of the Bell System in 1984.

   a. True

   b. False

2. A class 5 office is also known as a:
   a. Toll office
   b. End office
   c. Tandem office
3. Name the seven Regional Bell Operating Companies formed after the divestiture of 1984.
4. In the new switching hierarchy (post divestiture) the Access Tandem is used to:
   a. Connect to other end offices within the same LATA.
   b. Connect to toll offices within other LATAs.
   c. Connect to long distance carriers.
5. Prior to 1996, local telephone companies were allowed to carry calls from one LATA to another, bypassing the long distance carriers.
   a. True
   b. False
6. The Internet is the Information Highway.
   a. True
   b. False.
7. The first state-funded network modeled after the National Information Infrastructure was deployed in:
   a. California
   b. Oklahoma
   c. North Carolina
   d. Texas
8. __________ Division Multiplexing divides transmissions into different frequency blocks for transmission over an analog facility.
9. ASAI/SCAI are used to pass __________ parameters from the network through a PBX to an adjunct processor.

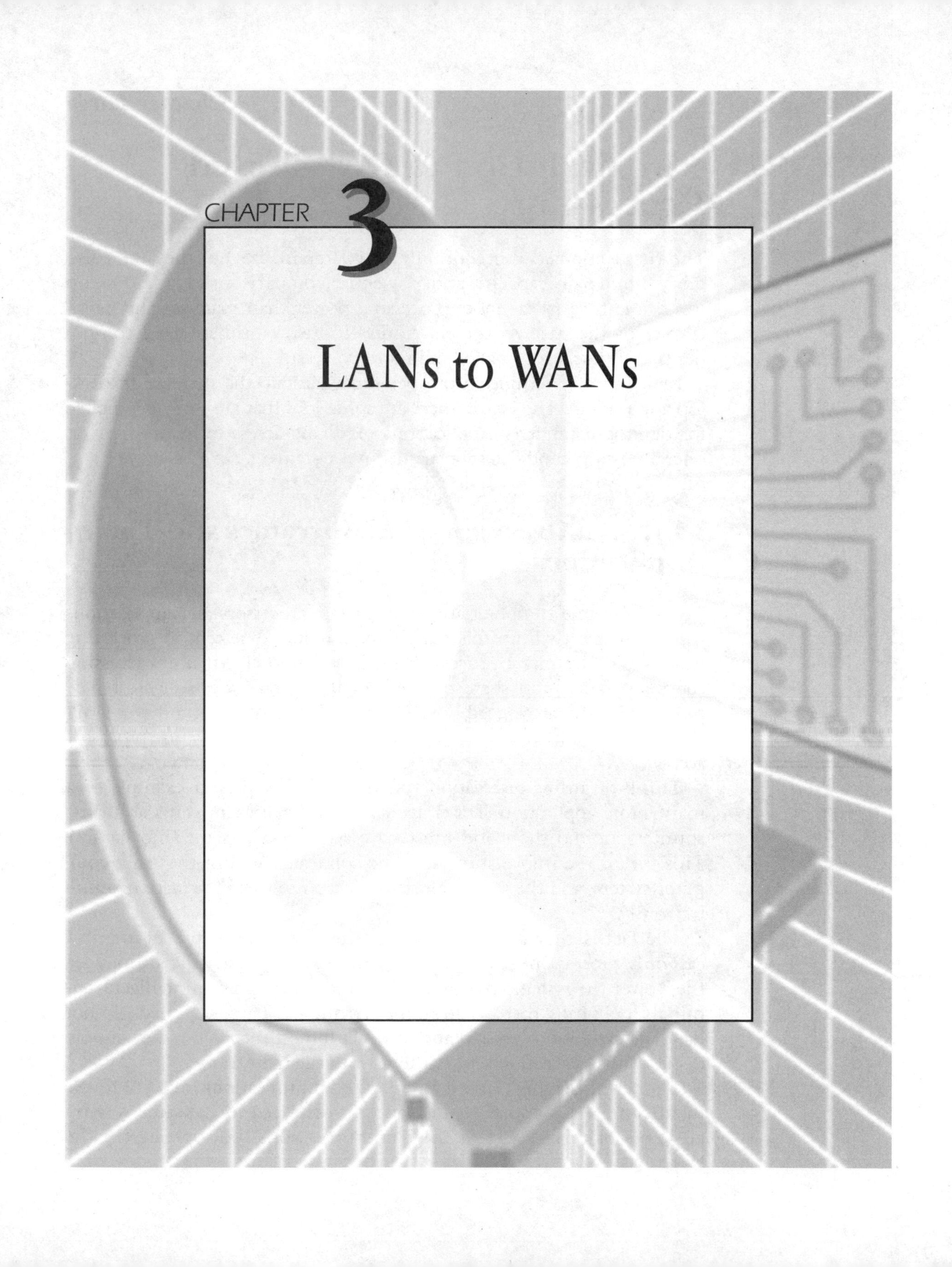

CHAPTER 3

LANs to WANs

3.1. Evolution to Distributed Processing

The first computers were not only very large in size, but they were also fairly limited in capacity and power. In comparison, today's desktop workstations carry far more processing power than many of the mainframes of the past. As the electronics industry miniaturized components, computers got smaller and more powerful.

Mainframes continue to prosper today, despite the increase in desktop applications. The fact is, there are some jobs that are just not suitable for desktop computers. Applications requiring access by many users at once are prime candidates for mainframe systems.

3.1.1. An Overview of Mainframes and Their Applications

The mainframe computer uses centralized processors and applications that are resident within the mainframe memory. The central processing unit is actually many processors working together. All users accessing the mainframe must share time with the central processing unit. The more users who are logged onto the mainframe system, the fewer available time slots. This means that users have to wait longer for processor access.

Think of airline reservation systems. This is a perfect example of a mainframe application. Travel agents as well as ticket agents at airline counters around the world must have access to the same information. This means the information must be collocated within the same computer system, and the agents must have access to this information simultaneously.

The fact is, they are really time-sharing. The central processing unit can only process one transaction at a time, so the busier the airlines get, the slower the system runs. Access is gained through nonintelligent terminals (politically correct for "dumb" terminals). These terminals do not have the capacity to process any information on their own; they simply display the information sent to them from the mainframe.

The terminals must have a common communications protocol to the mainframe. This is so the information is displayed in the same way on all types of terminals and the communications link operates in the

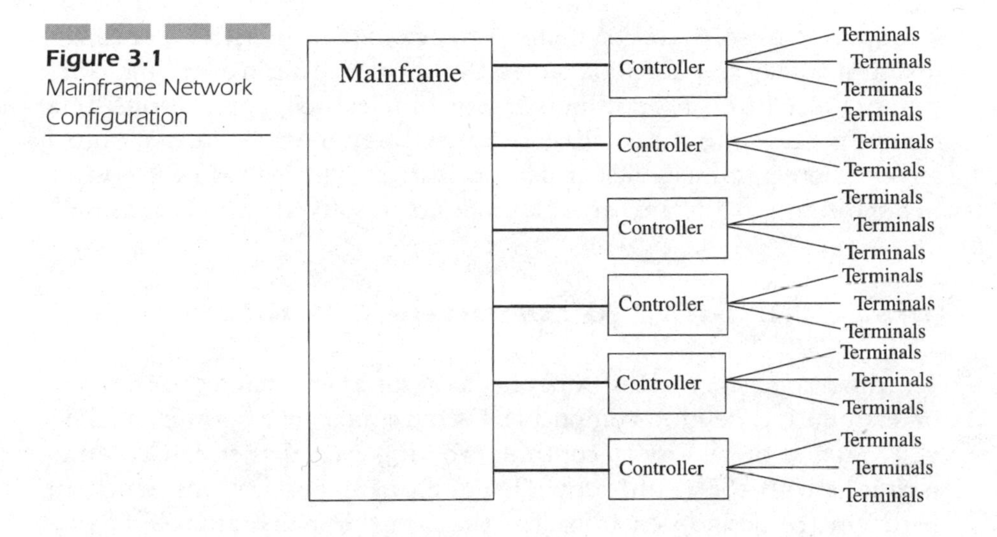

Figure 3.1
Mainframe Network Configuration

same way throughout the network. There are several protocols used within mainframe systems, the most popular being Systems Network Architecture (SNA), developed by IBM.

Mainframes can only provide so many ports for terminal connection, limiting the number of terminals they can support. Terminals, printers, and modems are connected to a cluster controller, which allows many terminals in one area to share the same communications link back to the mainframe. The cluster controller acts as a traffic cop for the central processing unit, allowing only one terminal at a time to communicate with the central processor (see Fig. 3.1).

This is where the communications protocol comes in. Each terminal is assigned a time slot within the communications protocol, and the cluster controller multiplexes these time slots into one serial stream over the communications channel to the mainframe.

There are some advantages and disadvantages to using mainframe computers. One of the advantages is the cost per user. Terminals are very inexpensive devices, especially when compared to today's desktop computers. This means that capital expenditures per employee are much lower in mainframe environments. However, when the mainframe goes down, productivity comes to a halt. Think of the reservation system where agents perform all of their transactions through the mainframe terminal. If they lose communications to the mainframe, they cannot make reservations. In fact, they cannot even check flight schedules or check baggage. This is one of the principal disadvantages to the mainframe system.

Applications software for mainframes cannot be purchased through discount catalogs or software stores. Companies using mainframe computers must hire programmers to customize their applications. If a change is needed to the database or a new entry must be entered onto a screen, a programmer must make the changes. This adds to the cost of maintaining a mainframe and causes delays in software upgrades as well.

3.1.2. The Move to Personal Computers

The disadvantages of mainframes helped spur a new market in the computer industry: desktop computing. Users wanted more power on their desktops and the ability to continue working even if they lost communications with the mainframe. There are many applications which fit better in the desktop environment than on the mainframe. Word processing is a good example. There is no need to use a mainframe computer for word processing. Since correspondence does not usually need to be stored in a central location, desktop word processors are more efficient.

There are many other examples of applications better suited for the desktop. Spread sheets, presentation software, and even desktop publishing is more efficient from the desktop. The miniaturization of electronics provided the ability to put powerful computers onto our desks and, today, into our laps. Palmtop computers are the next generation of computers, but they too will be tailored to specific applications.

As desktop computers became more popular in the 1980s, users began demanding more capabilities from their vendors. Exchanging files was difficult without sending the file to the mainframe, storing it on the mainframe, and then accessing it from another computer over the mainframe network. This proved to be slow and unreliable.

Users began seeking ways to interconnect their computers from one desktop to another. Computer manufacturers were quick to respond with networks designed specifically for the desktop computer. Xerox and DEC worked together to create a networking solution called Ethernet. Not to be outdone, IBM introduced their own solution based on a ring topology, called Token Ring.

As computers became more powerful, their popularity grew. This created more demands on the network, and soon a whole new industry was born: the Local Area Network (LAN) industry.

3.2. LAN Technology—Connecting to the Desktop

A LAN differs from the mainframe network in that there are no centralized processors. All processing is performed on stand-alone computers and workstations. The LAN interconnects these computers, allowing them to share files and communicate with one another.

All applications used by computer users are located on individual personal computers and workstations. This means that if the LAN should fail, the computers are still operational; they just cannot communicate with one another. The popularity of personal computers lead to the development of LANs, and today, the LAN industry has far outgrown the mainframe industry.

The disadvantage of a LAN is the cost. While the network costs are relatively low, the cost of individual computers can be rather high. In the mainframe environment, users communicated via terminals, which cost a few hundred dollars. In the LAN environment, users communicate through personal computers and high-powered workstations, some costing over $10,000 (depending on the applications they are used for). Even a simple computer with relatively low processing power can cost a few thousand dollars.

Another disadvantage is the support cost of personal computers. Applications continuously change, requiring upgrades. Oftentimes these upgrades require additional memory, forcing users to purchase additional memory chips for their computers. The cost of memory has not come down significantly, often costing as much as a new computer itself. Mass storage is fairly cheap, but memory is still rather expensive.

Still, when compared to the ongoing cost of mainframe equipment and productivity loss, the cost of LANs is really not much more than mainframes. In today's world, we have become so dependent on having computers that without one, we would be lost: no word processing, no spread sheets, no checkbook programs. This implies a shift in the computer industry from legacy systems to more personalized systems tailored to meet the needs of each individual user rather than a corporation.

Even this paradigm is changing. Many companies are now offering Network Personal Computers (NPCs) specifically designed for applications requiring central storage of information. The NPC is a computer, with memory and a processor. However, there are no disk drives in an NPC. This allows the operating system to be significantly smaller.

All applications must be run from a network server, reducing the cost of software upgrades. All files are stored on the server as well. This may introduce a bit of a problem in terms of throughput at the server if large workgroups are accessing it at the same time.

To alleviate the bottleneck at the server, workgroup servers will likely be used. All files and applications would be stored on the workgroup server, freeing the main server for other functions. Sound familiar? It should, since this is not much different than the mainframe environment.

3.2.1. Topologies and Basic Architecture

When connecting computers together, there are two forms of communications: point-to-point and multipoint. In point-to-point, two PCs are directly connected together and communicate with one another through this interface. In a LAN configuration, more than two computers are connected to this interface, but two computers are capable of addressing one another exclusively.

In point-to-multipoint connections, one computer is capable of connecting to many other computers at the same time. This is sometimes referred to as *broadcast* mode. The computer sends the same message to all the computers in the connection. This connection does not necessarily imply a physical connection. A logical connection can be obtained by addressing specific computer addresses in any mode, point-to-point or point-to-multipoint.

A topology is the logical layout of the network. It may not indicate actual physical connections, since many devices can be used to obtain the topology needed without physical wiring. For example, one may want a ring topology, which would require wiring all computers in a daisy chain from one to the next. The same can also be obtained by using a hub device where the ring topology is obtained through internal wiring in the hub itself. All of the computers are connected directly to the hub.

There are several different topologies used in LAN configurations. The star topology uses a hub device, with all devices connecting to the hub (see Fig. 3.2). The hub acts as a switch, allowing traffic from one computer to be switched to the addressed computer. The telephone network uses a star topology. This is not a very popular topology for LANs because the hub can introduce a bottleneck to data throughput.

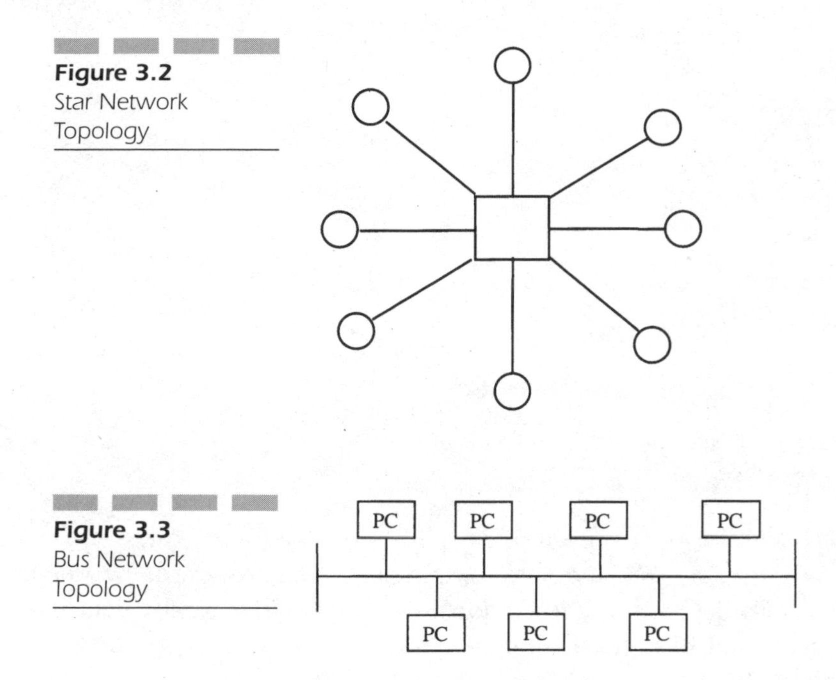

Figure 3.2 Star Network Topology

Figure 3.3 Bus Network Topology

Another more popular topology is the bus (see Fig. 3.3). In a bus topology, PCs are connected to one common bus. Transmissions from the PCs are placed on the bus for all other PCs to examine. Addressing is used to identify the destination of each data packet. As data is transmitted over the bus, each PC looks at the data packet header to determine if the address is its own. If it is, the PC then copies the data into memory. The original transmission stays on the bus, however, and must be absorbed at the end of the bus.

Data is transmitted in the form of electrical signals, which means the transmission is capable of reaching the end of the bus and reflecting back toward the originator. For this reason, the bus topology requires a terminator at each end, which acts like a sponge to absorb the electrical signal and prevent it from reflecting back onto the network.

While there are many different bus solutions, probably the most popular is Ethernet. Ethernet is a protocol designed specifically for LANs using a bus topology. We will discuss the Ethernet solution a little later.

A ring topology interconnects many computers together by daisy chaining one computer to the next. This means that every computer in the ring will have at least two connections: one for data coming in from its neighboring computer and one for data being sent to the next computer in the ring.

Figure 3.4 Ring Network Topology

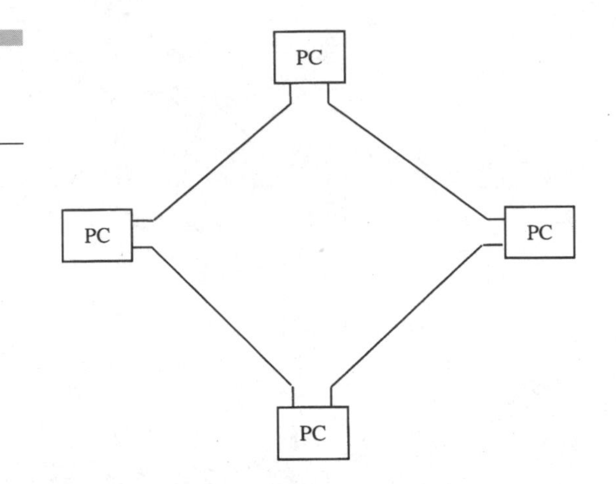

In this configuration, data is sent from one computer over the ring to the next computer on the ring. The computer then examines the header of the packet to determine if the address is its own. If not, the data is then retransmitted over the ring to the next computer. This process is repeated until the data packet reaches the final destination.

The fact that every computer in the ring retransmits the data implies that there is a built-in repeater function when using ring topologies. This means that repeaters are not needed; the electrical signal is regenerated every time the data packet is sent from one computer to the next. This is an inherent advantage when using ring topologies (see Fig. 3.4).

The downside to ring topologies is the fact that every computer must process the data packet as it makes its way through the network. Each computer introduces a delay in transmission while it determines the address and retransmits the data packet. This is different from bus topologies where the data is sent over the bus, and all computers examine the data without having to process the data.

Some ring networks use a dual ring topology. Fiber Distributed Data Interface (FDDI) is one example of a LAN solution which uses a dual ring. The idea is that one ring can be transmitting data in one direction, while the other ring is sending data in the opposite direction. If one ring breaks, the network is capable of healing itself by providing a connection between the two rings and forming one contiguous ring. This is discussed and illustrated in more detail in Sec. 3.2.5.

As shown in Fig. 3.5, in the dual ring configuration, each node on the ring has both an input and an output for each ring. Traffic is transmitted over both rings, but in opposite directions. By transmitting traffic

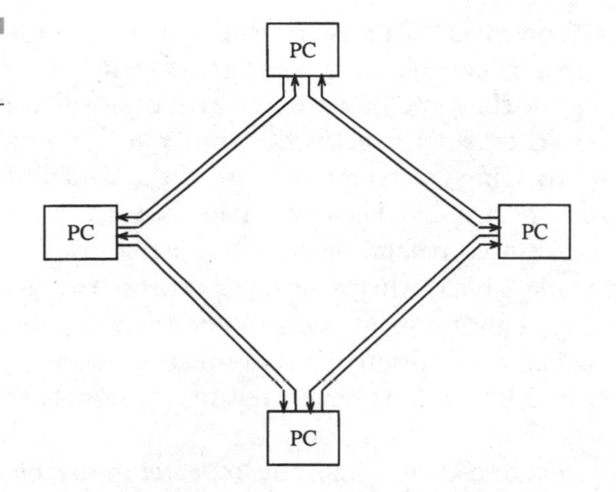

Figure 3.5
Dual Ring Topology

in opposite directions, any node can become a bridge between the two rings in the event there is a break.

Another feature of ring topologies is the ability to detect where a ring is broken. This is handled in a number of different ways, depending on which protocol is used. The protocol provides ring management, which usually entails nodes sending management messages to the neighboring node, constantly checking the status of their neighbor.

Dual ring topologies are also used in the Synchronous Optical Network (SONET), providing telephone companies with a more robust alternative to their existing digital transmission facilities. In the existing network, if there is a cable break, the central offices must determine which part of the network is out of service and route traffic around an entire office. With SONET, the ring is monitored by the protocol, and in the event of a break in the ring, traffic is automatically routed to the secondary ring.

Now that we understand the various topologies used in today's networks, let us look at the various devices used in a LAN and examine their functions.

3.2.2. LAN Devices

There are really only three basic network components in a LAN that are used to route data through the various parts of the network. Each one serves a specific function in the network. These devices follow a hierarchical approach in supporting the network.

3.2.2.1. Repeaters Data is transmitted through a LAN over copper wire. Copper is capable of transmitting data over a limited distance, depending on the type of wire and the protocol used in the network. Protocols can enhance a network's ability to send data over longer distances by changing the representation of the data and by using different compression techniques. However, there is still a limit as to how far data can be transmitted before the electrical signal used to represent the data begins to fade. This is when a repeater can be beneficial. Networks using bus topologies are frequent users of repeaters because the bus can only be so long before the original signal must be regenerated. It is important to understand here that repeaters regenerate signals; they do not amplify them.

To regenerate a data signal, the repeater must be able to determine what the original signal looked like. As data is received, the electrical signal is no longer as clean as it was. The waveform becomes more rounded rather than square. The repeater is capable of determining what the original waveform looked like, and it regenerates the signal based on an algorithm it performs.

Not all networks need repeaters. In ring topologies, the repeater function is provided by every node in the network. As the signal is sent from one node to the next, it is regenerated by each node. However, if there is a long distance between any two nodes, a repeater may become necessary.

Repeaters are considered layer 1 devices; they do not interpret any part of the data and are not capable of examining any of the addresses in the protocol header. They receive the data signal in the form of an electrical signal (or optical signal) and regenerate the entire signal based on the results of an algorithm.

3.2.2.2. Bridge A network bridge is used to join two different network segments (see Fig. 3.6). Understand that a bridge does not join two different networks together. Large networks must be divided into smaller segments to prevent too many nodes from creating congestion. The bridge is then used to interconnect the segments together.

When we begin discussing protocols, you will find that there are three addresses used in a protocol. The machine (node) address identifies the computer which generated data or is to receive a message. The network address identifies the network which sent data or is to receive the data. The third address type is used to determine which application in the computer sent or is to receive the data; this will be discussed later on.

The bridge is not capable of examining network addresses, which are

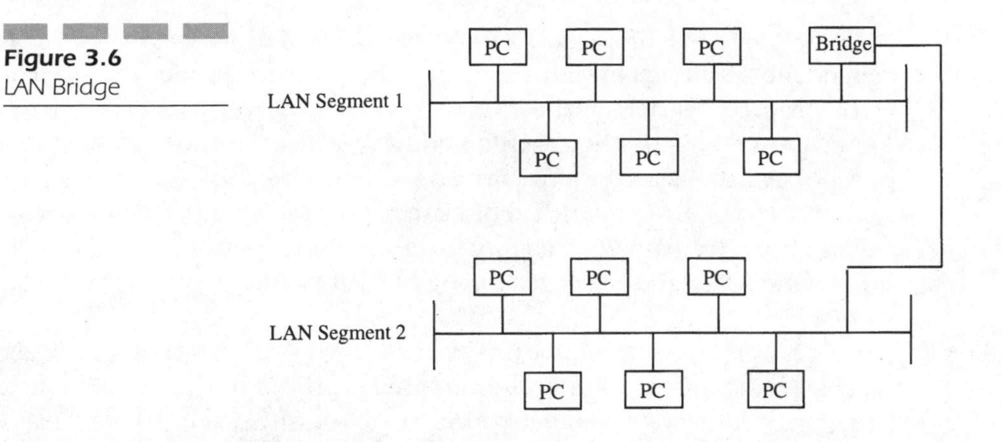

Figure 3.6
LAN Bridge

found in the protocol header. LAN protocols do not use network addressing because they only send data from one node to the next on the same network. Wide Area Network (WAN) protocols provide interconnectivity between LANs and use network addresses. The bridge is not used in WANs.

When data is sent over a network, all devices receive the data. Each node examines the machine address to determine if the data is addressed to it or not. The bridge also examines the machine address in the LAN protocol header to determine if the machine address is located on the other side of the bridge. If it is, the bridge copies the signal and sends it over the segment which it is attaching to.

The other traffic does not pass through the bridge and is simply ignored by it. The nodes on the other side of the bridge never see this data traffic. In today's complicated data industry, bridges often provide additional functionality, making it hard to draw a line between bridges, routers, and repeaters. For the sake of this discussion, we will stick to the basic functionality of a bridge.

As mentioned earlier, bridges only look at the machine address. For this reason, bridges are considered layer 2 devices. Again, as we discuss protocols and the various layers of a protocol further, this will become clearer. For now, just understand that these are layer 2 devices.

3.2.2.3. Routers A router is used to interconnect different networks together. A router may connect to many different networks, switching incoming data from one network out to another network. This is accomplished by providing multiple ports on the router. Each port is capable of receiving and transmitting data.

Routers depend on layer 3 (WAN) protocols. They too are protocol-dependent because they must examine the protocol header of the network protocol to determine the address of the destination network. They do not look at the machine address, which is found in another part of the data packet (and is part of another protocol).

Routers come in many different flavors, supporting multiple protocols and multiple transmission facilities (transmission facilities here refer to T-1, Frame Relay, and other technologies used to interconnect WAN segments).

Routers provide additional intelligence to the network and are often capable of making routing decisions based on the amount of traffic in a particular segment of the network and of routing around failed parts of the network. There are protocols designed specifically for routers, providing information to the router such as address tables and network status. We will discuss these further in Sec. 3.3.

3.2.2.4. Other Networking Devices There are other devices used in networks besides the three main components mentioned above. The Network Interface Card (NIC) is an integral part of any network. In order for a computer to connect to the network, it must have an NIC. The NIC provides the functionality of transmitting data over the copper or fiber facility. The NIC is protocol-specific and also provides the machine address used to identify each node in the network.

The machine address is "burned" into the NIC by the manufacturer, so it can never change. This ensures that each computer have its own unique machine address. Computers within a single network cannot have the same machine address; however, computers in different networks can have identical machine addresses (because there is also a network address to separate the two nodes). The general rule of thumb in any network is that every node within the same network must have a unique network "name," which serves as its address.

In Ethernet networks, a Media Access Unit (MAU) is used to connect the node to the network. The purpose of the MAU is discussed in detail in Sec. 3.2.3, but for now understand that the MAU is a connection point between the LAN and the computer in Ethernet networks.

A concentrator is often used to make an installation easier (see Fig. 3.7). For example, when installing a ring topology, it may become difficult to run wire from one computer to the next, maintaining a ring. If a concentrator is used, all computers can be wired directly to the concentrator, and the internal wiring of the concentrator forms the ring topology.

When connecting to a network using a protocol that is different from

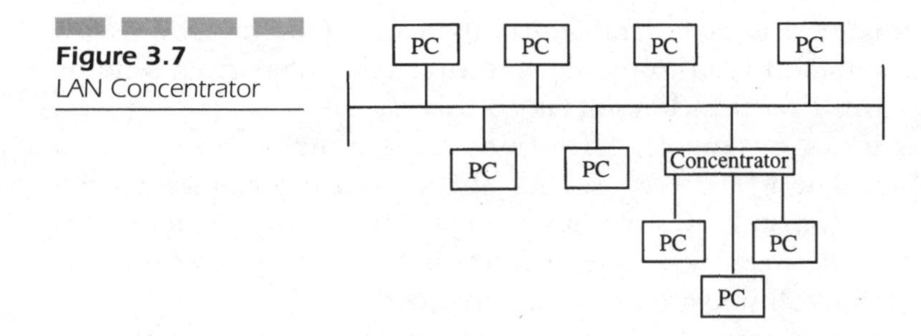

Figure 3.7
LAN Concentrator

that used by others, protocol conversion must be provided. For example, in connecting an Ethernet network to a Token Ring network, a device must be used to convert the protocol headers from Ethernet to Token Ring. This is so the devices in the network can interpret the data even though the various protocols place important information such as machine addresses in different parts of the data stream.

This conversion is often performed by gateways. The gateway function can be provided by many routers offered today, which makes it more confusing when trying to identify the various pieces of the network. Understand what the gateway function is rather than what devices perform the function.

Now that we have an understanding of the various devices in the network, let us look at the various protocols used in data networks. These are all LAN protocols, used in local networks. We will discuss protocols used to interconnect LANs a little later.

It is important to understand that the LAN protocol is used to interconnect computers located on the same bus or ring network. It does not communicate with nodes in different networks (at least not without the use of a WAN protocol).

3.2.3. An Overview of Ethernet

Ethernet was originally developed by Xerox and DEC as a way to interconnect their machines without the use of a mainframe network. The original Ethernet protocol was adapted by the IEEE, which made many improvements to the original design. The outcome was the Ethernet standard commonly used today, 802.3, which is called Carrier Sense Multiple Access with Collision Detection (CSMA/CD).

The IEEE also improved on the two protocols used in Ethernet and created another standard called 802.2. This standard deals with the actual packetizing of data and identifies the protocol structure, while 802.3 defines the standard used to prevent multiple computers from sending data at the same time (which results in collisions).

Ethernet uses a bus topology. To connect the computer to the bus, a MAU is used. The MAU serves several purposes. It determines when a node is able to transmit over the bus (using CSMA/CD), provides a standard interface between the node and the network, and determines if there has been a collision after transmitting data.

Not all Ethernet networks use a MAU to connect the computer to the network. A MAU is typically used with thick coaxial cable. This is commonly referred to as 10BASE5. Networks using thin coaxial cable have NICs with built-in MAUs. Thin coaxial is referred to as 10BASE2. In both cases, the 10 represents 10 Mbps, which is the bandwidth of Ethernet. The BASE indicates that the network is baseband, which means only one station can transmit at a time. The last number indicates the maximum distance supported by the medium. The 5 indicates 500 meters (m), while the 2 represents 200 m.

Many Ethernet networks today use twisted pair cable, which is designed to carry data at high speeds over the network. The difference between twisted pair for data networks and twisted pair for telephone networks is in the number of twists used in the copper. The tighter the twists, the faster data can be sent without interference.

In any bus topology, only one computer can send data at a time. If more than one computer is sending electrical signals over a copper wire, the electrical signals are going to "collide," causing the data to become corrupted and creating data errors. The Institute of Electrical and Electronics Engineers (IEEE) defined a means for preventing this from happening in 802.3. What it defines is a way for computers to listen to the network, looking for an electrical carrier signal before transmitting data over the bus. This is performed by the Ethernet MAU. If there are no carrier signals present, the MAU begins transmitting its traffic over the network. This is not a foolproof technique, but it works most of the time.

In the event that simultaneous transmission does occur between two nodes, the Collision Detection (CD) is capable of determining that there was an error; it waits to see if there is any carrier and then allows the node to retransmit its data. This is an improvement over the original Ethernet standard which did not have any collision detection or ability to determine if other nodes were transmitting.

In earlier Ethernet, a node would send a packet of data and wait for an acknowledgment. If an acknowledgment was not received within a

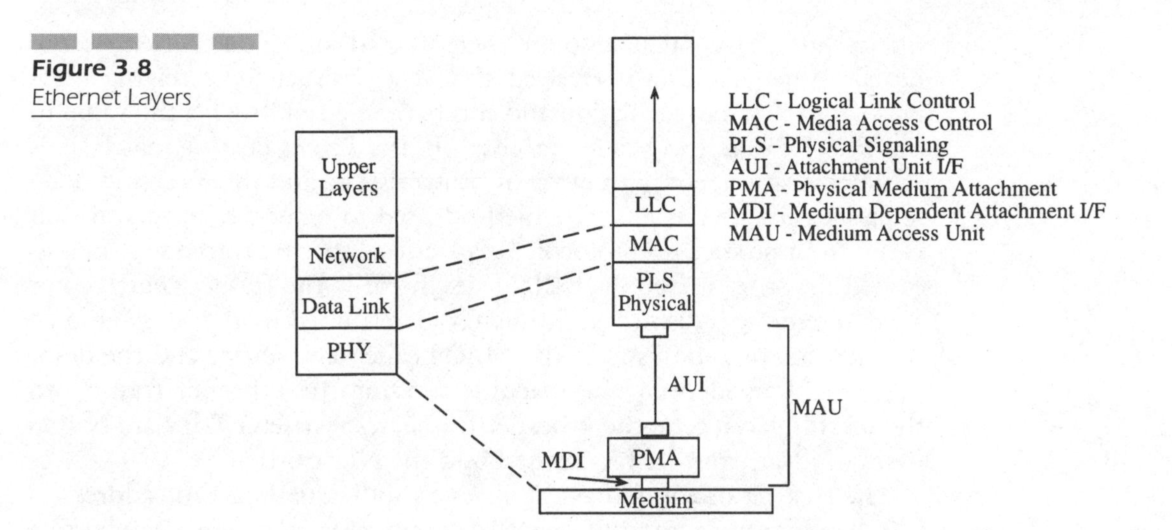

Figure 3.8
Ethernet Layers

few seconds, the node retransmitted the data. With CSMA/CD, the node will retransmit if collision is detected or if no acknowledgment is received.

Ethernet operates at the data link layer (as do all LAN protocols) (see Fig. 3.8). The functions defined by the Open Systems Interconnection reference model (OSI) for the data link layer do not sufficiently handle the needs of the LAN, so LAN protocols divide the data link layer into two layers, or sublayers as they are referred to. The sublayers used in Ethernet and most other LAN protocols are Logical Link Control (LLC) and Media Access Control (MAC).

3.2.3.1. Media Access Control The layer closest to the physical layer is the MAC sublayer. It deals with media access. It assembles the data into Ethernet frames when data is passed from the layer above (LLC) and prepares the data for transmission over the physical layer. The address of the node must be appended to the frame, and error detection/correction must be performed before transmitting the frame over the LAN.

When data is received by a node, the MAC sublayer reads the address from the protocol header and determines if the address is its own. If it is, the MAC sublayer strips off the protocol header, performs error detection/correction, and passes the data to the next sublayer (LLC).

Another function of the MAC layer is to add the Frame Check Sequence (FCS) to the header. This is used to determine if the data sent was received without error. As with many error detection/correction methods, an algorithm is used to check the data received from the LLC.

The results of the calculation are then entered into the FCS field of the header. When the data is received, the MAC layer at the receiving node performs the same calculation and compares the results with the value in the FCS field of the header received. If the values do not match, it is assumed that there was an error in transmission, and the receiving node requests a retransmission. The method used to request a retransmission varies from protocol to protocol. A particular Ethernet frame may be discarded for other reasons as well. For example, if the frame length is not valid, the frame is discarded. This is also determined by the MAC layer.

There are two addresses in the MAC header. The source and the destination address identify the machine sending the Ethernet frame and the machine to receive the Ethernet frame (respectively). They are both 6 bytes in length and are hard-coded into the NIC card.

The first bit of the address signifies an individual machine address (if it is 0) or a group address (if it is 1). A group address can be assigned to a number of machines on the same network. For example, a work group such as engineering may have a group address for hardware engineering and a group address for software engineering. This is determined by the system administrator, who must also configure group machines as part of a group. An address of all 1s is a broadcast address destined for all machines on the network.

Figure 3.9 shows the Ethernet frame and its parts at the MAC level. The LLC layer will of course add additional information, as we will see

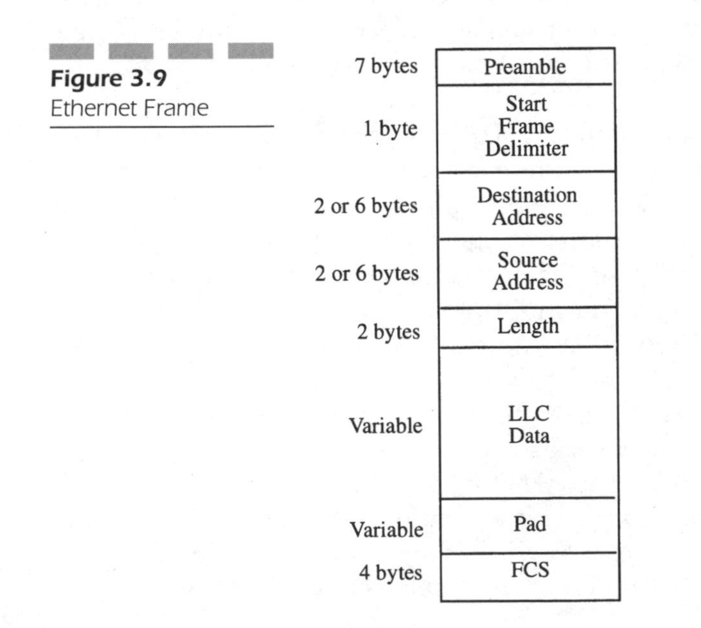

Figure 3.9
Ethernet Frame

later on. The preamble is a pattern of bits consisting of alternating 0s and 1s. The pattern is determined by the sending node and must not replicate any patterns sent within the data portion of the frame. This is then followed by the start frame delimiter, which is a predefined pattern which indicates the beginning of the Ethernet frame. This is used together with the preamble. The start frame delimiter is 1 byte in length.

Following the preamble and the start frame delimiter are the destination and source addresses. All machines on the network must have an Ethernet address. It is probably worth noting here that the Ethernet address is used only within Ethernet and is of no significance when a higher-layer protocol such as Transmission Control Protocol/Internet Protocol (TCP/IP) is used. TCP/IP uses its own addressing, but it does not identify the machine address of nodes within the destination network. We will discuss how protocols such as TCP/IP work with LAN protocols and their addressing later.

The length field of the frame identifies the length of the data field. This is necessary because the data field is a variable-length field. The data that is inserted in this part of the frame includes any data appended by protocols. For example, the LLC header can be found in the data portion of the MAC layer. You will find in the LLC header a data portion as well, which will include data appended from higher-layer protocols as well as the users data. This is an important concept to understand. As we look further and further into the protocol stack, you will see that each protocol used will append its own information into the data portions.

The pad field is used to maintain a consistent length to the overall Ethernet frame. The pad is nothing more than all 0s inserted after the variable-length data field to ensure consistency in Ethernet frame lengths despite variable lengths of data.

Now that we understand what is in the MAC layer and its function in Ethernet, let us look at the LLC layer, which is found in the data portion of the MAC header. There are many different types of LLC headers; their use depends on the type of service being provided by Ethernet. Three types of services are offered by Ethernet 802.3: unacknowledged connectionless service (type 1), connection mode service (type 2), and acknowledged connectionless service (type 3). The type of service provided is determined by the sending node.

This is the fundamental difference between the IEEE 802.3 standard and the Ethernet 2.0 standard. The Ethernet 2.0 standard only supports unacknowledged connectionless service and combines the MAC and LLC functions into one layer.

3.2.3.2. Logical Link Layer Unacknowledged connectionless service is the least reliable method of sending data. Connectionless means that prior to sending data, the protocol makes no attempt at establishing a logical session with the destination node. The data is placed in frames and transmitted to the destination node unannounced.

With this level of service, there is one basic message type (or data unit), unnumbered information (UI). Unnumbered means that there are no sequence numbers sent with each associated data unit. The frame must contain all of the information necessary to process the data at the destination node. Again, there is no guarantee that the destination node will ever receive the data because there are no data acknowledgments sent by the receiving node.

LLC management uses this type of service to exchange identification data with each node. Besides the machine address, the service access point (SAP), which identifies the various operations in the system, is also provided. This frame type is designated Exchange Identification (XID).

There is also a test message frame, used by protocol analyzers to send "loopback" frames over the Ethernet when testing. This frame type is also classified as unacknowledged connectionless service.

Connection mode service is a connection-oriented protocol. Connection-oriented protocols negotiate a session with the destination node prior to sending any data. A session is purely logical and does not represent a physical connection between nodes. The logical session guarantees delivery of all data by acknowledging every data frame received. Sequence numbers are commonly used in connection-oriented protocols, allowing the destination node to verify it has received the data units in the same sequence that they were transmitted, preventing out-of-sequence data errors.

The first phase in connection mode service is to establish a connection. Once the logical connection is established, data is transferred between the two nodes. Once all of the data has been transmitted, a disconnect is requested (by the originating node), and the resources allocated to the connection are released for another session.

There are several different frame formats used with connection mode service. The information frame is used for transmitting user data. Figure 3.10 shows the frame format for an information data unit.

The N(s) and N(r) fields are where the sequence numbers are found. The N(s) field contains the sequence number of the frame being sent, while the N(r) field contains the sequence number of the last received

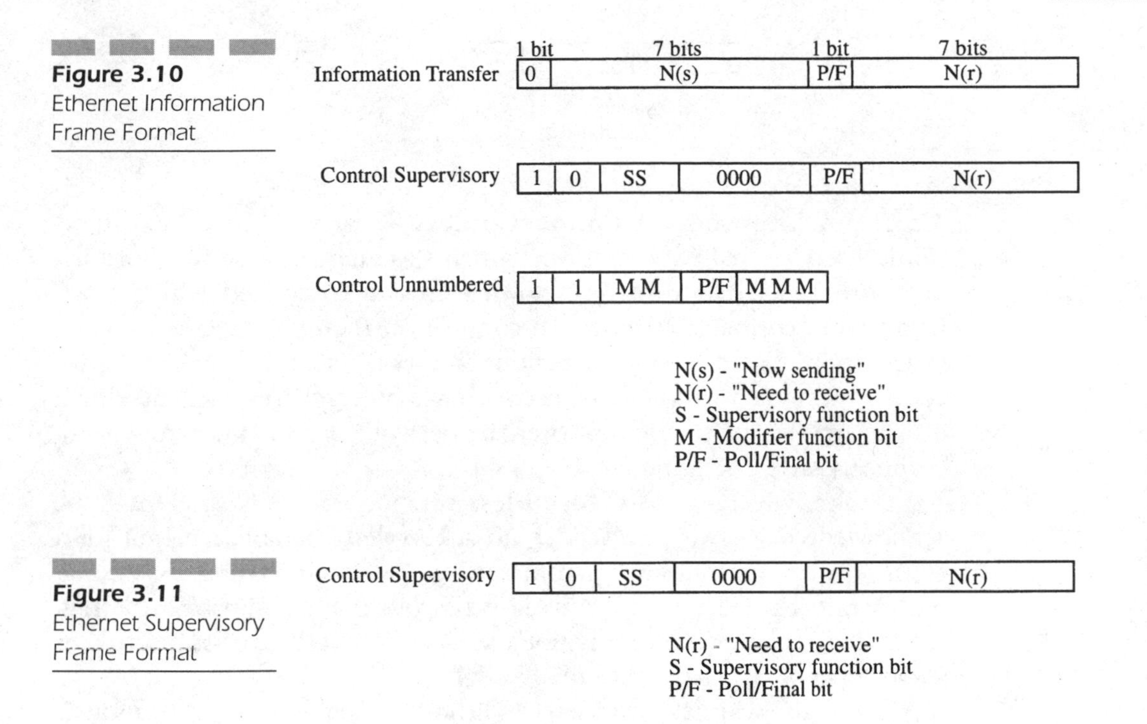

Figure 3.10 Ethernet Information Frame Format

Figure 3.11 Ethernet Supervisory Frame Format

frame. This allows any node to verify that data is received in the correct order and provides a means for acknowledging all received frames in one transmission.

The supervisory frame is used for sending acknowledgments when there is no data to send (see Fig. 3.11). If a node has received data but has no data to send in return, it can use the supervisory frame to send the acknowledgment (there is no user data field in the supervisory frame). This frame format is also used for flow control.

Flow control is used to stop data flow from the originating node. This consists of receiver ready and receiver not ready message types. The reject frame is used to request a retransmission in the event the frame is considered invalid. Invalid is different from a frame received with errors. The frame may have been of the wrong length, or it may be so corrupted that the header cannot be deciphered.

The unnumbered frame format is used when initially establishing a connection between two nodes. It is also used to send a disconnect message to a node, releasing a logical connection. The various message types used in unnumbered frames are shown in Fig. 3.12.

Control Unnumbered	1	1	M M	P/F	M M M

M - Modifier function bit
P/F - Poll/Final bit

Figure 3.12
Ethernet Unnumbered Frame Format

3.2.3.3. Acknowledged Connectionless Service This is a connectionless service that uses acknowledgments to guarantee delivery of each data unit. Every data unit is acknowledged as it is received, which is different from connection mode. In connection mode, the receiving node can acknowledge a whole contiguous series of data units. Not acknowledging every data unit as it is received saves on resources and cuts down on the amount of traffic sent over the network. This is known as asynchronous acknowledgment.

In acknowledged connectionless service, every frame must be acknowledged as it is received. If no acknowledgment is received for a frame, the originating node assumes it was lost during transmission and retransmits the data unit again. This of course adds traffic to the network. This type of service is not used often, but it can be useful for some types of data (such as e-mail).

This is just a quick overview of Ethernet, but it gives you enough information to understand how the protocol works. Use this information when learning about the other protocols in this book and compare functionality and protocol fields. You will find many similarities between protocols.

There is a new standard for "fast" Ethernet. Fast Ethernet supports transmission rates up to 100 Mbps over coaxial cable or category 5 twisted pair. The functionality of the protocol is somewhat the same, although some of the protocol has changed to support the higher data rates. Fast Ethernet is very competitive with FDDI and Token Ring networks because it can use existing network infrastructure.

3.2.4. An Overview of Token Ring

Token Ring was first proposed in 1969 by IBM. At the time, Ethernet was limited to 10 Mbps. Token Ring was capable of supporting 16 Mbps, which helped its popularity. Today, this is no longer an attraction since Ethernet is capable of supporting 100 Mbps (Fast Ethernet).

The IEEE later adopted Token Ring as a standard, numbered 802.5. There is also a Token Bus standard (802.4), which is similar to Token Ring except it uses a bus topology instead of a ring topology.

The principal difference between Token Ring and Ethernet is in the topology. While Ethernet uses a bus topology, Token Ring relies on a ring topology. Many of the same limitations of Ethernet apply to Token Ring, including the fact that only one node on the network can transmit at a time.

To resolve data collision, Token Ring uses a more reliable method than Ethernet does. A small data frame referred to as the token, is transmitted around the network. A node cannot transmit data until the token reaches the node. It can then hold the token and transmit its data. Once it has completed data transmission, it releases the token for the next node.

This is a very simplified explanation of how this protocol works. We will go into more details in a little bit. Another principle that is important to understand is that every node reads data frames as they are received from the network and then retransmits them.

This is different from Ethernet, where the data frames are transmitted over the bus, and each node reads the address of the frames. The data frame is really broadcast over the bus. In Token Ring, the data frame is passed from one node to the next, each time being re-created by the node.

Because the data frame is re-created at each node (or more accurately regenerated), there is usually no need for repeaters in Token Ring networks. Every node is a repeater, allowing for more distance in the network. However, this does not mean that Token Ring can handle large networks better. All networks become inefficient when too many nodes are attached to the same network segment.

Since the data frame must pass through each node in the network, delay becomes an inherent problem of Token Ring networks. The more nodes in the LAN segment, the more nodes the data must pass through. Each node can add as much as a 1-second (s) delay to data transmission. This is one disadvantage of Token Ring networks.

Like Ethernet, Token Ring divides layer 2 into two sublayers: the MAC and the LLC sublayers. Since LAN protocols tend to be topology-specific, there are obviously differences between the Ethernet MAC/LLC and the Token Ring MAC/LLC.

The physical layer is also very different from Ethernet. As mentioned earlier, each node is connected to a ring. This means that each node is connected to its neighboring node (rather than to a bus). There is only one connection, made using a connector. Earlier versions of Token Ring required that the nodes remain powered on at all times; otherwise they would cause the ring to fail (since they were responsible for regenerating

SD	AC	FC	DA	SA	INFO	FCS	ED	FS

SD - Start Delimiter
AC - Access Control
FC - Frame Control
DA - Destination Address
SA - Source Address
FCS - Frame Check Sequence
ED - Ending Delimiter
FS - Frame Status

Figure 3.13
Token Ring Frame Format

the data frame to the rest of the ring). This was later resolved through the NIC, where a shunt was placed to allow nodes to be powered down without interrupting the network.

If we look at the frame format of Token Ring (Fig. 3.13), we can see some similarities to Ethernet and other protocols. There is a delimiter to identify the start of the frame (much like the start frame delimiter in Ethernet) and an end-of-frame delimiter. The fields in the front of the frame are referred to as the start-of-frame sequence (SFS). The ending fields are referred to as the end-of-frame sequence (EFS).

The token that is passed around the ring consists of only the start-of-frame and end-of-frame sequence fields. There is nothing in between. As seen in Fig. 3.13, the only fields in the SFS and the EFS are the starting delimiter (SD), access control (AC), and the ending delimiter (ED).

The start-of-frame delimiter starts with 2 bits followed by a 0, followed by 2 bits and three 0s. This pattern cannot be repeated in the data field anywhere since it may then be confused as a new frame. This means that the pattern is variable, depending on the data contained in the frame.

The access control field is used to identify and control access to the ring. The first 3 bits are called primary bits, and they signify the priority level of the token. This is used by each node to determine whether or not they can hold the token or if they have to continue passing it along. Each node has a priority as well and cannot hold the token unless its priority is equal to or greater than the token priority.

This sounds complicated (and it can be), but the basic idea is to use a priority scheme to prevent nodes from taking the token out of turn. For example, let us say that node A is transmitting, and node C wishes to transmit next. While waiting for node A to finish transmitting and generate a token, node B suddenly decides it wants to begin transmitting. Since node A is connected to node B, node B would be next in line to receive the token, even though node C has been waiting longer. If node C could identify itself as having a higher priority than node B, the token would be passed through node B to node C. More about this later.

The next bit in the access control field is the token bit. This is used to identify a frame as a token. If the value in this field is a binary 0, the frame is a token frame. If the value is a binary 1, the frame is a data frame.

Following the token bit in the access frame is the monitor bit. This bit is used for managing the network. There must always be one node on the ring acting in monitor mode. This is the node that will be responsible for generating the first token when the network is first activated. The monitor node will also be responsible for removing from the ring data that never reached its destination. We will discuss the monitoring function in more detail later.

The next 3 bits are called reservation bits and are used along with the priority bits. This allows nodes on the network to reserve the next token for themselves. For now, know that as data frames are passed around the ring, these fields in the access control field are altered as they pass through each node on the ring.

The next field is the frame control field, which identifies the type of frame being sent (MAC or LLC). This is followed by an ending delimiter field, which carries a unique pattern of bits that is not duplicated in the data portion of the data field. This delimiter identifies the end of a frame.

The last field in the frame is the frame status. This field is used by the node identified in the destination address field. The receiving node of a frame identifies whether or not it recognized its address in the destination address field and also signifies whether or not it copied the frame into memory.

If there was an error of some sort, the destination node could signify that it recognized its address but did not copy the frame into memory. This is the way that Token Ring asks for retransmission of a frame. The originating node then reads this field when the frame comes back around to it again, and if the frame-copied bits are negative (did not copy the frame), it will retransmit the frame.

Now that we have identified the fields used in Token Ring, let us look at how data is passed around the ring. As a token passes around the ring, a node wishing to transmit data waits until the token passes by it and then seizes the token from the ring. The token bit is changed to indicate a data frame, and the remaining fields are appended by the MAC layer.

The node then transmits the data frame over the ring. If there is a lot of data to be sent, the originating node can continue transmitting data frames until it is either finished or a token holding timer expires. The token holding timer prevents any one node from "hogging" the network time by continuously transmitting and never releasing the token.

As the data frames are passed around the ring, each node on the ring reads the data frame, looking at the address fields. In addition to looking for an address, each node also performs error detection to determine if an error has occurred between it and its adjacent node. In the event an error is detected, the error bit found in the ending delimiter field is set.

The data frame continues its journey around the ring until the destination address is recognized by the destination node. The destination node must then determine if there is enough buffer space to copy the received frame. If there is enough room in its buffers, it changes the status bits to indicate that the address was recognized and that the frame was copied. The frame is then retransmitted over the ring.

The originating node is responsible for removing its data frame from the ring. When the data frame reaches the originating node, it is removed from the ring and the token is then re-created and transmitted. This sequence of events is repeated for every data frame sent on the network.

Each node on the network can also set a priority bit in a passing data frame, which is used in combination with the reservation bit to ensure that the node receives the next token transmitted. This mechanism is necessary because it is possible that a token could be seized by an upstream neighbor who has been waiting a shorter period of time to transmit data.

As data frames are passed around the ring, nodes wishing to transmit set the reservation bit to a priority higher than what is currently set in the priority field. When the originating node has completed transmission, it sets the priority bit one level higher and transmits a token. Stations which have indicated a lower priority cannot seize the token.

For example, let us say that node A is transmitting data. Node D reads the data frame and the priority bit, which is set to the value of 5. Node D then sets the reservation bit to a value of 5. When node A has finished transmission, it generates a token and sets the priority of the token at 5.

Other nodes on the ring cannot grab the token because they will have a lower priority than 5. When node D receives the token, it begins sending its data frames. Once node D has completed transmission, it repeats the cycle we just described, transmitting a token with a priority of 5 (or higher if another node has set the reservation bit higher). When the token passes by node A, node A decreases the priority back down to a lower priority so that it can be seized by upstream neighbors.

This is a very brief overview of the reservation system, which is actually a bit more complicated than described. The main point is each node has a fair shot at receiving the token and sending data. This requires the use of the priority and the reservation bits.

You may have figured out one flaw in this protocol. The originating node is responsible for removing its data frame from the network. But what happens if the originating node shut down shortly after transmission? The data frame can pass around the ring forever because the originating node is not present anymore. Even if the originating node were powered back up, it would not recognize the data frame as its own because all of its buffers would have been reset.

To prevent this from happening, there is always one station acting as the monitor station. The job of the monitor station is to read each data frame and determine if the frame has passed by it once already. The monitor station knows this because the first time a frame passes by the monitor station, it sets a monitor bit to 1. Any frame passing the monitor station with a monitor bit set to 1 has passed by the monitor station once already. The monitor station then removes the data frame from the network.

Any node can be the monitor station. When the ring is first initialized, the first station to be powered up sends a Active-Monitor-Present frame, indicating its wish to be the monitor station. If a node has already identified itself as the monitor station, when it receives the Active-Monitor-Present frame, it automatically goes into standby mode, relinquishing its status as network monitor.

This is a very high-level overview of the operations of Token Ring. The intent of this section is to familiarize you with the different protocols used in LANs so that you can better understand the specifics as you study any one protocol. As mentioned before, you will find many similarities between protocols. This makes it easier to learn new protocols. Once you have a firm understanding of what a protocol does and how it works, learning a new protocol takes less time.

3.2.5. An Overview of FDDI

FDDI was developed by the IEEE and adopted as an ANSI standard through the X3T9.5 committee. While this is a relatively new standard (first introduced in 1992), the work has been under way for some time.

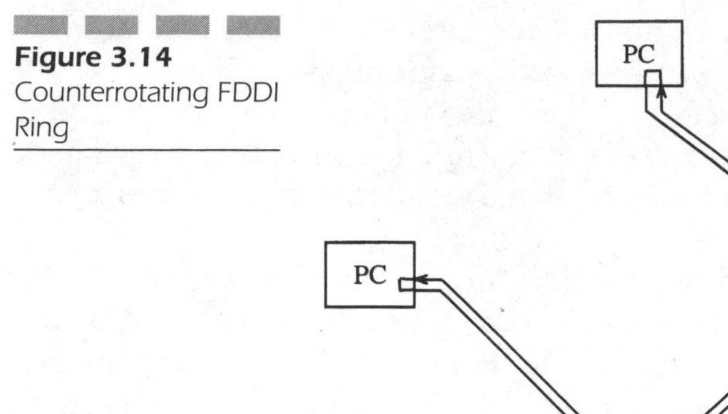

Figure 3.14 Counterrotating FDDI Ring

The concept for a fiber optics network utilizing a ring topology was first introduced in 1982.

FDDI is similar to Token Ring in that it uses a token to identify an idle network, eliminating the possibility of data collisions. Where FDDI differs is in its handling of the token. FDDI uses a timed token, allowing nodes to hold tokens for a predetermined time before they are forced to relinquish control of the token.

FDDI also uses a dual ring topology. This ring is counterrotating, which means that data is passed in both directions. This adds to the integrity of the network. As seen in Fig. 3.14, each node has two inputs and two outputs, each moving in opposite directions. If the ring should break between two nodes, the nodes pass data back to the other ring, forming one ring rather than two.

FDDI relies on other protocols above its own MAC layer, such as Ethernet MAC (802.2). In reality, where FDDI works best is as a backbone network used to move data from one LAN to another. FDDI provides 100-Mbps bandwidth, easily handling the capacity of Ethernet and Token Ring networks which may be attached. This is important for any backbone technology since the aggregate bandwidth from multiple networks may overwhelm the backbone.

Instead of going into the specifics of the FDDI protocol (such as frame types and protocol specifics), we will concentrate on how it operates. As mentioned earlier, FDDI uses a timed token mechanism. What this really means is that a node is allowed to hold a token for a limited amount of time and then it must release it. This is somewhat different from Token Ring since a node is allowed to hold the token for a longer period of time in Token Ring.

What is of significance in FDDI is the fact that the token is released immediately after the data frame is transmitted. The originating node does not wait for the data frame to return from its trip around the ring. This means that multiple nodes can be transmitting over the ring, as long as they wait until the token is received.

There is no danger of data collision because nodes must wait until a token has passed before transmitting. There is the possibility that multiple data frames will exist on the ring at the same time. As with Token Ring, the originating node is responsible for removing its own data frames from the network when it has passed one complete revolution.

FDDI also uses a distributed ring management function. Each node is responsible for communicating with its neighboring upstream nodes. Status information is constantly shared with adjacent nodes, allowing any node in the network to initiate ring initialization, isolate a fault, and recover from ring faults.

There are many other advantages to FDDI. Tables 3.1 and 3.2 compare FDDI with Ethernet and FDDI with Token Ring. As one can see, FDDI is capable of outperforming either protocol. Keep in mind, however, that there is now a 100-Mbps version of Ethernet which can be a stiff competitor with FDDI.

To understand more about FDDI, we will look at the various sublayers defined in the FDDI standard and will review their responsibilities. We

TABLE 3.1

Comparison between Ethernet and FDDI

	FDDI	Ethernet
Data rate	100 Mbps	10 Mbps
Maximum frame	4500 bytes	1518 bytes
Encoding	80% efficient	50% efficient
Distance between nodes	2 km	0.5 km
Maximum length	100 km	2.8 km
Maximum nodes	500	1024
Topology	Dual ring	Bus
Access	Token	CSMA/CD

TABLE 3.2

Comparison between Token Ring and FDDI

	FDDI	Token Ring
Data rate	100 Mbps	4 or 16 Mbps
Maximum frame	4500 bytes	No limit
Encoding	80% efficient	50% efficient
Clock	Each node	Master node
Maximum length	100 km	10 km
Maximum nodes	500	50
Topology	Dual ring	Single ring
Access	Token	Token

will not go into great detail about how the various sublayers work but will look at what they do.

The FDDI standard defines two layers which compare with layer 1 in the OSI model. The Physical Media Dependent (PMD) layer is the lowest layer, and it specifies all hardware and software interface operations. This includes the transmission of data over optical fiber, the connectors, services to the upper layers, and signal waveform and code requirements.

PMD supports single- and multimode fiber optic cable. The difference between the two lies in the way light passes through the fiber. In single-mode fiber, the inner core of the fiber is narrow, and the inner surface of the fiber is less reflective than in multimode. The actual core is hollow, allowing light to pass straight through.

A laser is used as a light source, shooting a light beam in a straight pattern through the inner core. Only a single beam can be used since other light beams would interfere with each other.

In multimode fiber, the light beams are aimed at angles. The inner core of the fiber is somewhat wider and more reflective. The light actually bounces off the inner walls of the fiber core at specific angles. Other transmissions can be represented by other beams at different angles, allowing for multiple beams to exist at the same time in the same fiber.

There are several different connector types used in FDDI. An MIC connector is used to connect two rings to one node. Each node has two MIC connectors and the transmissions are split between the two connectors for optimum redundancy. One pin on the MIC connector is for data coming in on one ring while the other pin is used for data going out over the other ring. With two connectors, a node can be ensured that it will always have a connection to one ring or the other even if there is a fiber cut.

An ST connector is used when a connection to only one ring is desirable. Each node utilizing an ST connector has two connectors, one for data coming in on the ring and the other for data going out on the ring. These are used for connecting nodes to concentrators.

Nodes can be connected in one of two ways. A node can be a Dual Attached Station (DAS) or a Single Attached Station (SAS). In a DAS configuration, a node is connected to both the primary and the secondary ring through an MIC connector. This is not always possible, depending on the area being cabled. In some cases, it may be more desirable to run a single set of fibers to a node and use a concentrator (to feed a number of nodes).

When a concentrator is used, a special cable is used to connect both rings to it. Each station is then connected using single fibers. Only the primary ring is connected to each node. This configuration is referred to as SAS.

The physical (PHY) layer protocol lies directly above PMD. PHY provides the encoding and decoding of data before it is transmitted over the fiber by the PMD sublayer. It is also responsible for transmitting and receiving of data, clock rate and synchronization, the various line states, and services to the upper layers.

Line states indicate what is happening over the fiber. There are several possible states.

- Quiet Line State (QLS)
- Master Line State (MLS)
- Halt Line State (HLS)
- Idle Line State (ILS)
- Active Line State (ALS)
- Noise Line State (NLS)

QLS indicates a loss of signal from PMD, which usually indicates a break in the fiber or a missing connection. MLS is used as part of the connection establishment phase. Before data can be transmitted over the fiber, it must first be established that all connectors are functioning and the ring is operational. The node will go into several line states during this process to ensure ring integrity before transmitting. HLS is also part of this process.

ILS is used to establish and maintain clock signals over the fiber. It is used as part of the connection establishment phase as well as between frames. ALS is used to indicate a connection has been established. NLS is used to indicate that there is noise on the fiber and there may be a faulty connection.

The MAC layer identifies the frame types being sent and also defines frame formats and sizes. When the network is idle, network management messages are passed around the ring to maintain integrity of the network. This includes status messages sent from one node to the next. Ring timing, fault isolation, and frame removal are also performed by the MAC layer.

There are several timers used in FDDI. They are used to determine how long it will take for a message to traverse one revolution around the ring. This information is also used to maximize the available bandwidth on the ring. The results from these various timers are used to calculate how long a node is allowed to transmit before it must surrender a token, allowing other nodes to transmit. This mechanism is a bit more sophisticated than the timer used in Token Ring.

The token rotation timer is used to time how long it takes a token to make one revolution around the ring. This is sent by station management (SMT) when the ring is first initialized during what is referred to as the claim process. Each node times how long it takes for a token to pass around the ring. A claim frame is then sent, with the value from the token rotation timer.

As the claim frame passes around the ring, each node examines the value of the timer and compares it with its own. If the node has a lower value than what is provided in the claim frame, it changes the value in the frame and regenerates the claim frame. This process is repeated at every node. When the claim frame has passed around the ring once, each station again looks at the value in the frame to determine if it matches its own timer value. If it does, the node "wins" the claim process and generates the first token.

The token holding timer is used to determine when a node must surrender the ring and generate a token. This means the node must stop transmission to allow another node to transmit data. If a token is received sooner than expected (according to the token rotation timer), the difference is then allocated to the token holding timer, allowing the node to continue transmitting.

The valid transmission timer is reset every time a valid frame is received. If the timer expires, the node then initializes ring recovery procedures. This is used to manage the integrity of the ring and provides a mechanism for ring error detection.

FDDI offers an alternative to broadband services for large networks such as campuses and hospitals. It makes an excellent backbone for interconnecting multiple networks to form one contiguous network. Based on fiber optics, it also offers a more reliable and error-free transmission medium than copper.

3.2.6. Client Server

First there was the mainframe, which used a centralized processor and centralized mass storage. All applications ran from the mainframe, accessed by terminals and later by personal computers. As the PC market flourished, the LAN proliferated. However, LANs do not address the need for centralized databases, applications which are shared, or centralized mass storage (for access by everyone on the network).

That is the purpose of client/server networks. A client is an application or program running on the desktop. It provides the protocols and interfaces necessary to interact with an application stored on a server. A server is nothing more than a high-powered desktop computer (although today we see dedicated computers designed for use as servers). This is located on the network and is accessible by all nodes on the network. Servers can be used to address a specific network application, or they can be multipurpose, dedicated to several network functions. Servers can be used as:

- Communications servers, providing access to modems and/or other network connections
- Print servers, controlling access to a number of printers located on the network
- Internet servers, controlling Internet applications such as File Transfer Protocol (FTP) and the World Wide Web (WWW)
- Application servers, providing access to common applications used enterprisewide
- Database servers, storing and managing enterprise databases

It is somewhat ironic that the networking industry has reverted to centralization of some functions, when the very reason for using LANs is to use distributed processing. The fact is, there are some network functions which must be centralized to be effective.

3.2.7. Network Operating Systems

Network operating systems (NOSs) provide diagnostics and management in corporate networks. There are a number of vendors who provide these, including Sun, Microsoft, and Novell. The NOS usually provides protocols which manage e-mail transmissions, disk allocation, mass storage access, and monitoring.

There are many network operating systems on the market today. Each is a proprietary solution with interfaces to more standardized solutions such as TCP/IP. In large enterprise networks (client/server), NOSs are necessary in order for network administrators to effectively monitor and manage the various resources on the network.

3.3. Bridging the Gap with Wide Area Networks

LANs remain within an enterprise, and while they may span several buildings, they cannot stretch across the nation without using some transport in between. This means there is another network used to interconnect various LANs within an enterprise and allow them to function as one contiguous network.

These networks are referred to as WANs. Previously, WANs were defined geographically, cited as networks whose boundaries spread across city limits. However, it becomes more difficult to delineate between LANs and WANs by boundaries alone. The easiest way to identify a WAN is by looking at its purpose.

If a network is connecting other LANs and the protocol uses network addresses rather than actual machine addresses, chances are it is a WAN. Let us look at some of the technologies used to deliver WAN connectivity and examine how they are used.

3.3.1. Basic Architecture and Options Available

A WAN uses routers to connect to LANs. The router is a layer 3 device which uses network addresses found in the header of the WAN protocol to route data through the WAN. Probably the most distinct difference between routing in a LAN and routing in a WAN is in the addressing. Routers do not use the machine address found in the header of the LAN protocol. In fact, routers have no knowledge of the end device, making routing decisions on the network address only.

There are a number of other technologies which are used to tie LANs together. Some of these are considered WAN technologies and some are not. By my definition, these are all components of a WAN.

3.3.2. X.25 Packet Switching

Introduced in the 1960s, X.25 has been used by many large corporations to link remote offices with headquarters' databases. Its popularity spawned the growth of many X.25 packet-switching carriers, providing connections across the nation. These network operators are still in operation today, despite the migration to newer and more robust technologies.

The concept of X.25 is to support asynchronous transfer of data without relying on the switched telephone network. When the switched telephone network is used for the transfer of data, a connection must be established and must remain connected throughout the transmission. If the data terminal is no longer sending data (idle), the connection remains until either data terminal requests that the connection be released.

This is an inefficient use of the network because facilities are being reserved during idle periods that have no data traffic. In the X.25 network, transmissions may use any available path, and while connections are still provided (connection-oriented services), they are logical, or virtual, connections, which means that facilities are not dedicated to the transmission and that any circuit can be used to route the data through the network. The logical connection (or virtual connection) is between the sender and the receiver of the data.

The X.25 protocol is the interface between the subscriber and the network provider. X.75 is the protocol used within the network. The two are similar, but the X.75 protocol provides services needed within a packet-switching network and does not worry about subscriber issues. X.75 can be considered a transport-only protocol, while X.25 supports retransmissions, segmentation, and reassembly.

The X.25 protocol uses a data link protocol called Link Access Protocol Bearer (LAPB). This protocol was developed to provide reliable transfer of data over a data link and, in the event of failure, retransmit the data. There is a lot of overhead presented in the LAPB protocol because each node in the network must examine each packet as it is received and determine whether the packet can be routed to the next node or if it must be retransmitted.

Another principle that must be understood about X.25 is that retransmissions are not typically from the originator of the data to the receiver. As the data is passed from one node to the next in the network, the packet is checked for errors. If an error is detected, the retransmission is from the last node to the node that detected the error. This means each node provides a lot of processing, driving up the cost of the equipment and introducing delays in the routing of data.

Despite the drawbacks of this network, X.25 works well in areas where highly reliable circuits cannot be provided. In areas where fiber optic networks have been deployed, X.25 is no longer a good choice. Technologies such as Frame Relay provide the same level of service with a fraction of the overhead found in X.25.

3.3.3. Using T-1 for Connectivity

When connecting LANs between two sites, a T-1 can be used. T-1 provides 1.544 Mbps of bandwidth, which is typically ample, but another protocol is needed to transfer data between the two locations. TCP/IP is a good solution for this type of application. Frame Relay can be used as well, especially if a point-to-point connection is configured.

If a full T-1 is not needed, a fractional T-1 can be provided as an alternative. Telephone companies divide a full T-1 between multiple subscribers, depending on the amount of bandwidth required. In some areas, you can order just a few channels (each at 64 kbps) or many channels, depending on your needs and what is available.

3.3.4. Switched 56

Telephone companies provide a 56-kbps facility from the subscriber to their switched network. The facilities are not packet-switching networks, but circuit-switched ones. Data can be switched through the network to any location (but the location must be configured first). This has been a popular solution for many data users because switched 56 is affordable.

3.3.5. Frame Relay

Originally developed as an interim solution and stepping stone to ATM technology, Frame Relay has quickly become the first choice for many corporations faced with connecting their LANs to remote offices. As file transfer and remote access to large databases become more critical to businesses, Frame Relay is an excellent choice for many reasons.

Frame Relay is a layer 2 protocol. It encapsulates data (including protocol headers from other network technologies) into a frame and sends it over a Frame Relay network to its destination. If a message originated in a TCP/IP network, the TCP/IP header information would be encapsulated into a Frame Relay frame and sent to its destination. If the destination was also a TCP/IP network, the header information could then be used to continue routing the data over the TCP/IP connection.

The unique feature of Frame Relay is that it does not provide any error detection/correction or real flow control (between devices). The only flow control provided in Frame Relay is used by the network to prevent network congestion. Even then the procedures are very simple, requiring little or no processing on the part of the network devices.

That is the real intent of Frame Relay. Remove all of the requirements for processing within the network and let the upper-layer protocols (such as TCP/IP or ISDN) provide those procedures. This means the end devices must provide the flow control procedures and error detection/correction outside of the network.

The end result is a very efficient network protocol that introduces very little delay because there is so little processing required to deliver messages.

3.3.6. ISDN

Integrated Services Digital Network (ISDN) is more a concept than a technology. The idea is to provide one facility which can support voice, data, and video. ISDN specifications define the protocols to be used to interface the subscriber equipment to the network.

There are two types of interfaces supported from the subscriber to the network, basic rate and primary rate. Basic rate interface (BRI) provides two 64-kilobits per second (kbps) channels and one 16-kbps signaling channel over one circuit. BRI is usually used for residential services or from an ISDN Private Branch Exchange (PBX) to the desktop.

Primary rate interface (PRI) provides twenty-three 64-kbps channels and one 64-kbps signaling channel. The facility is a T-1 facility (DS1). PRI is typically used to connect commercial subscribers with digital PBXs capable of supporting a direct ISDN connection.

There are protocols defined for ISDN from the subscriber to the network. The network-to-network protocol is not needed because the voice and other "bearer" traffic can be passed over existing digital facilities. Signaling information is passed from the ISDN signaling channel to the Signaling System #7 (SS7) network, which is separate from the rest of the switched telephone network.

ISDN and SS7 were designed for each other. To be more specific, when the ISDN concept was first conceived, it was determined that there would need to be a network to support the transfer of signaling information and provide access to intelligent network elements (such as databases). This network was developed before the ISDN interface to the subscriber. ISDN services cannot be provided end to end without full SS7 deployment between all of the participating networks.

3.3.7. TCP/IP

TCP/IP started as a robust network used by the government, called ARPANET. It became popular among scientists and researchers and soon found its way onto U.S. campuses involved in research projects for the government. Naturally, students who had access to the ARPANET looked for ways to continue using it. When the ARPANET was later made public and became the Internet, many companies began looking for ways to get connected.

The reason for all of the interest is because TCP/IP provides a number of protocols and solutions that meet just about every corporate networking need. To top it off, TCP/IP is a very robust yet efficient network solution, making it cost effective for businesses who need to link various locations with corporate headquarters.

TCP/IP is not just for linking locations with one another. Its about linking millions of users with one another. This network solution allows businesses to communicate with one another, share e-mail across the globe, send files from one city to another, and exchange ideas through special forums called newsgroups.

The concept of the Internet has moved indoors. Many are now finding that the same suite of protocols used to connect the world work just as well within the walls of the corporate office. For this reason, companies are now looking to TCP/IP to support their own Intranets: TCP/IP networks that are used internally to share files, e-mail, and data exchange.

3.4. Internet as a Model

The Internet has become one of the best models for determining the effects of global networking. Most companies today have some type of connection to the Internet, using it for e-mail, access to the World Wide Web (WWW), and even access to newsgroups. The way the Internet is used in business is evolving and is serving as a learning experience for all involved.

How the Internet will be used in business long term is still uncertain. One thing is certain. The Internet is not the Information Highway touted by the White House. The Internet is a piece of what will become the Information Highway and is probably the best model we have of what the Information Highway will mean to enterprise networks worldwide.

3.4.1. Lessons to Be Learned from the Internet

There are many lessons we can learn today from the Internet. The most pressing problem many are faced with today is managing information glut. There is so much information at our fingertips, managing that information has become a challenge.

Servers are becoming pressed for storage capacity. Networks are being stretched to their limits as workers download large files from one location to another. Safeguarding data on enterprise networks has created a new industry as companies struggle to prevent unauthorized access to their corporate databases.

While these are the network issues which the Internet has introduced us to, there are also social issues to deal with. Employees now spend a lot of time looking for information on the vast Internet, in some cases spending hours of valuable company time "surfing the net." This has become a concern for human resource managers, who are learning how to deal with productivity issues and network misuse.

Despite all of the problems the Internet introduces to companies wishing to participate, there is a good side. All of the experience we gain from dealing with these issues today will make migration to the Information Highway much smoother. The Internet is preparing us for a new way of information gathering, sharing, and socializing.

3.4.2. Issues to Resolve—Corporate Policies and Legislature

There are many issues to resolve when one connects to a global network such as the Internet. Company policies need updating to make the rules clear and concise. Company secrets can easily make their way over the Internet, in the form of documents and specifications, with little effort. This makes security an important issue for companies with connections to the Internet.

Access from external connections can also breach security in corporate networks. Already there have been many computer "break-ins" where computer hackers have made their way into government and corporate networks, accessed confidential information, and even destroyed database records. This should be of major concern to any corporation that is providing access to their network through the Internet or any other public network.

Firewalls, dedicated computers with security software designed to screen incoming access to the network by examining the addresses of incoming messages, can be effective. However, even the best firewall cannot prevent an experienced and seasoned professional from gaining access to any network.

One of the biggest issues yet to be solved is that of productivity. Employees with access to the Internet may spend hours upon hours surfing the Internet on company time. There are ways to monitor a station's access to outside networks, and there are other measures which can be taken to limit the kind of access employees have. These solutions restrict and censor access to the Internet, when the whole intent of the Internet is to bring a wealth of knowledge and opinion to the desktop.

Maybe the best solution is one of education. Teaching employees how to use the Internet in their everyday work may be a better approach, educating them on how to search out information and use the Internet responsibly. Policies and procedures should clearly outline acceptable activities, while also making it very clear the consequences of abusing Internet privileges while at work.

This will not solve all corporate issues with Internet access, but it will certainly lay a foundation which can be adopted when a much more global network is put into place, the Information Highway. If you think we have problems now with a network which provides plenty of text and pictures, wait until videos and interactive games become available right on every employee's desktop.

3.4.3. Corporate Solutions—The Intranet

The WWW has introduced a new philosophy about how we read and distribute information. Programs called browsers allow us to read and view text and pictures on our computer screens as they were formatted, but what makes these browsers especially unique is their ability to connect to other files, simply by clicking on a section of text or a part of a picture.

What provides this functionality is a language called Hypertext Markup Language (HTML), which was derived from an older language called Standard Generic Markup Language (SGML). SGML is still in use today for formatting documentation used in large companies. Many companies have replaced SGML with desktop publishing applications, which are easier to use and provide complete automation for documentation developers. However, these applications use proprietary code, which makes it difficult to pass files from one application to the next.

With HTML, the same file can be read by any computer using any operating system. HTML is not a proprietary code (although it is not standardized either), and as long as the browser used can interpret HTML, the file can be read. There are many browsers available on the market today, making it an attractive alternative for companies wishing to distribute their documentation over an internal network.

The concept is to make all company documentation available over an internal network, providing links between the various documents. This allows employees to find virtually anything they need to do their jobs without wasting valuable time searching through vast directories on corporate servers. It also provides a more interesting method for distributing employee newsletters, company directories, and human resource information.

Intranets have become the latest rave in Internet technology and have spurned a whole new industry. Many predict that in the next few years interest in Intranets will far surpass interest in the Internet.

3.5. The Internet Infrastructure—Worldwide Networking

The Internet consists of many different networks, all attached through a common backbone network. This backbone network was formally owned and operated by the National Science Foundation (NSF). The purpose of this network was to allow universities and research centers involved in government research to share information. This was later expanded to include the defense network and then commercial access.

It was not until commercial access was granted that the Internet began to flourish. Companies wishing to connect to the Internet could send e-mail to anyone with a connection, making their e-mail networks far more global. This became an important business tool, opening communications with vendors and customers alike. Today, we see the value of e-mail over the Internet whenever we receive a business card. Chances are, the individual's e-mail address will be on the card (right next to the pager, cellular, and voice mail numbers).

Corporations thrive on communications, especially where downsizing and consolidation have forced employees to play multiple roles within their companies. Busy executives and managers are hard to reach in today's work environment as they become more involved in the daily activities of their business. By using the Internet for e-mail, employees can be reached no matter where they are (providing they have a means for connecting to the Internet).

The WWW has opened a whole new concept in advertising, with the ability to reach millions around the world. Virtually every major corporation has a presence somewhere on the WWW, which they use for advertising their products, recruiting new employees, and sharing financial reports with their shareholders (and potential shareholders).

Companies with large databases use the Internet to give remote employees access to their companies resources. Remote offices no longer have to rely on dedicated lines to their corporate headquarters to access these databases. They can access them through Internet connections using a specialized protocol (TELNET) from their office or even from home.

The wide selection of newsgroups found on the Internet can be a valuable resource for those involved in research. Newsgroups are special interest groups that form a forum on the Internet. Anyone can read the contributions to these bulletin-board-like newsgroups, and anyone can make contributions. The newsgroups cover a wide variety of topics and are usually uncensored, which raises an issue for companies.

While the Internet introduces new challenges in the area of company policy, the advantages far outweigh the issues for Internet access. Companies will find the benefits of connecting to such a global network economical, and because the Internet is a global network, they will find they will be able to reach places they never thought of as markets.

3.5.1. Who Is in Control?—Supercomputer Centers

The first intention of the Internet backbone was to provide a network which would interconnect the various supercomputers located at research centers. These supercomputers were being used for research and development in government projects as well as by the defense department. There are six supercomputer centers in the United States which use powerful Cray computers. They are located at universities and non-profit research firms throughout the country.

Today, these computers are still used to maintain the Internet, but the government has handed over all aspects of the Internet to the public. The Internet Engineering Task Force (IETF) is responsible for ongoing development of the Internet protocols (TCP/IP). It makes decisions about the evolution of these protocols, supervises the testing and deployment of new protocols, and represents the users of the Internet in the industry.

The NSF was the original keeper of the Internet and still has a role to play in its development and upkeep. However, its role is a small one in comparison with the users of the Internet and the Internet Service Providers (ISPs).

So, who really is in control of the Internet? The answer is a difficult one because there is no clear owner of this network. The Internet is not just one global network but is millions of smaller networks all interconnected by the backbone created by the NSF. So, the Internet is really controlled by the user and the provider.

3.5.2. Direct or Indirect—Getting Connected

There are several ways to get access to the Internet. For individuals wishing to have personal accounts on the Internet, the most economical way is to get an account through an ISP. An ISP is not the same as services such as CompuServe and America Online. These are content providers, which offer their own services as well Internet access. In many cases, the services they provide are in direct competition with the services found on the Internet.

An ISP simply provides connections to the Internet. It may provide additional services as well, such as creation of home pages used on the WWW. It may also provide network services for larger companies who want to use the Internet as part of their corporate network.

In the near future, you will see the local telephone company providing the services of an ISP. The telephone companies have finally seen the value of the Internet and have suddenly taken an interest in providing access to it. They are busy building their own network, with connections to the Internet backbone, so that they can support high-speed access to both residential and business customers.

Regardless of who one gets a connection from, there are a handful of choices. For the individual, a dial-up account is the most economical (although many are looking at ISDN in their residence to get high-speed access to the Internet). A Point-to-Point (PPP) connection is the fastest way to get access over a dial-up account, and it provides full capability.

A simple dial-up account using a modem does not allow you to use the many protocols of the Internet (such as the WWW). These simpler accounts only allow you to dial into a large computer and access their services remotely. This is not as favorable as a PPP account because you rely on the applications made available over the connection rather than on your own software. With a PPP account, you can use whatever Internet software you want, provided it supports PPP access.

Larger corporate users will undoubtedly want to look at high-speed pipes to the Internet. ISDN is a logical choice, despite the problems in configuration and ordering. Frame Relay provides another attractive alternative. Whatever is chosen, corporate accounts can be purchased through ISPs and many telephone companies as well.

3.6. Internet Services

There are many different services provided over the Internet. All of them use specialized protocols (all or part of TCP/IP, which is discussed in a later chapter). Remember that the intent of the Internet has always been information sharing. All of the protocols developed for the Internet are designed for information distribution. There are several ways to distribute information, hence the many faces of the Internet. Here we will discuss the various ways the Internet distributes information to its users.

3.6.1. E-Mail—Global Delivery

The many advantages to using the Internet for e-mail have already been mentioned. The Internet gives users access to the entire networking community. For the first time, it is possible to send mail to someone outside of the company network. Businesses have obviously found this to be an important part of their networking strategy since today's business card is considered incomplete without an Internet e-mail address right under the telephone number (or several telephone numbers, as pointed out earlier).

E-mail requires a specialized protocol. The Internet set of protocols is TCP/IP, which includes over 1000 different special-purpose protocols used to provide access to various types of servers and routers. E-mail uses the Simple Mail Transfer Protocol (SMTP) to transfer mail from one server to the next. Once the mail message has arrived at the destination mail server, the Point-of-Presence (POP) protocol is used to transfer it to the desktop.

The Multipurpose Internet Mail Extensions (MIME) protocol allows data other than text e-mail messages to be sent using the SMTP protocol. For example, a photograph can be sent as an attachment to e-mail using the MIME protocol to handle the attachment.

3.6.2. Information Exchange—File Transfer

FTP allows files to be transferred from one server to another over the Internet. These files are treated as binary files rather than as formatted text files. This allows binary programs, pictures, text, and almost any other form of data to be transferred over the Internet.

While FTP was initially used to download free software, many companies have found it to be an economical distribution tool for their software updates and application patches. Many computer manufacturers offer free upgrades and software fixes through an FTP server. This allows them to distribute software to their customers without the expense of shipping materials.

3.6.3. Cheap Remote Access—Terminal Emulation

TELNET allows users to access mainframe computers and minicomputers (or most any remote computer) using a dial-up or direct Internet connection. For example, a company may have a customer database located at their headquarters in California. Their offices across the nation can access that database through their Internet connections, saving the cost of dedicated circuits back to headquarters.

This is much more economical for a number of reasons. With a dedicated circuit, they would be paying for a connection which is only used for one application, database access. Naturally, this facility would not be utilized all of the time. By using the Internet connection, they can use the facility for e-mail, remote database access, and any other Internet application.

TELNET is also used for accessing university databases and databases in state and federal government. This can be a useful tool for businesses that deal with the government or rely on university resources for research.

3.6.4. Blessing or Curse?—Newsgroups

Newsgroups are a lot like bulletin board systems. They are made up of special interest groups of virtually any topic. There is one central newsgroup server, which is responsible for collecting all contributions sent to the billboard service and then distributing them to all who subscribe.

The way it works is actually simpler than it may sound. A large mainframe in Virginia is used to poll news servers from around the world. These servers are specialized servers which collect "postings" from local connections and forward them to the main storehouse in Virginia. While sending postings, the news server also downloads any new postings which may be in the storehouse computer. However, the news server only downloads postings for newsgroups to which it has subscribed.

Subscription to a newsgroup is free; all it takes is configuring the server to query the main computer for all postings within that newsgroup, which is an autonomous function. The local server then stores those postings for its own subscribers. If the news server is owned by America Online, all AOL customers would connect to the AOL news server, which gets its postings from the main server.

The subscribers can only receive postings from newsgroups which their service subscribes to, which allows service providers to pick and choose the newsgroups it wants to provide access to. This is important to content providers such as CompuServe and AOL, who both offer their own versions of competing newsgroups.

The newsgroups use the Network News Transport Protocol (NNTP), which was developed by two students in North Carolina who wanted to exchange research data between their two universities over the Internet. Today, over 25,000 newsgroups are available over the Internet.

3.6.5. Commercialized Internet—World Wide Web

The WWW has changed advertising from printed media to interactive multimedia. The combination of text with video, animation, and photographs has created an exciting new way to distribute information about products, companies, or most any form of information. Specialized browsers decode the files, created using HTML, and display them according to rules defined in the browser application.

Newer versions of these browsers have the ability to access mail servers to retrieve e-mail, access news servers to retrieve newsgroup postings, and even access FTP servers to download files from remote computers. All of the functions of the Internet can now be provided through one application program, the WWW browser (Netscape is the dominant browser in the market).

Companies are now looking at these sophisticated browsers for their own internal networks, to view documents stored on corporate servers. This allows companies to distribute all forms of documentation electronically, without the cost of printed copies and without the worry of version control. With only one master file located on the corporate server, changes and updates are guaranteed to get distributed since all who have access will be viewing the same file.

New products aimed at making the WWW more powerful are flooding the market. One of these products is quickly becoming a de facto standard. Adobe has developed a new type of format, called PDF, for all types of files. With PDF files, all of the formatting information is stored in the file, and using their specialized viewer, you can view the document as it was created.

This new file format is much like Postscript, which is used to print files created in word processors and desktop publishing programs. PDF can also be used for illustrations and virtually every type of file used to convey information. Already, corporations are converting their documentation into PDF to be shared over the Internet, over an internal Intranet, and even on CD-ROM.

The WWW is still maturing. There are a lot of changes taking place in terms of how the WWW is used and what is placed on the WWW. One thing is for certain: The WWW will continue to be the highlight of the Internet well into the future.

3.6.6. Fad or Reality—Voice on the Internet

There have been many attempts to use the Internet to place telephone calls. They have been somewhat successful, depending on what one would call successful. The Internet is capable of transporting voice; however, delays in the routing cause delays in the delivery of the voice packets. This means conversations are full of pauses, which can be frustrating to those of us used to full-duplex communications.

New developments in the TCP/IP protocol suite include a new protocol designed for real-time applications such as voice transmission over the Internet. Once these protocols have been completed and deployed, it will be more feasible to use the Internet for voice transmission, competing directly with long distance telephone companies. However, in the meantime, voice over the Internet remains a fad and will not be serious competition for some years to come.

3.7. Chapter Test

1. What is the name of the topology that interconnects terminals in a ring configuration?
   a. Ring
   b. Bus
   c. Star
   d. FDDI
2. What suite of protocols is used in the Internet?
   a. ISDN
   b. TCP/IP

c. SS7
d. X.25

3. In a client/server application, what is the client?
4. What topology is used in FDDI?
5. What topology is used in Ethernet?
6. What topology is used in Token Ring?
7. Frame Relay is a good solution as long as the facilities used for transmission are reliable.
   a. True
   b. False
8. In mainframe environments, what is the device that terminals connect to and that connects directly to the mainframe?
9. What is an Intranet?
10. Who owns the Internet?
11. Which of the applications on the Internet supports special interest groups and provides them the ability to share "postings" with one another?
12. Can voice be transmitted over the Internet reliably?

CHAPTER 4

TCP/IP—Protocol of the Internet

4.1. Introduction

Transmission Control Protocol/Internet Protocol (TCP/IP) is a family of many protocols, each tailored to address specific applications within an internet. The fact that it has gained such popularity in the last few years can be attributed mostly to the success of the Internet. With so many corporate networks connecting to the Internet, vendors are busy developing products for these corporate users.

In addition to the explosive growth of the Internet, many corporations are seeing the value in the TCP/IP technology. TCP/IP offers a robust suite of services and applications, and it is standardized (which resolves a number of interoperability issues). No other technology offers the services of TCP/IP, the ability to interconnect with almost any vendors equipment, and the maturity of the TCP/IP protocols (TCP and IP have been standardized and deployed for over 12 years).

TCP/IP is not a single protocol but is a suite of over 100 protocols, each addressing a specific application within an internet. This is one of the factors which makes TCP/IP so flexible; each protocol can be used independently of the others, making them compatible with other transport technologies.

The fact that TCP/IP was developed in the 1960s does not make it obsolete. Development on TCP/IP protocols continues, as new demands are placed on internets. IP addressing has been strained to the point that new addresses are distributed on a limited basis. Old protocols have been updated and in some cases replaced by newer more robust protocols. There is no doubt that TCP/IP has a long future.

4.1.1. History of TCP/IP

During the 1950s, the U.S. defense network consisted of several mainframe computers linked by point-to-point transmission links. The concern at the time was the survivability of such a network during wartime. The Department of Defense (DOD) was looking for a replacement.

What was needed was a network which could heal itself. If any node, or worse, any section of the network were to suddenly become unreachable, the rest of the network must be able to continue operation. Such a technology did not exist at the time, so the Defense Communications Agency (DCA) began development in the 1960s on behalf of the DOD.

The result was TCP/IP. The first efforts concentrated on the network layer (IP) and the transport layer (TCP). Later, other protocols were added to provide additional functionality to the government internet. The Advanced Research Projects Agency (ARPA) network was first commissioned in 1968, but without TCP/IP. It was several more years before TCP/IP was commissioned in the defense network.

The group was disbanded in 1971, and work was resumed by the Defense Advanced Research Projects Agency (DARPA). DARPA began deploying TCP/IP in all ARPANET computers in 1983. The network was later split into two separate networks, the Military Network (MILNET) and the ARPANET. The network remained in use until the last original node was decommissioned in 1990.

Today, the ARPANET is known as the Internet, and it links millions of nodes that are connected via subnetworks located around the world. The Internet is no longer limited to military use, and the TCP/IP protocol suite continues to evolve through contributions from commercial, scientific, and educational institutions.

One of the smartest decisions made by the DOD was granting the right to distribute TCP/IP code to the University of California in Berkeley. It did not take long for the code to spread to other universities, which were already using the UNIX operating system. Universities began developing additional protocols, providing many of the applications we enjoy today on the Internet.

TCP/IP is a nonproprietary networking protocol, it works on any type platform, and it is flexible enough to use with any other type of technology (TCP/IP can be run over X.25, Frame Relay, or even the Integrated Services Digital Network, or ISDN). This is one of the reasons it has grown in popularity to become one of the most widely deployed networking technologies today.

4.1.2. Overview of Internets

Before we begin our discussion of TCP/IP, an overview of internets is in order. An internet is two or more networks linked together, forming one ubiquitous network. The internet should be transparent to the end user and should interoperate with all other subnetworks. This presents a challenge when mixing equipment from many different vendors. It is for this reason that standards are so important.

Throughout our discussions, we will use the term *internet* to identify any combination of networks interconnected with one another. The term *Internet* (capitalized) is used to reference the worldwide Internet that is used by many corporations and individuals today to link their individual networks.

Subnetworks are the individual networks within a larger network deployed by different corporations, service providers, and other large network users. A subnet may have a small number of nodes, or it may consist of other subnets (as is the case with many service providers). As a result, hierarchically, an internet may be several layers deep.

It is important to understand the relationships various subnetworks have with one another to understand the inner workings of TCP/IP. The devices used to interconnect the various subnetworks are routers and gateways.

A router receives data packets and forwards them through a port to another network or another part of its own network. A gateway works the same way as a router (from an TCP/IP perspective) but provides access to another network. A gateway can be considered an entry/exit point from one network to another. In other chapters, we discuss other functions for gateways, which will differ from our discussions in this chapter.

4.1.2.1. Autonomous Systems Internets are grouped into autonomous systems. An autonomous system is a group of networks joined together and maintained by a single authority. Autonomous systems are then linked to other autonomous systems by gateways. This provides a hierarchical approach to internets and simplifies the task of routing within an internet.

We will discuss the concept of autonomous systems later when we discuss routing in an internet. There you will see the role of this concept and the devices which are involved. Hospitals and universities are typical examples of where autonomous systems can be found.

4.1.3. Description of TCP/IP

As mentioned earlier, TCP/IP is a suite of protocols. These protocols fall within various layers of the protocol stack. To understand the protocol stack, one should understand the interactions between layers.

Subnetworks are managed at the physical and data link layers. While these layers are not part of the TCP/IP protocol, they do interact with the protocol stack. For example, to reach an IP address, the IP address must first be translated (or resolved) into a Local Area Network (LAN) machine address. If the IP address is part of a subnetwork, the subnet mask must be determined and the address resolution based on the results of the subnet mask.

Internetworking is managed by the IP protocol at the network layer. IP does not support error control, so it relies on another protocol for this function. The Internet Control Message Protocol (ICMP) provides error correction and flow control for IP. ICMP, although a user of IP, is still considered part of the network layer (ICMP information is encapsulated in IP packets, making it a user of the IP protocol).

TCP and the User Datagram Protocol (UDP) are considered service provider protocols and reside at the transport layer. TCP is a connection-oriented protocol, while UDP is a connectionless protocol. Both rely on the services of IP but do not require IP. For example, TCP and UDP can be transported over X.25 or Frame Relay services (both of which reside at the network layer).

The applications service layer consists of a number of protocols such as File Transfer Protocol (FTP), TELNET, and Network News Transport Protocol (NNTP). These are not really applications themselves but are protocols which interface to the various applications that are necessary to use these services. They provide the communications to remote devices but do not provide the user interface to interact with the various remote services.

All of these protocols encapsulate data into envelopes referred to as protocol data units (PDUs). There are many different labels used for these PDUs at various layers. In this chapter, *segment* will be used to describe PDUs from the transport layer (such as TCP) down to the network layer. In other words, when a protocol passes data from TCP to IP, the data unit is referred to as a segment.

A datagram is used to refer to PDUs passed from the network layer down to the data link layer (as in from IP to Ethernet). Datagrams sometimes refer to packets in connectionless protocols, such as UDP, but in this chapter they will refer to data units at the IP to data link layer.

Once a data unit has passed through the various layers and is sent to the physical layer, it is considered a frame. Once the data unit has been passed over the network, it is referred to as a pocket. These labels are pretty much consistent with other technologies as well and will be used throughout this book as described here, unless otherwise noted.

To interact with other parts of a host (software modules), protocols must interact with interfaces provided by the operating system. The operating system provides ports as entries to applications. As a data unit is passed to the application layer, the operating system provides a connection to the application by way of a logical port. The connection is established and maintained throughout the transaction period until data segments have been terminated.

A socket identifies an endpoint communications process. In order for communications to pass from one machine to another, a port must be connected and a socket defined. Internet ports are usually predefined (0 to 255) for well-known applications (such as FTP and NNTP). Undefined ports are provided as well, allowing operating systems to define their own ports when necessary.

Now that we understand some the terminology used with this technology, let us look at some of the advantages of TCP/IP. When data is sent from an application down to the transport layer, the data may be too big to fit into one data unit. TCP provides a service called fragmentation and reassembly, to handle this problem. The data is divided into evenly sized data units and is then passed to the network layer for further processing.

At the network layer, the individual data units may require further fragmenting. There is nothing wrong with this practice since protocols work within a peer-to-peer relationship. Data fragments created at the transport layer cannot be processed at the network layer; they must be passed to the transport layer before the fragments can be reassembled and processed. So fragmenting at various layers does not pose a problem.

When the data unit is passed over the network, it must pass through routers and gateways to reach its destination. It is possible that a data unit may pass through a network that will not accept the size of the IP fragments, and the data units will have to be fragmented further. This is not a problem for IP or TCP, which manages fragmented data at both the network and the transport layers.

When fragments are received by a host, the IP layer looks for the other parts of the data. Timers are used to determine when fragments are considered as lost, and when a time-out occurs, the received data is thrown out and an error message is sent to the source to generate retransmission of all of the fragments.

In addition to this handling of data fragments, multiple addressing conventions can be supported between various subnets. Different routing methods can be used within the various subnets with absolutely no effect on the end-to-end transmission itself. For example, a message may be passed to a subnet using source routing (the source defines the path to take to reach a destination), even though the source originated the message using nonsource routing (this is discussed in more detail in Sec. 4.3.4). In short, a message does not have to follow one method of routing all the way through the network. Each portion of the network can use any routing mechanism it wants without affecting the delivery of the original data unit.

By the same token, various services can be provided from one subnet to the next. A data unit can be originated using the connection-oriented services of TCP, but along the way a subnet can use connectionless UDP to pass the message through its own network as long as TCP services are used at the destination.

There are many other advantages to using TCP/IP in an internetwork environment. We will examine a number of them as we discuss the other functionality of these protocols. In short, TCP/IP was designed to support data communications through a number of nonrelated networks, ensuring reliable delivery of data even when networks fail along the way.

4.2. TCP/IP Standards

TCP/IP is a national standard, developed and standardized in the United States. Standardization is important because it ensures that various vendors' equipment will be compatible within the same network (provided the vendors followed the standards). TCP/IP standards are a little harder to track than other standards, partly because there is no central authority responsible for developing and writing the standards [as in the American National Standards Institute (ANSI)] and partly because of the way standards evolve.

In this section we will discuss how TCP/IP standards are written and how they evolve from contributions to standards.

4.2.1. Standards Documentation

TCP/IP standards are submitted in the form of a Request for Comments (RFCs). An RFC can be submitted by anyone and does not become a standard right away. In fact, there are thousands of RFCs available on various subjects regarding TCP/IP, but not all of them are standards. Many of them have not been implemented. This makes it very difficult to determine which are approved standards and which are not.

The first step in the standards process is the submission of a preliminary draft RFC. The draft is made available to anyone wishing to add to or comment on it (which is why it is called a Request for Comments).

Drafts can be found on the Internet at ds.internic.net (an AT&T server) and can be freely downloaded. Many network providers also provide access to these RFCs through their own servers.

Once an RFC has been submitted, the Internet Engineering Task Force (IETF) reviews it and makes a recommendation to make the RFC a standard. The document than becomes a draft standard. It takes about 6 months to move from preliminary draft to draft standard.

After another 4 months of review, and actual implementation, the draft standard can be moved to a published standard. This is again decided by the IETF, as well as the Internet Advisory Board (IAB). These committees and their activities are discussed in more detail below.

The IAB also publishes the IAB Official Protocol Standards Document List, which provides the status (preliminary draft, draft standard, or standard) of all RFCs. This is the best source for tracking actual TCP/IP standards. Before implementing any RFC, check this document first, or you may be implementing a nonstandard technology.

This process is far quicker than those used by other standard organizations. The cycle of ANSI or the International Telecommunications Union (ITU) standards may take 4 to 10 years before a standard is published. This is due in part to the use of a committee, rather than the actual user community as in the Internet.

4.2.2. Standards Groups

The Internet provides an excellent proving ground for TCP/IP standards. With its many variations and millions of users, the Internet is the best source for testing and validating new technologies, as long as the preliminary implementation is isolated to a subnet and does not affect the entire Internet.

There are several organizations which oversee the activities of the Internet and the standards process used to implement new technologies in the Internet. These organizations look after all TCP/IP standards.

The IAB oversees the entire development process. Using the resources of the IETF, they determine which RFCs will actually become standards and oversee the activities of the IETF.

The IETF evaluates RFCs and provides technical expertise to the IAB. Consisting of engineers and networking professionals, the IETF is the technical arm of the IAB. They also design and implement new standards under the direction of the Internet Engineering Steering Committee (IESC).

The IESC consists of IETF leadership and provides direction for the IETF staff. It is chaired by committee leaders and works under the direct input of the IAB.

The Internet Research Task Force (IRTF) oversees long-term issues of TCP/IP network architecture. They work under the direction of another steering committee, the Internet Research Steering Group (IRSG).

One other organization which works apart from those mentioned above is the InterNIC. The InterNIC provides services to the Internet community, such as IP address administration. Domain names are also issued by the InterNIC. The InterNIC is sponsored by the National Science Foundation (NSF). They are active in governmental issues as well and represent the Internet community in regulation issues. Now that we understand the standards and the various organizations responsible for the standards, let us look at the protocols themselves.

4.3. Internet Protocol

The IP resides within layer 3 (network layer) of the OSI Model. It provides end-to-end transport of data units through internets using connectionless services. Being connectionless, IP does not provide reliable data transfer, but this is not an issue if the upper layers provide reliability and error control.

An IP host must encapsulate data into IP headers, which are then passed to the data link (such as Ethernet). The protocol at the data link layer then encapsulates the IP header with the data into its own data unit (the datagram). The datagram is then passed down to the physical layer, where it is passed over the network as a serial bit stream (with possible encapsulation again, depending on the technology used).

For data to leave the local network, it must be sent to a router. Routers are network layer devices and are capable of processing the Ethernet and the IP headers. If the data is to be passed to another network, the Ethernet (or data link header) is stripped from the data, and the IP header is then processed.

Before transmitting the data over a port to the next network, the router must create a new IP header and place the data (consisting of the TCP header, possibly application header, and user data) into the IP header. The datagram is then given to the data link layer (which may now be X.25, ISDN, Frame Relay, or even Switched 56), and the whole process is repeated.

IP does have its limitations, the biggest being the number of addresses available. As you will see when we discuss IP addresses, there is a severe limitation in the number of addresses that IP can support. This issue has brought about the need for a replacement to IP. Internet Protocol next generation (IPng) provides a 16-byte address rather than a 4-byte one.

The primary function of IP is to provide routing information for data being transported through internets. Any error control is provided by the Internet Control Message Protocol (ICMP), which resides at layer 3 as well. This protocol does not provide error control but merely reports errors to the originating hosts.

IP is not a requirement for TCP. The TCP protocol can use almost any network layer protocol for delivery as long as the protocol is capable of providing routing services and supports the interfaces between the two layers. Remember that the concept of layering was to allow various layers of a protocol to be changed without affecting the layers above or below it.

4.3.1. IP Header

Figure 4.1 depicts the IP header and its fields. The first field in the IP header is the *version* field. This is used to identify which version of IP was used to create the header. This is important in internets because not every network is running the same version of a protocol. If the IP header was created in a network using the latest version of IP, it may contain information not recognizable by an older version of IP.

When this occurs, the receiving network (running the older version of IP) knows to ignore unrecognizable fields because the version field indicates a version newer than its own. This is valuable information for an internet and can be found in a number of the TCP/IP protocols.

Figure 4.1
IP Header Format

IP Header

<table>
<tr><td>Version</td><td>Length</td><td>Service Type</td><td colspan="2">Total Length</td></tr>
<tr><td colspan="3">Identification</td><td>Flags</td><td>Fragment Offset</td></tr>
<tr><td colspan="2">Time to Live</td><td>Protocol</td><td colspan="2">Header Checksum</td></tr>
<tr><td colspan="5">Source IP Address</td></tr>
<tr><td colspan="5">Destination IP Address</td></tr>
<tr><td colspan="4">IP Options (optional)</td><td>Padding</td></tr>
<tr><td colspan="5">Data</td></tr>
</table>

Figure 4.2
Type of Service Field

PPP	D	T	R	00

PPP = Precedence
D = Delay Attributes
T = Throughput Attributes
R = Reliability Attributes

The version field is followed by the *header length* field, which provides the length of the header itself. The data portion is not indicated here. One would think that the IP header would always be the same length, but there are some variable options which can be included in the header. The length is measured in 32-bit units (or words). A value of 1 indicates a header length of 32 bits, while a value of 2 indicates a header length of 64 bits.

Following the header length field is the *type of service* field (see Fig. 4.2). In other protocols, this is often referred to as Quality of Service (QoS). It stipulates the level of service the data requires. This field contains the four values discussed below.

Precedence is used to assign a level of priority to a data unit. It can be used for congestion management (lower-priority data units could be discarded while higher-priority data units are allowed to pass) and flow control. Not all networks implement this parameter. The precedence field consists of a 3-bit code indicating the type of precedence assigned. The values are shown in Table 4.1.

The *delay* field indicates whether a delay should be applied when sending the packet. The standard does not indicate how much of a delay

TABLE 4.1

Type of Service Values

Precedence		Reliability Attributes		Throughput Attributes		Delay Attributes	
111	Network control	0	Normal	0	Normal	0	Normal
101	Critic/ECP	1	Best possible	1	Best possible	1	Best possible
011	Flash						
001	Priority						
110	Internetwork control						
100	Flash override						
010	Immediate						
000	Routine						

should be used when this bit is set to 1. There are two possible values for this field, normal and low delay. A value of 1 indicates low delay.

The *throughput* field can be used to indicate that high throughput should be used with the particular data unit. For example, if the data unit was generated by a real-time application (such as an interactive game), the application may request a speedy delivery of the data units, requiring high throughput. The possible values are normal or high.

The *reliability* field is used in a similar fashion, indicating whether or not this data unit requires high reliability or normal service. If high reliability is indicated, it may be necessary to apply additional services at the upper layers to provide a high-reliability transmission, or it may mean that the data unit should not be routed over certain routes which may not provide a particular level of reliability.

The standard does not define how these bits should be implemented within an internet. It is left to the network provider to determine how these should be implemented within its own network. The protocol does provide the mechanisms for various service types if network providers choose to use them.

Following the type of service field is the *total length* field, which provides the length of the header and the data field (which would consist of the TCP/UDP header and the user data). The maximum length data unit at the IP level is 65,535 octets. This parameter allows nodes to determine the length of the data field by subtracting this value from the header length.

The *identifier* field is a 16-bit field used to correlate data unit fragments. When a data unit is fragmented, a number is assigned by the source to the fragments so that the receiver can match the IDs and reassemble the packet. The ID for associated fragments is the same, so the receiver can determine which fragments belong to each other.

A 3-bit *flag* field is used to qualify data units for fragmentation and to identify the last fragment in a series of fragmented data units. The first bit in the 3-bit field is always set to zero. The second bit is used to identify whether or not fragmentation is allowed for a data unit.

This bit can be used when routing a data unit through a network and you want to prevent intermediate networks from fragmenting the data unit. For example, a data unit may be sent in whole but may be too large for an intermediate network to pass. The intermediate network would then fragment the data unit. This bit would prevent the intermediate network from fragmenting the data unit and would possibly force the data unit to take a different route.

When data units get fragmented, the protocol must identify where each particular fragment belongs in the reassembled data unit. There are a number of ways protocols accomplish this. In IP, the fragment offset field is used to identify where a fragment should be placed in relationship to the whole data unit. This is then used at reassembly by the receiving node.

It is important to note here that IP does not "hand off" the single fragments to TCP and expect TCP to provide all reassembly of fragments. IP may have fragmented data units several times during its route to its final destination, making it impossible for TCP to reassemble the data unit. IP provides reassembly of fragmented data units at the IP level, and TCP provides reassembly of fragmented data units at the TCP level.

The *time-to-live* field is actually a hop counter. Each time the data unit traverses through a router, the router decrements this field by 1. The field originates with some value (maximum is 15, since 16 is considered as unreachable). The value is determined by the originating network, depending on the quality of service required by the data unit.

When the time-to-live field reaches a value of zero, the data unit is discarded. This is used to prevent circular routes with an internet. For example, a cluster of routers may be passing a particular data unit around in a circle because of some network failure. When a router reads the time-to-live parameter and sees the value has reached zero, the router will immediately delete the data unit and pass an error message (using the ICMP protocol) to the originator.

The *protocol* field identifies the protocol contained in the data field. Each protocol related to TCP/IP is identified by a number in the standards. Table 4.2 contains the numbers assigned by the standards for well-known protocols.

Error detection is provided at the IP level, but the user data is not

TABLE 4.2

Protocol Field Values

No.	Protocol
1	Internet Control Message Protocol (ICMP)
2	Internet Group Management Protocol (IGMP)
3	Gateway-to-Gateway Protocol (GGP)
6	Transmission Control Protocol (TCP)
8	Exterior Gateway Protocol (EGP)
9	Interior Gateway Protocol (IGP)
17	User Datagram Protocol (UDP)

checked for accuracy, only the IP header. This provides some streamlining of the protocol processing, while still ensuring reliability of the routing information. The *header checksum* checks only the header data, which includes the source and destination IP addresses.

When checking the IP header, the checksum validates the IP version number and validates that the time-to-live field does not equal zero. It also checks that the IP header is not corrupted and the message length is acceptable.

The *source* and the *destination IP addresses* are used by the routers and gateways within an internet to route the data unit. These addresses remain the same throughout the life of the data unit and are not altered by intermediate networks. We will discuss the format of IP addresses in greater detail in the following section.

An optional *option* field is provided for specific applications but is not always used. The options field is usually used by network control or for debugging purposes. When the *record route* feature is used, for example, the option field will indicate this. The record route data then follows the option field.

The data provided by IP options is variable and will depend on the actual application using it. For this reason, *padding* is provided. Padding is the placement of all zeroes to maintain a 32-bit boundary within a protocol. If the variable data causes the information to be less than a 32-bit segment, the rest of the space is filled by zeroes (as filler) and ignored.

Figure 4.3 shows an entire IP header, including the option field with the *record route* option enabled. The record route data fields are shown after the option field. Record route is described in more detail in Sec. 4.3.4.

The data field contains the layer 4 header as well as the customer data. This field may consist of a fragment instead of an entire data unit, depending on the size of the data unit when it was received by layer 4

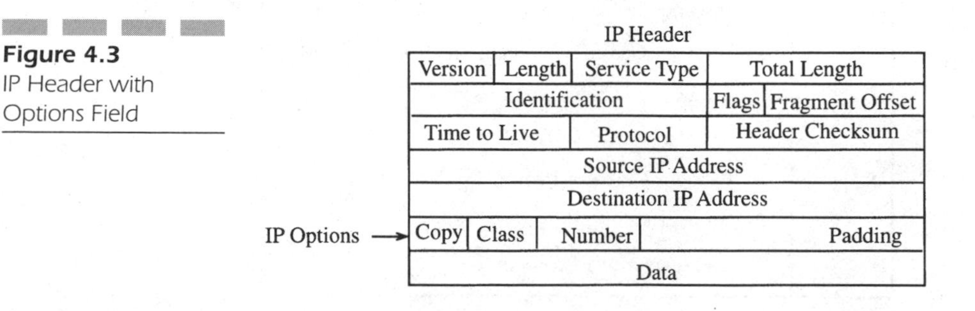

Figure 4.3 IP Header with Options Field

and depending on the services being provided by the upper layer. IP does not care about the contents of this field and has no visibility to its data.

4.3.2. IP Addressing

Addressing occurs at several layers within any protocol. For data to flow from a computer application, through the computer to an external interface, over a LAN, and out through a router to an internetwork, there must be addresses which identify all of the various interfaces used along the way. Let us begin this discussion by looking at the various levels of addresses relevant to IP and where they are generated.

The application will create a username. This is the first level of addressing and identifies a specific user to receive data. For example, if you are sending e-mail to someone, you do not know his or her machine address or individual IP address. All you know is the e-mail address, which can be translated by the receiving mail server into a username and machine name.

The internet application will identify a port to be used for the transmission. Consider the port as an internal logical address, not a physical address. A socket is assigned by the operating system and is a combination of the port number and the IP address.

The transport layer will identify an address for the protocol to be used to translate the data and present it to the application. For example, data generated using a TCP application must then be received by TCP at the destination, so that it may be processed and presented in its original form to the application.

The network layer will depend on IP addresses to route data units through various networks to reach the destination. Routers will read the IP address to determine which physical port the data unit should be transmitted through. The addresses we have already mentioned are transparent to the IP layer and are processed by host-resident software only.

Service access points (SAPs) are used by the LAN protocol for addressing within the Logical Link Layer (LLC). This is not part of the IP protocol but is part of many LAN protocols such as Ethernet and Token Ring. The LLC is defined in the IEEE standard 802.3.

Media Access Control (MAC) addresses are defined in 802.3 as well and can be found in the lower portion of the data link layer. Ethernet and Token Ring (as well as many other LAN protocols) divide the data link layer into two sublayers, called the MAC layer and the LLC layer. Refer to the discussion on the OSI Model in Chap. 1 for more information on their functions.

4.3.2.1. Sockets and Ports IP addresses and port numbers combined create a unique socket address, maintained and monitored by the operating system. The socket identifies a logical entity above the LLC layer. Remember that ports identify an application and usually have predefined numbers from the IP standard.

A socket is the combination of the originating IP address and the port number. The operating system provides the socket and maintains the logical connection established by the protocol. The socket concept allows multiple users (identified by the IP addresses) to address the same application (identified by the port address). This could be compared with the session established by a user in a mainframe environment. This concept was derived from the University of California, Berkeley version of UNIX back in the 1960s.

It is important that ports be standardized and well defined in the IP standards. This allows host-resident applications to identify ports without conflict. There is room for proprietary ports or ports not identified by the standards. These can be assigned within a private network for use within a corporate internet.

The concept of sockets and ports is an important one to understand. Earlier we discussed the various layers of addressing used within a protocol. Together, these two addresses are used internally (within a computer) to route data units to the proper application.

4.3.2.2. IP Addresses IP addresses are used at the network layer (layer 3) to route data units through the internet. There are four classes of addresses supported, although the standards are moving away from class addressing to classless addressing. This provides much more flexibility in the addressing scheme and allows for more IP addresses to be assigned. Currently there is a limitation to the number of addresses that IP can support, which has become an immediate issue because of the explosive growth of the Internet's popularity.

Figure 4.4 shows the structure of the four classes of addressing. Class A addresses are used most commonly within private "closed" networks. A closed network is one in which there is no external connection. Class B addresses are also used within closed networks. Class C addresses are the most commonly used address and are used for communications to external networks.

Figure 4.4 shows that a Class A address supports 126 network IDs and a total of 16,777,124 hosts. This class of address would not work for large internets because the number of networks outweighs the number of hosts. In a large internet, the nodes within the internet (routers and gate-

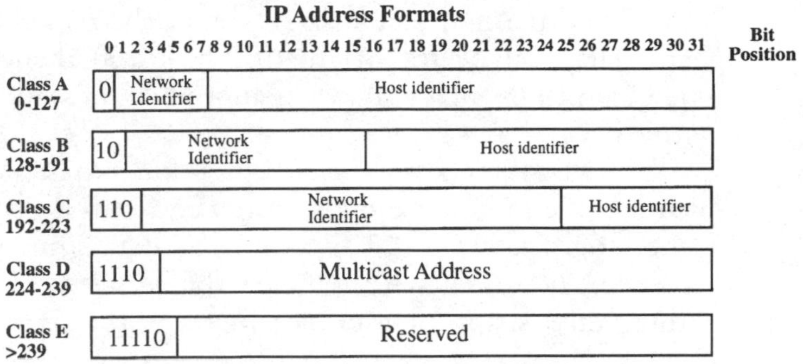

Figure 4.4
IP Addresses

IP Address Formats

	Bit Position 0–31		
Class A 0-127	0	Network Identifier	Host identifier
Class B 128-191	10	Network Identifier	Host identifier
Class C 192-223	110	Network Identifier	Host identifier
Class D 224-239	1110	Multicast Address	
Class E >239	11110	Reserved	

ways) do not care about the individual hosts. They only care about the network ID. So Class A addresses can only be used within closed networks.

A Class B address supports 16,384 networks and 65,534 hosts. This is designed for medium-sized networks and is not suitable for large internets. A private internet used to interconnect smaller networks within a large corporation would use Class B addresses.

Class C addresses are used for connections to the Internet itself. The Class C address supports 2,097,152 networks and only 254 hosts within a network. This is well suited for connecting to the Internet because it supports a large number of network connections. However, we are already running out of Class C addresses because of the popularity of the Internet. A new version of IP (IPng, due out sometime before 1998) will allow addressing up to a trillion networks.

Class D addresses are considered multicast addresses. Within a network there can be a multicast address, which is used to reach a group of individual hosts. Each host is assigned a multicast address and can be addressed as a group through the Class D address assignment. Routers manage the multicasting of data units to these addresses and maintain the table of hosts which are assigned these addresses.

The IP address does not identify the physical address of a particular machine. Remember that the actual computer may be connected through a LAN, which requires an address of its own (machine address). The machine address used by LANs is not compatible with IP because of the address structure. The IP address identifies a machine's connection to the IP network.

Consider this scenario. A computer is connected the LAN using an Ethernet card. The Ethernet card has a machine address hard coded into the card (nonchangeable) which is used to deliver data units over the LAN. As a new data unit is created, the IP address is embedded into the data unit by the originating host and then passed down to the LLC, which in turn will assign another address for use at the LAN level (the SAP).

The LLC then passes the data unit down to the MAC layer, which will embed the machine address (from the Network Interface Card, or NIC) into the data unit and transmit the data unit over the LAN to the router. The router will not send the MAC or LLC addresses over the Internet; it strips them off, leaving only the IP address at the network layer.

This again shows the flexibility offered by layering protocols. The addressing required at each layer can be included without interfering with the other layers. Without layering, this would be a difficult task. If a host is moved within a network to a new connection, the machine address does not change, nor does the IP address. If the machine address changes (because a new NIC card was inserted), the IP address may need to change, but only within the server that provides mapping from IP addresses to machine addresses.

The machine can move within a network without requiring the IP address to change. It makes no difference where a computer is as long as it has an IP address to identify its connection to the Internet. Some argue that this is not efficient use of IP addresses since a machine does not stay connected to the Internet all of the time. Because of this, there are some new standards which identify dynamic IP addressing.

Dynamic IP addressing requires a machine to log into a server and obtain its IP address from the server. The server will then provide the next available IP address to the machine. The server is able to keep track of the assignments by machine address because the request for an IP address is encapsulated in a LAN data unit, including the machine address. This is a more efficient use of the IP addresses because they are only used when a connection to the Internet is needed. Ideally, there would be lower demand for IP addresses using this scheme.

There are cases when the IP address does not sufficiently meet the needs of a large corporation. For example, ACME corporation may need to identify more than one network. They may have a need to address their various offices independently. This would require a Class C assignment for each location, rather than a Class C assignment for the corporation. This is when subnetting is used.

The Class C address identifies the corporation's series of networks. By using the subnet mask explained below, the corporation can further define which of its networks should receive a particular data unit. This is extremely useful in addressing several networks within a Class C assignment and has been extended for use in classless addressing.

4.3.2.3. Subnet Masking As mentioned above, gateways only address networks and do not look at the host portion of an IP address. Large corporations may need to address different segments of their network using different network IDs for each segment. This would require Class C assignments for each network segment if there were no mechanism within IP to accommodate this requirement.

Fortunately, IP provides for subnetting of an IP address by using a subnet mask (see Fig. 4.5). This allows an IP address to be divided into various subnets. The subnet address remains transparent to gateways and is only processed by the local router or gateway providing access into the corporate network.

The subnet mask uses portions of the host ID for identifying the subnets. The subnet mask identifies which portions of the host address should be used to identify the subnet. This of course will further limit the number of actual hosts which can be addressed within a network, but it allows for the division of larger autonomous networks.

The subnet mask is known only to the local gateways, which provide access to the various subnets. The main gateway providing access to the corporate network looks at the network ID portion of the IP address and determines that the data unit is destined to its own network. It then sends it to a routing table where the subnet mask is determined.

Once the subnet address is determined, the data unit can be sent to the appropriate router (or gateway) for processing. The subnet router then checks to verify that the subnet address is the same as its own and processes the host portion of the IP address for routing to the appropriate host within the subnet.

This is an hierarchical approach to routing within an internet and has many advantages. The biggest advantage is that intermediate networks used to reach the final destination do not need to know about the various subnets within the corporate network. The corporate network can then be maintained outside the realm of an internet.

This allows network administrators to change their addressing schemes within the autonomous network without affecting the rest of their internet. Changes made within the subnets are transparent to external networks, and routing tables in intermediate networks do not have to be updated (again, because subnets are transparent to the rest of the internet).

The routing table within the gateway must be aware of the subnet

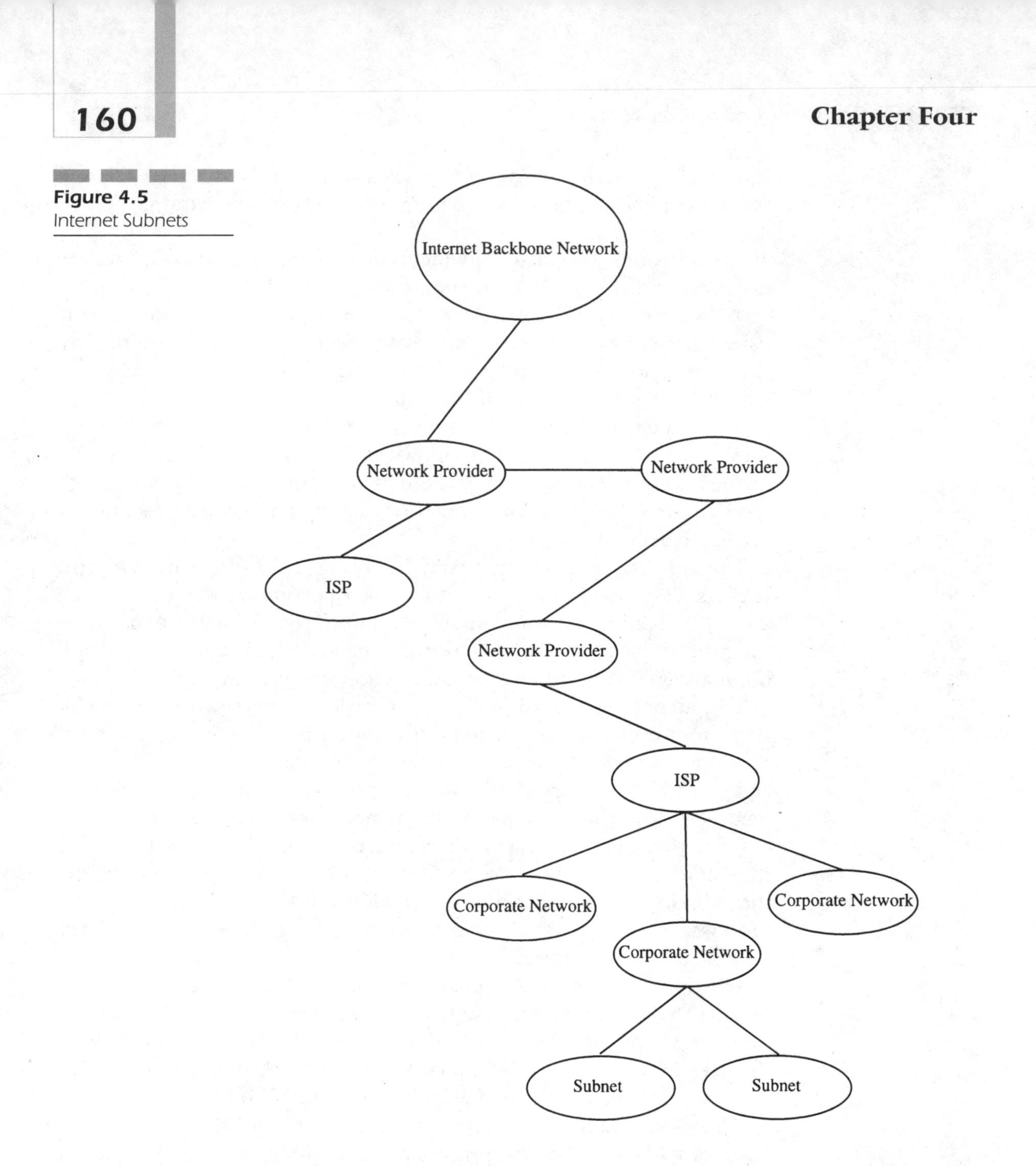

Figure 4.5
Internet Subnets

mask and know how to resolve addressing within its own network. This is a far better approach than assigning individual addresses to every subnet and updating routers and gateways throughout the internet every time these addresses are changed. It also prevents intermediate routers and gateways from having huge routing tables by minimizing the number of addresses which they need to be aware of.

The subnet mask is the same length as the IP address. When converted to binary, the subnet mask will usually consist of a string of binary 1s, followed by a string of binary 0s. The subnet mask is then ANDed with the IP address. The end result is the subnet address.

For example, we will take the IP address of 199.72.6.100 with a subnet of 255.255.246.0. The router will perform the AND operation against these two numbers. The result would look like the example below:

IP Address=199.72.6.100	11000111	01001000	00000110	01100100
Subnet Mask=255.255.246.0	11111111	11111111	11110110	00000000
Subnet Address=199.72.14	11000111	01001000	00001110	

The entry within the routers tables would then look something like:

Destination	Subnet Mask	Next Hop	Port #
199.72.0.0	255.255.246.0	199.72.14	A

Any combination of 1s and 0s can be used in a subnet mask. In short, a subnet mask can be very simple or very complex, depending on the needs of the individual network. This method of routing to subnets is critical where classless addressing is used. There are no classes of addresses defined in the newer IP standards because masks can be used to determine the network ID and host portions of an IP address.

There are a number of RFCs addressing possible replacements to IP. We will not look at these alternatives in detail because they have not been identified as completed standards at this writing. The alternatives do support a much larger IP address, as well as classless addresses.

So far, we have identified how IP addresses work. IP addresses are of no value to the LAN, however. The machine address resident on each computer's NIC is the only way to route data units from an internet over the LAN and to the host. Consequently, an IP address must be converted, or "resolved," into a machine address. This function is performed by a dedicated server.

When a data unit is finally routed to a LAN, the IP address is no longer usable for delivery to the destination node. LANs depend on machine addresses, which are resident in firmware on a computer's NIC. The LAN protocol is not able to process IP addresses and must rely on machine addresses to deliver the data unit.

Routers at the end of the delivery path must be responsible for changing the IP address into a machine address which can be used by the LAN protocol to deliver the data. When a node powers up on a LAN, it sends its machine address to the router to notify it of its existence. The router stores this information in a routing table for later use. The router may send messages to the node periodically to make sure it is still accessible. This is accomplished through routing protocols, which we will discuss a little later.

The important thing to remember here is that the IP address is only good for getting to a local router. Once it reaches the proper network, the router must resolve the address into a machine address for delivery over the LAN.

4.3.3. Domain Name System

No one likes to use numbers when sending e-mail or when trying to connect to another network. The *Domain Name System* (DNS) supports the use of a naming convention for addresses rather than IP addresses requiring numbers. The address must be converted at some point to an actual IP address, but this process is transparent to the user.

DNS was developed by the Internet Network Information Center (InterNIC) in 1983 and is still administered by InterNIC today. All users of the Internet must apply for a domain name which corresponds to their IP address. The IP address must also be registered by the InterNIC (actually, you have no choice of which IP address you receive; it is assigned to you). Domain names are a popular possession and have become the trendy identity over the Internet.

DNS uses an hierarchical approach, which fits well within the IP addressing scheme. A dedicated server must be used to resolve the domain name into an IP address. This same resolver also provides the machine addresses of all the local hosts (with a direct connection to the server).

The name resolver is a server application, which must then rely on the services of a name server. The name server performs the actual lookup of the domain name and provides the actual IP and machine addresses of the host. The name server and name resolver only know about local hosts and do not concern themselves with addresses outside of their own autonomous networks.

The name server can reference other name servers. This is often the case when an IP address for an entity outside of the local network must be resolved. Name servers are deployed in such a fashion that they know of all other name servers to which they have a direct connection. Requests may be sent through a number of name servers before they are actually resolved and an IP address provided.

Once resolved, the name server (and the name resolver) can maintain the IP address in cache memory. This eliminates the need to query name servers for domain names used frequently. The domain names do not stay in cache forever and are eventually removed from the cache. The time-to-live is determined by the network administrator and stored in the name server.

Name servers perform in one of two modes, *recursive* and *nonrecursive*. When in recursive mode, the name resolver sends a request to the name server. If the name server cannot resolve the address, it sends a query to another name server within its domain. If that name server does not know how to resolve the address, it in turn can send a request to name servers within its domain. The end result is that the address is eventually resolved and the results returned to the name resolver.

In nonrecursive mode, the name resolver sends a request to the name server. If the name server cannot provide resolution, it cannot send a query to other name servers. Instead, it sends an error message along with the address of another name server which can resolve the address. It is then up to the name resolver to send the query to the other name server.

There must be one name server labeled as the authoritative name server for each network. This is the name server administered by the local network administrator. The name server is responsible for a defined zone (such as within a corporate network), and the various "subtrees" within the domain (subnets, for example) are defined as zones. For large corporations with several subnets, the name server may be located within each zone, administered by the local network administrator.

The primary name server for a network exchanges databases with secondary name servers within the same network (or outside of the network) to provide redundancy. There is usually a primary and a secondary name server within each network. The name servers use the UDP protocol (or they can use the TCP protocol) to exchange database information. Likewise, queries to the name servers use the UDP or TCP protocol (usually UDP).

The domain itself identifies the user, the subnet (if applicable), and the domain in which they are located. The actual structure of the domain name address follows the convention of:

TABLE 4.3 Domain Name Extensions

Extension	Description
.com	Commercial
.edu	Educational institution
.gov	Government institution
.mil	Military branches
.net	Major network providers
.org	Nonprofit organizations
.int	International organization

```
user.subdomain.domain.domain
```

with the subdomain identifying the subnet. A typical domain name address might look like:

```
trussell@interpath.com
```

The company name is almost always part of the domain name, unless the e-mail services are being provided through an Internet access provider, in which case the provider's name would be in the domain name (interpath in the above example). The extension identifies what type of organization the domain name belongs to. Table 4.3 shows the extensions defined by the InterNIC.

The domain name system adds a level of user friendliness to the Internet and has been embraced by all of the TCP/IP community. Today, domain names can be seen everywhere, from newspaper articles to television commercials. It provides a means for identifying not only a user within the Internet community but also the organization and type of organization that they represent.

4.3.4. Routing in an Internet

Routing decisions can be based on two criteria: state of the various nodes and links or distance to the destination. When the distance to a destination is the prime criteria for basing routing decisions, delay, throughout, and the ability to reach various gateways and routers along the path are the only determining factors when a route is chosen.

When the state of the network is the prime consideration, there are a number of criteria used to determine the best route. Link capacity, the number of packets in queue for any particular link in the route, link security requirements, and the number of hops to reach the destination (distance) must be considered.

Hosts within a network receive routing information from routers within the same network and use this information to update their own routing tables, but they do not send their own routing tables to other nodes. Only routers and gateways propagate routing information throughout the network. Hosts use the routing information for a variety of applications, which will be explained below.

4.3.4.1. Source Routing With source routing, the host determines the route to be used for a particular data unit. Within the IP portion of the data unit, the host will include the IP addresses of all intermediate nodes, which determine the route for the data unit. This is based on information received from adjacent routers regarding routes. Hosts calculate the best route based on distance to reach the destination rather than state of the various nodes within the route. This technique is commonly used in large LANs but is not efficient for an internet. There are too many variables involved with large internets for this method to be efficient.

Nonsource routing is the most commonly used in internets. Routing decisions are based on the destination address in the IP header and the routing information maintained by each gateway and router along the route. As each intermediate node receives a data unit, the destination address is examined, and a route is determined by the receiving node. As the data unit moves from one gateway to the next, routing decisions must be made at each node. This is a better method of routing because it allows intermediate nodes to make decisions based on the dynamics of the network, which may not be known by the originating host.

Within a LAN, the routers do not process the IP portion of the data unit. Only the machine address is used for routing within the LAN. If a data unit is destined for an outside host, the machine address of the gateway or router is used. The router then examines the IP address to determine the route to take to the destination.

4.3.4.2. Time Stamping Time stamping allows each router to store information regarding round-trip delay for each route. This information is gathered using a routing protocol. As the IP header is processed by each gateway/router, the time the data unit was received is inserted (in milliseconds) into the IP header, along with the IP address of the node recording the time.

This continues until the data unit reaches its destination. Once the destination is reached, the time stamp information can be extracted and added to the host's routing tables. Round-trip delay can then be calculated for any number of routes. However, round-trip delay is not a perfect science. There are a lot of variables involved when calculating delay. For example, delay may be introduced because of the type of facility or because of a more temporary condition, such as congestion. To be efficient, these would have to be recorded in real time. This is not possible with most routing protocols.

Time stamping uses the IP header and attaches an option field for the recording of the IP addresses and time stamp of each node encountered over a particular route. Time stamping is not highly accurate since each node is running an independent clock. A clock synchronization network, much like those found in telephone companies, is necessary to make this method highly accurate.

An IP protocol, the Network Time Protocol (NTP), is used specifically for clock synchronization and addresses the issue of accuracy to some degree. It allows nodes to exchange time information for synchronization. One node must be designated as the root node and is responsible for obtaining time from a reliable source. There are a number of sources which broadcast time via radio waves for networks.

Once the root node has obtained the accurate time and updated its own clock, it must then send the time information to all of the other nodes in its area using NTP. Obviously, the further away the root node is from the receiving node, the less accurate the time information becomes because of delays within the network. For this reason, it is more advantageous to have many root nodes reporting via NTP to gateways and routers within short distances.

4.3.4.3. Circular Routing Circular routing is a condition any mesh network needs to deal with. When routes suddenly become unavailable, it is highly possible that data units may be sent to a node and returned over the same link to the sending node. This happens because the receiving node thinks that the link is a better route, based on conditions it knows about on other links. The problem is that the adjacent node may not know those conditions, and when the data unit is received again, it will end up sending the data unit back over the same link. This continues until the time-to-live parameter in the IP header expires or until the routing tables are updated by a routing protocol.

There are other methods of preventing circular routing, discussed below.

4.3.4.3.1. Split Horizon When routing information is received over a particular link, the receiving node updates its routing tables but does not send any routing advertisements from its own routing tables over the same link. For example, say node A was connected to node B. A sends routing information concerning a new route over a link to node B. Node B updates its tables and then sends routing information back to node A.

The result is A may think there is a path to a destination through node B, when in actuality, the path to the destination from node B is back through A. If a data unit were to choose this route, it would pass through A to node B, which would in turn send it back to node A (based on its routing information). Node A would then send the data unit back to node B (once again, because of its routing information).

To prevent this, a split horizon prevents node B from sending its routing updates to node A. Node A would then select another route for data units rather than node B. The scenario is a bit more complicated, but for brevity sake we will keep the explanation simple.

4.3.4.3.2. Poison Reverse With poison reverse, the same principle is used except for one thing. Updates are allowed in the backward direction, but with a weighting factor of infinity. This is interpreted as an unreachable route and remains as such until the next routing update. The problem with poison reverse is it increases the size of the routing tables, which in turn take more bandwidth when sending routing table updates to other nodes. If bandwidth is of major concern, this method is not recommended.

4.3.4.3.3. Triggered Updates Probably the best solution is a triggered update. Routing updates only occur when conditions change. There is no need to send routing information if there has been no change in the routing table. The more routing exchanges occur, the more the likelihood for circular routing.

When a change does occur (such as a link failing at a particular router), the router sends an update to its adjacent nodes, showing the route with the failed link as unreachable. The receiving nodes in turn update their tables and send routing advertisements to their adjacent nodes. This creates a cascade of data units through the network, and in itself could introduce problems with bandwidth. To counter the problem, random time delays are set at each router so they do not all send their routing advertisements at the same time.

4.3.5. IP Routing Protocols

Now that we understand the addressing conventions used within TCP/IP networks, let us look at how data units are routed through internets using TCP/IP. There are a number of protocols which are used within a network and within an internet by routers and gateways. Remember that TCP/IP is capable of rerouting data units around failed nodes, making it "self-healing." This requires a level of intelligence within the internet itself.

Routers and gateways must be able to exchange routing information with one another autonomously. Otherwise, a network administrator would have to continuously update the routing tables manually. This would be a formidable task given the size of many of these routing tables and the frequency at which changes occur within an internet.

By allowing routers and gateways to exchange routing information with one another, the networks can maintain the status of all of their neighboring nodes and alert other nodes of any status changes. This must follow an hierarchical approach, however, to prevent a flood of broadcast messages throughout the network.

Routing protocols differ depending on the geography of the network. Gateways attached to the backbone of the network do not need to exchange routing information with routers located within an autonomous network. They only need to communicate with adjacent gateways and only concern themselves with network addresses, not host addresses.

There are also two types of routing protocols. Link state metric protocols send link and node information so that routers can make routing decisions based on the state of the network. Distance vector protocols include information regarding the number of hops a data unit must travel to reach a destination and the amount of delay encountered. Routers then make routing decisions based on the distance and delay factors.

Gateways using distance-vectored protocols will periodically send test messages to adjacent gateways to ensure that the routes are still reachable. Should a route test message fail, the gateway must then update its routing tables and share the updates with its adjacent gateways. This is not real-time processing and results in slow updates to routing tables.

Routing protocols allow devices to communicate with other devices within a geographical area. It is not efficient for all devices to communicate with all other devices. For this reason, routers and gateways are placed into groups, and they communicate accordingly.

Routers are placed into areas and communicate with other routers within the same areas. This prevents routing tables from becoming too large. The network administrators are responsible for the deployment of routers and gateways and should always configure their networks in such a fashion.

A gateway provides access to other gateways outside of the area. Gateways then communicate with other gateways, exchanging network addresses and ignoring host addresses. The intent of the gateway is to provide a path from one area (or network) to another.

Gateways and routers are classified four ways:

Interior gateway

Border gateway

Exterior gateway

Gateway-to-gateway

For each of these classifications, there is also a routing protocol. This allows gateways to exchange routing information with other gateways at the same level.

Interior gateways exchange routing information within an autonomous system with other gateways or routers using the Interior Gateway Protocol (IGP). These are locally maintained routers which provide access to other subnets within a network or to an exterior network through a connection to an exterior gateway.

The Exterior Gateway Protocol (EGP) exchanges routing tables with other routers and gateways connecting other networks. For example, company XYZ may have an exterior gateway connecting them to the Internet. Their gateway would exchange routing information with gateways from other networks (with which it has direct connections) but would not share routing information about destinations within its own network.

The EGP may be replaced with the Border Gateway Protocol (BGP). This protocol was first deployed in 1989. However, some feel this is a more robust protocol, and if it is embraced throughout the Internet community, it may indeed replace the EGP.

For gateways outside of any network (part of the internet backbone) the Gateway-to-Gateway Protocol (GGP) is used. These gateways are used by service providers to connect autonomous networks to the backbone of the Internet or to a smaller private internet. They exchange routing information about how to route from one network to the next.

It is important to remember the scope of these routing tables. A gateway has information about how to reach all destinations within an internet. The destinations are not individual nodes, but individual networks. Routers and gateways associate routes with physical ports, so it is likely that many destinations will use the same port.

The only portion of the IP address these nodes are concerned with is the network ID portion of the Class C address. It seems that routing tables would be enormous because of the number of users on the Internet itself, but in actuality, routers only need to know about the networks themselves and not the nodes within those networks. Exterior and border gateways maintain information about interior gateways and routers, and these routers in turn maintain routing information about the individual nodes within their own networks.

Also remember that routers and gateways do not communicate with every other router and gateway within an internet. They only exchange routing information with routers and gateways to which they have a direct connection. This helps prevent a flood of messages from congesting the network. It also provides a check and balances mechanism with which routers and gateways can keep track of the status of neighboring nodes.

All routing protocols use the TCP protocol to exchange routing information. The frequency at which this information is exchanged depends on the protocol used and the configuration of the network itself. It may also depend on network activity. If activity is high, the routers and gateways will not communicate as frequently as when the activity is low.

4.3.5.1. Address Resolution Protocol The Address Resolution Protocol (ARP) is part of the IP protocol stack and is an IGP. ARP provides translation of IP addresses into machine addresses. This requires communications with end nodes or computers on a LAN.

The router within a LAN will use mapping tables, which map the IP addresses received to the proper machine address. Of course, before this can be done, the router must know of the machine address. The router communicates with all nodes on the LAN to determine which machine addresses are reachable and what their IP addresses are.

When the router receives a data unit, it checks its mapping table to see if there is an entry for the IP address received. If there is, the router can create a packet (such as an Ethernet packet) and transmit the data over the LAN.

If there is no entry for the IP address, the router sends an ARP broadcast message over the LAN. All active nodes on the LAN will see the broadcast message and process it. If a node recognizes the IP address in the broadcast message as its own, it will reply with its machine address. This is then placed in Random Access Memory (RAM) cache for temporary storage.

The cache entry is not a permanent entry. There is a parameter associated with every entry which determines how long the particular entry shall remain in cache. When the entry expires, the router has to send a broadcast message again to receive that node's machine address.

The purpose of the ARP cache is not to provide a permanent record of all hosts on a LAN (since these hosts are not always on and accessible) but to provide a temporary mapping table so that the router does not have to continuously send broadcast messages over the LAN.

The ARP cache can also identify the hardware type (such as Ethernet or Token Ring), which is important for multiprotocol routers. These routers may have connections to several different types of LANs, and they need to know what type of protocol packet needs to be created to route the IP packet to the host.

Also part of this table is the type of protocol used to obtain the routing information (such as ARP or RIP), the routing age (how long the information has been in cache), and the subnet mask for the destination host.

ARP is used by routers when addressing hosts which know their IP address. This requires a workstation (or PC) with some form of permanent storage (such as a hard drive). The IP address must be configured into the host (through IP client software) so that the PC knows its identity. Not all workstations know their IP address because they may not have disks, in which case a different protocol is required.

4.3.5.2. Reverse Address Resolution Protocol Some workstations may not have hard drives (diskless workstations) and are not capable of storing their IP configuration. In such cases, the workstation must send a query to the server to determine what its IP address is.

The Reverse Address Resolution Protocol (RARP) allows diskless workstations to send queries over the network (in broadcast mode). Servers on the network read the machine address from the query and send a reply with the workstation's IP address.

4.3.5.3. Routing Information Protocol The Routing Information Protocol (RIP) is an interior gateway protocol used by routers to exchange routing information within an autonomous network. RIP is based on two Xerox protocols, PUP and XNS.

RIP is a distance-vectored protocol. It is the most popular of the routing protocols. The routes are determined based on distance, with no consideration of the state of the links and/or nodes which must be used to reach a destination.

There are two modes of communications used with RIP, passive and active. Passive mode is used by workstations, which read RIP messages and update their routing tables but do not exchange routing information with other nodes. In active mode, devices read and send routing information to adjacent nodes. Routers and gateways are the only devices which operate in active mode. When a router or gateway receives a RIP message, it updates its own routing table and then generates its own RIP message to send the updates to their adjacent routers/gateways.

If no updates are received for a given route, the route is marked unreachable. The assumption is made that the route has either failed or is unreachable and should therefore be removed from the routing choices.

RIP is not an efficient routing protocol because it sends routing advertisements on a periodical basis rather than on a select one. Consequently, the network is burdened with extra traffic which may not be necessary.

4.3.5.4. Open Shortest Path First Open Shortest Path First (OSPF) is a link state metric protocol. It makes routing decisions based on link capacity, delay and throughput requirements, the number of data units presently in queue for transmission over a particular link, the number of hops required to reach a destination, and the ability to reach gateways and routers along the route.

This makes OSPF more robust than RIP, which bases routing decisions on the number of hops to reach a destination. In RIP, there is no consideration of capacity or link state. In OSPF routing tables are created based on the above criteria, and a weighting factor is applied to each route. This allows for routing based on the dynamics of the network. It also allows routing decisions to be made based on the type of facility. For example, a satellite link may be given a higher weighting factor even though it is the shortest path to a destination. Satellite links introduce additional delay, and even though there may be fewer hops through a satellite link, there could be longer delays by taking such a route.

Another reason for the popularity of OSPF over RIP is the frequency at which routing advertisements are sent. With OSPF, these advertisements can be made more selectively, rather than being based on time. The only time advertisements need to be made is when changes to existing routes are made (such as the case when a router or link should fail).

The advertisements are then sent in response to a condition rather than in response to a clock.

OSPF is not one protocol, but a series of routing protocols consisting of test messages and routing exchange messages. It allows routing based on the dynamic state of the network rather than distance.

4.3.5.4.1. HELLO Protocol HELLO is used by OSPF for communicating with adjacent nodes. This protocol first establishes an adjacency with another node and then continuously monitors that node (through test messages) to ensure it is still reachable.

There are several states in which an adjacent node can be in. When a node is first placed onto the network, it is considered to be in the DOWN state. This indicates no adjacency has been established. Once a HELLO packet has been received, the node is placed in the INIT state by the sending router.

When a response is sent to the HELLO packet, the node is moved to the two-way, or EXSTART state, which implies that communications have been established, but adjacency is still being negotiated. Routing information cannot be exchanged until the node has reached a FULL state. Once adjacency has been established, the two routers can begin exchanging routing information.

Rather than have all routers within an autonomous network share their routing information with a gateway, one router can be designated as the prime for exchanging routing tables with a local gateway. This prevents routers from creating congestion by advertising their routes to the same gateway.

4.3.6. IP Services

We now have a clear understanding of how IP routes data units through an internet. IP provides many services other than routing (although these other services are associated with routing of data units). In this next section, we will look at the services provided by IP.

4.3.6.1. Fragmentation and Reassembly Fragmentation is the division of data units into smaller data units. This is an important feature in any network and can be a complex issue as well. IP provides fragmentation across networks, which means a data unit can be fragmented into several data units at the source, and then the various fragments can be fragmented even further by intermediate networks.

There are two fields within the IP header which support fragmentation, the "more" bit and the fragmentation offset. The more bit indicates that there are more fragments to be received, and the receiver should not reassemble the data unit until the last fragment is received.

The fragmentation offset is used by the receiver to reassemble the data unit. It identifies where the fragment belongs in the entire data unit. The offset is always in relation to the original data unit as it existed at the source and not after fragmentation by an intermediate network.

It is possible that a fragment could be received by an intermediate network and fragmented further because the received data unit is too large to pass through the intermediate network. In fact, data unit fragments can take different paths and arrive out of order. IP ensures that all of the fragments are received and can be reassembled properly.

To ensure that all fragments have been received, IP uses a timer. All fragments must be received before the timer expires, or an error message will be returned to the originator, resulting in a retransmission of all of the fragments.

IP cannot retransmit one fragment of a data unit because there is no way for IP to determine which fragment was lost. In the event a fragment is lost, all of the fragments are retransmitted, and any received fragments are thrown away.

4.3.7. Internet Control Message Protocol

IP has no error control and relies on the Internet Control Message Protocol (ICMP) for the delivery of error messages. ICMP resides in hosts and gateways, and it provides administrative and status messages. Routers cannot generate ICMP messages.

Typically, gateways generate an ICMP message with the originating host as the recipient. This means that the ICMP software residing in gateways is more complex than that found in hosts. Since gateways are usually the message generators, there is no need for complexity in the host, which saves in memory and processing.

An ICMP message is generated if a destination is deemed unreachable by a gateway, if the time-to-live field expires for a data unit, or if a gateway determines that a header is in error. If an error occurs with an ICMP message, no report is generated.

ICMP reports errors with fragments, but it only reports the first fragment in error, not all subsequent fragments. The information field of

Information (variable)	Parameters (optional)	Checksum	Code	Type	IP Hdr

Figure 4.6
ICMP Header within IP Header

the ICMP message contains the first 64 bits of the fragment in error, so the host can determine how to handle the error.

It is important to understand that ICMP does not provide error detection for IP. It is simply a reporting mechanism used by IP to report errors to originating hosts. IP is still a connectionless transport, which means delivery is unreliable and not guaranteed. IP relies on upper-layer protocols (such as TCP) to provide robust error detection and correction.

ICMP is considered part of the IP suite, but it uses the services of IP for delivery of error messages. It is carried in the data portion of the IP header. The IP header field "protocol" will indicate the message type as ICMP (see Fig. 4.6).

Within the ICMP header is a type field, which indicates why the ICMP message was generated, for instance, "destination unreachable." There are 13 types defined in the protocol. The code field provides additional information about the error.

As mentioned earlier, the first 64 bits of the data unit are provided in the information field of the ICMP header. This allows the originating host to correlate the data unit with one already in its transmit queue waiting for acknowledgment.

4.4. Transport Control Protocol

The TCP protocol is a layer 4 protocol and is a user of IP services. It does not require IP, however, and can use almost any other layer 3 protocol for transport services. The software resides in hosts, but not in routers. A router does not need to know about TCP because it does not process the TCP header. Only hosts (both originating and receiving) process this header.

There is one exception to this rule. Sometimes routers will use TCP software for network management purposes, in which case they must be able to generate an TCP header to encapsulate the network management message. User data is never processed in routers.

Likewise, gateways do not use TCP (other than for the generation and processing of network management messages). We will discuss network management later on in this chapter, but for now understand that network management is considered an application, and it relies on TCP and IP for the delivery of network management data units.

TCP provides end-to-end session control between two hosts. Remember that TCP is connection-oriented, which requires a session to be established before data can be exchanged. TCP also provides error detection and correction for applications requiring connection-oriented services.

As with any connection-oriented protocol, once the session is established between two logical entities, the receiving host must acknowledge all received data units. If a data unit is not acknowledged, there are no error messages (negative acknowledgments) sent by TCP. Instead, the sending host sets a timer, and if no positive acknowledgment is received before the expiration of the timer, then the data unit is automatically retransmitted.

Before discussing more TCP services, we need to examine the TCP header and understand the various fields in it. The TCP header can be found in the data portion of the IP header.

4.4.1. TCP Header

The data units sent between two entities are called a segment. The segment is sent down to the IP stack where it is encapsulated by an IP header (see Fig. 4.7) and becomes a packet (recall the discussion at the beginning of this chapter).

The addresses contained in TCP are somewhat different from the IP addresses we discussed in the previous section. Addressing at the TCP

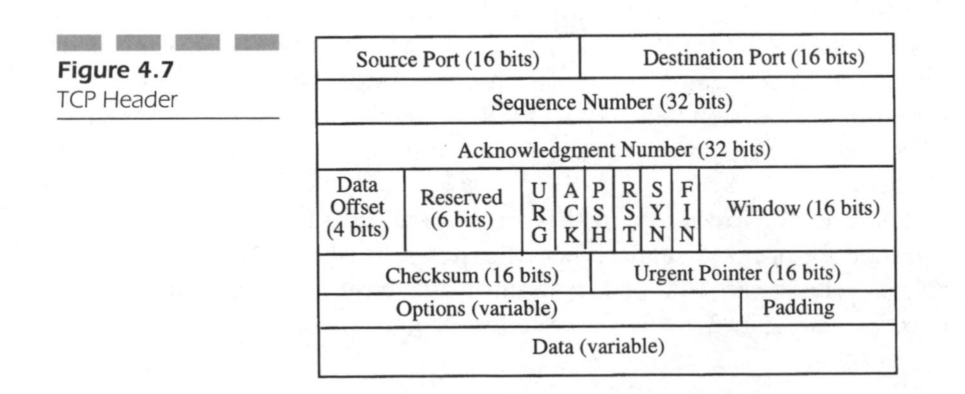

Figure 4.7
TCP Header

level is of logical entities within a host rather than actual user connections to the network. Recall that the IP address is not a physical address; it indicates a connection to the network and identifies a user.

TCP uses a destination and source port number. A port is a predefined number identifying the application using the TCP services. This may be FTP, TELNET, or Simple Mail Transfer Protocol (SMTP; e-mail). These applications will be discussed later in this chapter. The port number is 16 bits long.

A 32-bit sequence number is used to verify that all data units have been received. This is a number sent in serial fashion with every data unit. Sequence numbers can be received out of order as long as they are received within a time constraint (determined by a TCP timer). If a sequence number is received out of order and the timer expires, all unacknowledged sequences must be retransmitted.

An acknowledgment number follows the sequence number and identifies the next expected sequence number. The acknowledgment number is 32 bits long. As mentioned earlier, there are no negative acknowledgments, only positive ones. This field is considered a positive acknowledgment. It identifies which sequences have been received properly and which sequence number is expected next.

If the originating host has already sent the sequence number identified in the acknowledgment number field, the host will not retransmit until a timer expires. The originating host allows a certain amount of time, which is determined by calculating the round-trip time for the route used, before retransmitting the sequence again. Since this is a dynamic value, the process can be somewhat complex. We will not go into detail here as to how round-trip time is calculated, but if you need additional information, see the Request for Comments (RFC).

The data offset field identifies where the data portion of the TCP header begins. It identifies how many 32-bit words are in the header, preceding the user data field. This field is 4 bits long and is necessary only because there is an optional field at the end of the header which can be of variable length.

Several 1-bit fields follow the data offset field and are used for processing the TCP data unit. The urgent bit identifies that the urgent pointer contains data. The urgent pointer is a 16-bit field which identifies the offset in the user data field containing data considered to be urgent. We will discuss how urgent data is processed in more detail a little later, but for now know that this field is used to identify the presence of urgent data and its location in the data unit.

The acknowledgment bit identifies that an acknowledgment is present in the acknowledgment number field, and it alerts the receiver that this number is acknowledging previously received sequences. The receiving host knows then to delete the sequence numbers being acknowledged from its transmit queue.

The push bit is similar to the urgent bit. It notifies the receiving host that the data unit received should be processed immediately and causes the data unit to be processed when received rather than being placed in a receive queue. We will discuss the processing of push data later.

The reset bit causes a session (logical connection) to be reset. This usually means that all queues associated with the session are flushed and all associated counters and timers are reset to zero. This is used when an error occurs with a connection and the connection must be reestablished.

The synchronized bit is used when establishing a logical connection, and it indicates that sequence numbers need to be synchronized. Synchronizing sequence numbers allow both hosts to identify where the sequence numbers will begin so that each entity knows what sequence numbers to expect. This bit is used during the handshaking process, which takes place during the connection establishment phase.

The finish bit is the same as end of transmission. It indicates that there is no more data to be sent, and a session can be closed. The session will then be terminated and the resources released for another session.

The window field is used during session establishment. Each host must negotiate how many data units can be sent before an acknowledgment. This is considered the window size and is determined by the size of the queue and the amount of processing already occurring from other sessions. The window size cannot be changed once the session has been established.

The checksum field is used to check for errors in the header as well as in the user data. If you remember in our discussion on IP, the checksum in IP does not verify the user data in IP; it only verifies the header itself. The checksum in the TCP header does verify the user data.

The options field is a variable field designed for future TCP implementation. There have been no uses for this field at the time of this publication. Padding follows the options field to maintain a 32-bit boundary.

4.4.1.1. Processing of Urgent Data Urgent data can be an interrupt, control information sent from a terminal, or almost any other data requiring real-time processing (or near real time, since TCP/IP is not a real-time protocol). The method used in TCP to identify urgent data is sometimes referred to as out-of-band notification.

Out of band means that data must be sent outside of the data stream. For example, if a terminal emulation application (such as TELNET) is running, the normal data stream would contain data being transferred to and from the remote terminal. Urgent data (control characters on the keyboard) is processed outside of this data since it is not part of the information being sent or received.

When a host receives data marked urgent, it processes the data identified by the header (the urgent pointer field). The data is not placed in a buffer as the normal data stream would be. This ensures the data is processed when received without delay.

Urgent data will also receive a sequence number, but because this data is processed immediately, it leaves a gap in the sequence numbers received in the buffer. Software must manage this gap so that when the data in the receive buffer is processed, the host knows that the missing sequence numbers have already been processed.

The urgent pointer in the TCP header identifies where the urgent data ends rather than where it begins. This allows other normal data to be received in the same data unit as urgent data. The host processes all data up to the pointer and then leaves the urgent mode and begins its normal data processing.

If the window size has been reduced to zero, urgent data is still processed. Again, this ensures that all data has been processed and is not left in a queue or ignored because of processing capacity.

4.4.1.2. Processing of Push Data The push bit also provides a means for bypassing the receive buffers. Push data can be buffered if the data unit is fragmented. This is different from urgent data, which is passed to the application without buffering. The difference is that urgent data is normally small data units and does not require fragmentation, while push data may be larger.

When multiple data units are used for push data, the data stream is buffered until all of the other fragments have been received. If all of the data is not received within a specified time frame (managed by a TCP timer), the data already received is passed to the application.

If there are missing fragments, the application may return an error, indicating that not all data has been received. It is the responsibility of the application, rather than TCP, to report this.

4.4.2. TCP Ports and Sockets

We discussed ports earlier and identified a port as a predefined number used by the protocol to indicate the application using the data unit. This tells the host which upper-layer application to pass the data unit to.

A socket is something managed by the operating system and is made up of both the port number and the originating IP address. Together, the port number and the source IP address form a socket identifier, which is used to identify logical connections established by TCP.

All sockets must be unique since a pair of sockets (originator and receiver) identify a session. The socket is of significance only within the host. The originating host will know the socket number but will not manage the resources required to support the socket at the destination host. Each host is responsible for managing resources allocated to each socket within its own entity.

The defined port numbers commonly defined by the TCP protocol are:

20 = FTP data

21 = FTP control

23 = TELNET

25 = SMTP

42 = Nameserv (host nameserver)

53 = Domain (domain name server)

109 = Post Office Protocol (POP) 2 (used by e-mail applications)

Many other port numbers have been defined, but these are the most common. The port numbers 0 through 255 are predefined (cannot be defined by network operators), while any port number above 255 can be defined by network operators.

A port is capable of supporting multiple sessions. The operating system manages the resources needed to support multiple users. The actual number of sessions that can be supported depends on the platform and the operating system.

4.4.3. TCP Services

TCP provides a number of services to hosts. Remember that TCP is a connection-oriented protocol and must provide session management as well as reliable transfer of data units. IP is a connectionless protocol and relies on TCP to provide reliable data transfer.

In this section we will discuss two important services provided by TCP, error control and flow control. These are typical in any connection-oriented protocol, but TCP uses mechanisms that are somewhat unique.

4.4.3.1. TCP Error and Flow Control We have already discussed error detection and correction to some degree. In this section we will look at TCP error management and flow control in more detail. One unique factor with TCP is that there are no negative acknowledgments. Only positive acknowledgments are sent.

If a sequence is received in error (detected by the checksum), the data unit is deleted from the buffer. No reply is returned to the originating host by TCP. Remember that IP checks only the checksum of the IP header, not that of the user data (which includes the TCP header and the user data).

The originating host sets an acknowledgment timer when data units are sent. In the case of an error (either checksum or other) the timer will expire and the originating host will retransmit the data unit. The same applies to data units received out of sequence.

The timer is not a fixed timer, but a variable one based on round-trip time. Round-trip time includes delay in the forward direction (or time to reach destination), processing time at the destination, and delay time in the backward direction (or time for acknowledgments to reach the originating host).

This is not a very accurate mechanism because round-trip time is very difficult to determine. There are many variables in the network which effect round-trip time, including network congestion and node failures. There are a number of new techniques being researched to improve the accuracy of round-trip time calculations.

TCP uses both acknowledgments and windowing for flow control. The TCP header contains fields for both the acknowledgment number and the window size. The actual window (which is the number of data units which can be received without acknowledgment) is determined by both the acknowledgments outstanding and the window size value in the TCP header. For example, if the acknowledgment number sent is 6 and the window value is 12, the actual window size is 18 (acknowledgment number+window value). This is more efficient than using positive acknowledgments only because it allows the host to adjust its window based on resources available, as well as buffer size.

The window size is negotiated when a session is established by TCP. Once the window size has been established, the value cannot be changed by the host without negotiation. This prevents the host from arbitrarily changing its window size at any time.

TCP also maintains an internal window, referred to as the congestion window. This window is not advertised to other hosts. It allows for the throttling of data units transmitted based on retransmissions. When the acknowledgment timer maintained by the originating host expires, the originating host assumes that the destination is congested and enters into the congestion mode.

While in congestion mode, the congestion window is decreased. The value is derived by taking the window size sent in the TCP header and the number of retransmissions sent. If retransmission continues to occur, the congestion window continues to decrease using what is called multiplicative decrease.

What this means is that the congestion window is decreased using a multiplicative formula, with the number of retransmissions being the factor. When there are no retransmissions being sent, the originating host assumes there is no congestion at the destination, and the congestion window will equal the same value as the window size advertised in the TCP header.

When congestion subsides, the congestion window does not immediately change to the window size value. There is a slow restart implemented so that the destination is not flooded with data units, causing it to become congested again. During the restoration process, the congestion window steadily increases until it reaches the same value as the advertised window size.

If a retransmission to the congested destination occurs during the restoration process, congestion mode is entered again, and the congestion window is decreased using the multiplicative decrease algorithm.

4.4.3.2. TCP Management Besides error detection/correction and flow control, TCP is also responsible for maintaining established sessions. Software resident in each host is responsible for establishing and maintaining sessions. Remember that a session is not a physical connection between two entities, but a logical one controlled by software (both TCP and the operating system).

TCP keeps information about every session in what is referred to as a Transmission Control Block (TCB). The information kept by the TCB includes the source and destination addresses, ports assigned to the session, round-trip time (calculated by the individual host), and data units (sequences) sent and received.

In addition, the TCB at the originating host maintains records on acknowledgments and retransmissions sent to the destination as well as statistics gathered regarding the connection (such as the number of retransmissions and number of bytes transmitted).

This data is maintained for every session established by a host. Remember that there may be several sessions to one host (especially in the case of a server), which means TCP must maintain many sessions at one time. Some operating systems are better at maintaining multiple sessions than others (such as UNIX). This is why some platforms may be better for TCP/IP applications than others.

4.5. User Datagram Protocol

UDP is called a connectionless protocol, but it really is not. The operating system maintains information about each active UDP socket, which implies a connection-oriented service. In a true connectionless service (such as IP) there are no sockets maintained. Data units are sent to a specific destination, with a port address. The socket is not needed because the data unit is processed when it is received by the application, and there is no further action to take with the originating host.

What UDP does not provide is error correction and flow control. There are no acknowledgments sent for received UDP data units; they are assumed received. The application using UDP services is responsible for determining if there have been errors or if data units are missing.

This makes UDP an unreliable service, but reliability is not always a concern with applications such as e-mail and some network management functions. Upper-layer protocols can make up for what UDP is lacking. This provides a streamlined protocol which does not require a lot of processing at the originating or destination hosts.

If there is not a lot of processing required, data units can be sent and received with very little delay. Many newer protocols operating at this layer (layer 4) and the lower layers use this philosophy.

Figure 4.8
UDP Header

16 bits	16 bits
UDP Source Format	UDP Destination Port
UDP Message Length	UDP Checksum
Data	

4.5.1. UDP Header

The UDP header is much simpler than that used in TCP (see Fig. 4.8). You will find there are no sequence numbers, which means no acknowledgments. Likewise, there is not an acknowledgment number field or urgent data processing capability.

There is a source and destination port number and length and checksum fields. Nothing else is needed. Obviously, this allows receiving hosts to process UDP data units rather quickly since all that is needed is to send the received data unit to the proper application (identified by the port number).

Applications which do not need the services of a true connection-oriented protocol use UDP rather than TCP. Some of these applications are identified below; others we have already discussed.

4.6. Internet Application Protocols

We have discussed applications which use the services of both TCP and UDP. Now we will look at the most commonly used applications. These applications also use a protocol which is processed internally within hosts and is used for communications between the destination and originating hosts.

You will remember in our discussions about protocols that we described peer-to-peer protocols. These are protocols that communicate with the same protocol at the other end of a communication (whether it be a router or a host). Where the processing takes place depends on which layer the protocol resides on.

Internet applications are processed within hosts and are transparent to the network devices (routers and gateways). The network devices have no visibility to these and treat them as user data.

4.6.1. TELNET

The TELNET protocol allows terminals or PCSs acting as terminals (terminal emulation) to communicate with remote hosts. The remote host may be a server running an application or a mainframe. Terminals are proprietary and are usually not compatible with one another. DEC terminals use a different terminal protocol than IBM terminals do. These differences must be resolved when connections are established with remote hosts. This is the task of TELNET. This protocol provides a means for all terminal types to communicate with remote hosts regardless of the terminal type. It allows the host to determine the characteristics of various terminals.

This is done using a series of commands to negotiate services the host can support. The host must agree to each of the services before the session can begin. All commands are sent using a format defined by TELNET. One example of services to be provided by the remote host is echo, where all data is echoed by the remote host back to the terminal. Other services include control character mapping (defining which keyboard keys will emulate specific control characters) and how flow control is to be managed.

The terminal characteristics supported by the TELNET protocol depend on the terminal emulation being used. However, the main purpose of TELNET is to allow a mechanism by which the remote host can determine what terminal characteristics will need to be supported.

There are many uses for TELNET, and it has many advantages. Without TELNET, terminals rely on modem communications or dedicated leased lines to reach remote hosts. With a connection to an internet, virtually any remote host running the TELNET client software can provide remote access services to all hosts on the same internet.

Corporate users of the Internet have found it very cost effective to connect their corporate servers to the Internet, allowing employees in remote locations to access the company servers without a dedicated connection. This means they can connect from anywhere that they can get a connection to the Internet.

Once a connection has been established, TELNET operates as any other terminal emulator. What the users see on their screens is the data provided by the remote host, in the format supported by the remote host.

4.6.2. File Transfer Protocol

FTP is used to establish a session with remote hosts to download (or upload) files. Downloading files means transferring a file resident on a remote host to the local computer. Uploading files means transferring files from a local computer to a remote host.

FTP uses some of the services of TELNET to establish a session. TELNET services are needed to allow users to view directories on remote hosts and navigate around the host to find files. However, users cannot run applications on remote hosts using FTP.

Two sessions are needed to run FTP. One session is used for sending control information (such as TELNET commands). The other session is used for the actual transfer of data files. This allows files to be transferred without interruption, while still allowing control information to be sent to the remote host.

There are several data types supported by FTP. Files can be transferred in ASCII or EBCDIC format. These file types are used when transferring text files. Files from word processors and desktop publishing applications cannot be sent using these formats because they include proprietary characters not supported by ASCII or EBCDIC.

If files of this nature are transferred using ASCII or EBCDIC, they will be unintelligible because the formatting characters are only recognized by the application which created them. All formatting characters will be translated into ASCII text, which will result in a lot of garbage.

For all other file types (including word processed or desktop publishing files) FTP supports binary file types as well. When transferred as a binary file, the file is not converted at all but is sent as is. Binary type is used for all files other than plain text files; this includes image files.

FTP uses the services of TCP and supports authentication and security functions. Some remote hosts may require a login and password (supported by TELNET), while others allow an anonymous login, which is accessible by anyone.

4.6.3. Trivial File Transfer Protocol

Trivial File Transfer Protocol (TFTP) is less complex than FTP. It does not support any authentication or security features, which means it does not need all of the TELNET features. It does still rely on TELNET for establishing a connection and negotiating the session.

TFTP does not require the same level of services FTP does, so it can use UDP rather than TCP. This means that TFTP is not as reliable as FTP because UDP does not provide acknowledgments or error control of any type. For this reason many network operators do not use TFTP.

When transferring data, TFTP sends a block of data using UDP and then waits for an acknowledgment from TFTP at the distant end. Once an acknowledgment is received, another block of data can be sent. This increases the amount of processing to some degree. TFTP also provides some limited error handling.

Sessions are tracked using a random source transfer ID (similar to the socket we talked about in TFTP). The destination transfer ID is always 69, which is a well-known port identifier for TFTP. The source transfer identifier is unique for every session established.

There are five different data unit types used by TFTP. A read request asks to read a file from the destination directory. This is used to transfer a file from the selected directory to the remote directory.

A write request is used when modifying a destination directory. Modifications include deleting (provided the operating system provides the necessary user permissions), renaming, or moving files to other directories on the same host.

A data type is used for the actual transfer of blocks of data. An acknowledgment type is used to acknowledge receipt of data and is sent after each block of data. If an error occurs, an error type is used to indicate the type of error that occurred.

FTP is a very popular application within the Internet. There are many hosts supporting FTP capability loaded with files which can be downloaded at no cost. The server must have FTP host software, and the remote host must have client software supporting the access of FTP servers.

4.6.4. Simple Mail Transport Protocol

SMTP is used to deliver e-mail in an internet. This protocol uses a spooling methodology, where a data unit is sent to a mail server, which then holds the data unit until the user accesses the server using an SMTP client application. The data unit is then downloaded to the user.

SMTP is used between servers. The host which generates the e-mail uses a different protocol—POP, explained later. SMTP provides the functions necessary for an e-mail server to transfer mail to the destination mail server using TCP services.

The servers can be configured to either save the data unit or discard it when it is downloaded. This of course oversimplifies the whole process for sake of brevity. The mail server tries periodically to send the mail to the user, usually until a counter expires, at which time the mail is either discarded or saved on disk.

If the mail cannot be delivered within a certain time, the server can be configured to send an error message back to the sender. This is a function of SMTP and not the underlying TCP or IP protocols. SMTP provides its own error control at the application level.

Clients can be connected continuously to the mail server or can connect intermittently (as is usually the case). When users connect intermittently, it is up to the server to maintain the mail until the user actually connects, and then it can download the mail to the user upon request.

When sending mail to a mail server, SMTP first establishes a connection with the mail server. It then waits until it receives an acknowledgment from the server to send a receipt message. The receipt message identifies the recipient of the upcoming mail. The server can then acknowledge that the username identified in the receipt message is valid (by sending an acknowledgment), or it can return an error message indicating the username is not valid.

When an acknowledgment is received, SMTP begins sending the actual data (mail message). When the transmission is complete, SMTP sends an end-of-transmission followed by a QUIT command. This terminates the session and closes the connection.

Addresses used in SMTP utilize the Domain Name System (DNS) convention. If the user is not connected to a network using SMTP, the actual address may follow a different convention than standard DNS addresses. This is because the address must contain additional information about the non-SMTP hosts used to gain access to the Internet. For example, the address user%remote-host@gateway-host indicates that the user is located outside of the SMTP environment. The standard convention would look more like user@host.com.

Some e-mail applications support Multipurpose Internet Mail Extensions (MIME), which allows nontext information such as pictures, movies, and audio to be attached to the e-mail as binary files. The files can then be saved on disk for viewing with a specialized application that is capable of viewing such file types.

4.6.4.1. Post Office Protocol POP supports remote access of e-mail. It stores the mail message until the remote user accesses the mail server and requests mail delivery. Desktop hosts use POP, rather than fully functional SMTP, to access mail servers. With POP, desktop hosts do not need to provide the processing resources required by SMTP. POP is not as complex as SMTP and is only used by hosts accessing mail servers.

4.6.5. Network News Transport Protocol

Two students from Duke University developed NNTP so that they could share research notes with other students from North Carolina State University. The protocol uses a store and forward methodology, much like SMTP.

Host software resident on servers collects "postings" and forwards them to a central server, which collects these postings and then forwards them to other servers upon request. This is what makes the Usenet community on the Internet work today.

Directories are configured on the various servers, and the postings placed in the directories from which they were generated. The postings are nothing more than simple text messages; they can be image files as well. The various servers on the Internet which download postings and upload postings are called news servers.

News servers each subscribe to certain newsgroups (or directories), which are topical groupings of postings. On a periodic basis, the various news servers connect to a central news server owned by UUnet in Virginia and download postings from the various groups to which they subscribe.

There are over 20,000 newsgroups in existence today, making NNTP a crucial part of communications over the Internet. The various newsgroup subscribers use news readers on their computers to download the postings of interest from their local news server to their desktops. The news reader allows them to view the directory contents and select specific postings for viewing.

The local news server can subscribe to all of the newsgroups on the UUnet server or to selected newsgroups. The network administrators choose which newsgroups they subscribe to. Users who connect to the local news server can only download postings from newsgroups which their local server subscribes to. This allows corporations to control which newsgroups their employees contribute to.

The NNTP protocol uses the services of TCP. It is capable of sending text or images, as well as other binary files (usually as attachments). The binary files are converted to ASCII text by a utility program so they can be transferred by NNTP.

When users receive the files, they must use a conversion program to convert them back to binary files. NNTP does not support the transfer of binary files in their native form. One of the most popular encoding/decoding utilities is uuencode, provided by the UNIX operating system and supported on almost every computer platform.

NNTP is a unique way of supporting special interest groups without requiring a dial-up connection to some central server. The newsgroups are similar to bulletin board systems with one exception: Individuals can connect to all of the newsgroups and view all of the postings on multiple newsgroups rather than connecting to one server and then having to connect to another. By using NNTP, special interest groups are accessible to everyone on the Internet, and postings can be distributed throughout the Internet.

4.6.6. Hypertext Transport Protocol

The Hypertext Transport Protocol (HTTP) supports the transfer of files stored on dedicated servers. These files are unlike text or data files. The files are coded using Hypertext Markup Language (HTML), which is derived from Standardized Generic Markup Language (SGML). Text files are coded with special characters which identify which characters are headings and which characters are underlined, etc.

Images can be referenced as well, without being a part of the file. A reference is inserted into the file to point to the location of another file. When the file is opened by a special-purpose viewer application, the user sees the text displayed according to the coding and the images referenced in the source file.

Images and text files do not have to be collocated. In fact, text and image files can be located anywhere on the Internet and linked to the main file by references in the text. This allows files to be maintained locally rather than downloaded to a central server somewhere which must be maintained by a central authority.

The HTTP server (commonly referred to as a Web server) is accessed with an application which interprets the HTML file and displays the information according to the HTML code. The file is stored in the local computer when it is being viewed. HTTP uses TELNET services when downloading HTML files for viewing.

To access a HTTP server, a Uniform Resource Locator (URL) is accessed (part of the client), and an address is typed in. The address uses a format that is similar to that of e-mail. The protocol must be identified first (such as HTTP or FTP), followed by a colon, and then the address. The address is in the DNS format, and may look something like:

```
http://www.tekelec.com
```

The address will vary depending on where the files are located. Most companies which are using HTTP servers to distribute information use this format with their own company identity, making it simpler to find them even if their address is not advertised (an educated guess will often get you to the proper HTTP server).

HTTP is more commonly found on the Internet and is used for the distribution of marketing information. However, many companies are setting up internal HTTP networks for the distribution of internal documentation, such as human resource information and specifications. These networks are referred to as intranets.

4.6.7. SLIP and PPP

Serial Line Interface Protocol (SLIP) is typically used by those on a dial-up connection. It is a protocol that supports TCP/IP over serial communications lines where routers and gateways are not used. If you are using a modem and need to send IP packets over a dial-up line, the transmission will not work because you must connect to another modem, and, of course, modems do not understand IP.

SLIP encapsulates the IP information or information from layers above IP and transmits it over a serial line. Addressing is not used because a serial line is a point-to-point communications link. When it is received on the other end, the SLIP data unit is deleted, and the IP data unit can then be sent over a conventional TCP/IP network. The termination point for SLIP connections is usually a gateway computer.

Point-to-point Protocol (PPP) is a newer version of SLIP that provides faster, more efficient communications. PPP uses a frame format such as High-level Data Link Control (HDLC), with an information field containing the IP header. PPP uses another protocol, the Link Control Protocol (LCP), to establish a connection between either end. Once the connection has been established by LCP, the PPP transmission can begin.

LCP negotiates with the gateway for link configuration, quality of service, network layer configuration, and link termination. The protocol residing in the PPP data unit is identified by LCP at the time of negotiation.

Those using modems at home who want to communicate with company computers or to connect to the World Wide Web (WWW) on the Internet must use either a SLIP or PPP connection. A simple dial-up connection without SLIP or PPP can be used, but access to many Internet services (such as WWW) cannot be supported on dial-up connections.

4.7. Network Management

The most important part of any large network is the network management tool used to monitor and evaluate the performance of the network and all of its nodes. There are many ways of accomplishing this, anything from placing devices at every node which report back to a central location to protocols which are capable of providing information about various nodes autonomously.

In TCP/IP networks, there are a couple of different such protocols which report to network management clients about the status of the various nodes within the network, the links attaching to those nodes, and the performance of each of the nodes. The next section discusses the most popular of those network management protocols to clarify the role of network management and provide some insight into the possibilities of such protocols.

4.7.1. Simple Network Management Protocol

SNMP was developed by the IAB for use on the Internet. Prior to SNMP, vendors were implementing proprietary solutions, which were not capable of interconnecting to other vendors' equipment. Special training was required to teach users how to use the devices which interact with the protocol since they were not standardized.

The IAB solution provides a standard protocol which can be used on any platform. Vendors have already implemented SNMP support in their products, which resolves the issues of interoperability. Modems, routers, gateways, and host computers from different vendors can all share network management data by using SNMP.

SNMP uses simple code, allowing it to be incorporated into devices which do not typically have a lot of processing resources. Small routers and modems can be equipped with SNMP software without affecting the performance of the device. Yet SNMP is also flexible enough to allow changes to be made easily.

SNMP uses the services of UDP to share data between the various nodes on the network. Network administrators then use client software to access SNMP host software which is responsible for the gathering of network data. The server uses a store and forward methodology to maintain information about network entities.

The client application used by the network administrator may provide a simple interface with a Graphical User Interface (GUI), which is usually the case, or with something simpler and less user friendly. This of course is transparent to SNMP, which does not define the user interface.

It is important to understand that SNMP defines the means by which devices communicate network data to one another and not how the data is to be displayed to the user. This is the job of the client application, which collects the data from an SNMP server and displays it in a variety of formats.

The various devices on the network use an agent. The agent reports to managers, which are communications software which query devices for network data. The data includes status of the device, the links attached to the device, and maybe even measurements data used for statistics. This information is then stored in a database on the server. The database is called a Management Information Base (MIB).

SNMP can be used on any type of network and is not limited to TCP/IP. The only requirements is that the network must provide a simple transport service (such as X.25 or ISDN). More complex transports can be used but require additional overhead because of their complexity.

4.7.1.1. Management Information Base The MIB is stored within the various devices which provide data to the management server. The device stores status and performance data about itself in the MIB, which when queried, provides the data to the manager. The manager then stores the information for later retrieval by the client, which is the user's interface.

The client will then display the information to the user (usually the network administrator) in a variety of formats, depending on the capabilities of the application. Remember that the client software used to display the SNMP data is not standardized (nor should it be).

The information displayed for each device includes the system identification, the number of interfaces and the status of those interfaces, routing table information (such as available routes and status of those routes), and traffic measurements. Statistics such as the number of Cyclic Redundancy Check (CRC) errors encountered and the number of retransmissions are also common.

The MIB is used to define the data which is to be stored by each entity. This is configured by the user of the device, such as a network administrator. Keep in mind the types of devices we are talking about; routers, gateways, modems, and network servers.

SNMP is one of the more popular network management protocols used today, but there are others. The debate about which one is the best continues. Many networks are implementing different solutions, which does not complicate things in a private network. In public networks, standardization is critical, and SNMP is the standard of choice today in many internets.

4.8. Chapter Test

1. What is the protocol used to support e-mail in an internet called?
2. TCP/IP protocols are documented by using ______ and are stored on a number of servers for retrieval over the Internet.
3. What is the name of a network or a group of networks which are maintained by a single authority?
4. A port identifies the interface used by an application. What is a socket?
5. Which organization is responsible for overseeing the development of protocols used in TCP/IP networks?
6. In the IP header, what does the flag field indicate?
7. What portions of the data unit does the header checksum in the IP header validate?
8. How many network addresses can be identified using a Class C IP address?
9. If the IP address of 128.1.17.1 has a subnet mask of 255.255.240.0, what is the subnet address?
10. A name server used in DNS reports the IP address of a domain name to name ______.
11. In the DNS, which type of operation is allowed to forward a query to another name server for resolution?

12. Name one type of routing protocol that is considered an Interior Gateway Protocol (IGP).
13. Diskless workstations use which routing protocol to learn their IP address?
14. A distance-vectored routing protocol determines the best route based on what?
15. A link state metric routing protocol determines the best route based on what?
16. What protocol does OSPF use to establish an adjacency with other routers and gateways?
17. When using source routing, intermediate nodes determine the next hop based on what?
18. What are split horizon, poison reverse, and trigger updates used to prevent?
19. What is time stamping used to determine?
20. Which protocol uses a primary source for clock synchronization and propagates that information to other nodes?
21. How large of a boundary does IP use for fragmentation?
22. What protocol provides error messaging to IP?
23. What happens to data units sent in a TCP header with the "push" bit set to 1?
24. What is the port number for the data being transferred using the FTP protocol?
25. What does TCP use to throttle message transmission when congestion is detected at another node?
26. Congestion is determined by TCP by what event?
27. How does TCP indicate that a data unit was not delivered?
28. What is the Post Office Protocol (POP) and where is it used?
29. What protocol is used to support newsgroups on the Internet?
30. What kind of files are stored on servers on the World Wide Web?
31. What is used to establish a link between two devices communicating via modem with the Point-to-Point Protocol (PPP)?
32. Where is network management data stored before being sent to servers upon query in an SNMP network?

CHAPTER 5

Signaling System #7

5.1. From Signaling to Control

Signaling System #7 (SS7) is the backbone of today's communications network. At the heart of every telephone company, long distance provider, and cellular provider is a network used by the providers' switching equipment to send control and signaling information between switches. This network is the communications network for communications equipment.

Signaling is certainly not new technology. In fact, signaling has been used in one form or another since the days of Alexander Graham Bell's first telephone device. The industry has refined signaling methods and since the digital revolution, has found new ways to communicate circuit- and network-related information over a separate network, SS7.

Today, the entire communications industry depends on the SS7 network. For example, making an 800 call would be difficult if not for the SS7 network. Remember that telephone switches route calls based on the digits dialed. The digits dialed represent an area, an office providing service in that area, and the subscriber's number. If we see the area code 212, most of us know that area code is for New York City. This analogy was recently made invalid with the FCC mandate for Local Number Portability.

The area code 800 can be used nationwide, and soon around the world. However, the area code 800 cannot be routed by telephone switches. The number must first be converted into a routing number that the telephone switches can understand. That is where SS7 comes in.

The telephone companies store these translation tables in databases that are centrally located within their networks. When a telephone switch receives digits beginning with the area code 800, it knows it must first query the 800 database to retrieve the routing number before proceeding with the call connection. The switch generates an SS7 message, which is sent over the SS7 network (not the voice network) to the database. The database then sends the routing number back over the SS7 database to the switch.

The use of the SS7 network does not stop here. The switch then sends a call setup message over the SS7 network to the switch at the other end of the circuit which will be used to connect the call. The two switches then exchange several call setup messages before the call is actually connected. When the call is released, messages are exchanged over the SS7 network again to release the circuits used for the call.

This is an oversimplified overview of how the SS7 network is used.

Later on we will get into more specifics about SS7 networks and the various applications used. To understand what SS7 is and why it is so important, we need to first look at what signaling is and how it is used in the nation's networks.

5.1.1. Signaling Methods—How They Evolved

As mentioned earlier, signaling has been used to set up circuits for telephone calls since the early days of Bell's inventions. The methods of signaling have changed dramatically over time, from electrical current to analog signals to digital messages over a packet-switched network.

There are several types of signals used by telephone switches to manage telephone calls. Circuit-related signaling is called line signaling and includes:

Call progress signals

Supervisory signals

Control signals

Address signals

Alerting signals

Call progress signals are used to notify the caller of the call progress. This includes a ringback tone, sent to the calling party after the digits dialed are processed and the called party is determined as available. Busy signals, all circuits busy, and intercept tones are all considered call progress tones.

Supervisory signals are usually electrical signals (on the subscriber loop) used to indicate an on- or off-hook status. When the receiver is lifted off of the cradle on the telephone, current is sent from one side of the telephone line, through the telephone circuitry, and back to the central office switch. This current is considered a supervisory signal and indicates an off-hook condition (which is interpreted by the switch as a request for service or circuit busy if a call is routed to the subscriber number while it is off hook).

Control signals are used by auxiliary equipment in the telephone network. They are tones sent from one device to another over the voice circuit (in non-SS7 applications). In today's network, almost all of these signals are carried over the SS7 network as data messages rather than as audible tones.

Address signals are the tones generated when a telephone is dialed (or the electrical pulses generated by a rotary dial). Once received by the central office switch, these address signals are converted into data messages and sent over the SS7 network. This was not always the case, however, as we are about to learn.

Alerting signals are used to notify parties of an incoming call. This is the ringing of the line, which causes the telephone to ring (or buzz or make some other audible tone in the case of digital or electronic telephones). Alerting signals can also be sent as data messages.

Now that we have learned the types of line signals used, we will learn about the different methods of signaling. The earliest form of signaling was inband signaling, which was actually sent over the voice circuit along with the voice call. A telephone circuit was established between switches, and tones were sent over the circuit. The frequency of these tones was within the same band allocated for the audible voice transmission.

Voice transmission over analog facilities is sent within 0 to 4000 hertz (Hz). Voice does not require that much of the bandwidth and is actually transmitted between 300 and 3800 Hz. The frequencies below and above the voice frequencies are used as a buffer to filter out cross-transmissions from adjacent circuits.

The signaling tones used the 300 to 3800 voice band and single- or dual-frequency tones. This did not work well because the voice transmission could interfere with the signaling information, and in some cases, the audio sent over the circuit could be mistaken for a signaling tone, resulting in disconnected calls.

This method was eventually phased out and replaced with out-of-band signaling. This form of signaling uses the buffer zones of the voice band, which are 0 to 300 Hz and 3800 to 4000 Hz. This method is better, but it still poses problems. Interference can still cause signals to be distorted.

Besides the problems with interference and inefficiency, analog tones limit the types of signaling information that can be conveyed. Querying a database for routing information would be next to impossible using audible tones. Out-of-band signaling also causes delays in the call setup and release of circuits.

Work began on common channel signaling in the 1960s to create a separate path for signaling information (apart from the voice circuit) and develop a standard for digital messages to signal and control telephone company switches and databases. Today, this network is the nucleus of many services offered by the telephone companies. SS7 is the future of telecommunications, supporting everything from Plain Old Telephone Service (POTS) to customized calling services such as custom routing and now local number portability.

5.1.2. Common Channel Signaling—The Advantages

There are many advantages to SS7. Because this is a digital network, all signaling and control information can be sent in the form of data messages. There are endless possibilities for the types of messages that can be represented. The SS7 standard itself provides a standard set of protocols and message types to support virtually all forms of signaling and control information, supporting telephone company needs for now and the future.

The links that are used within the SS7 network are also digital facilities. The most commonly used data link in SS7 networks in the United States is DS0A, providing 64 kbps to each node (called signaling points in SS7). This allows signaling information to be sent quickly through the network with little delay. When using the voice circuit to send analog signaling information, there were long delays in call setups and releases.

To understand this better, let us look how a call is connected through the telephone network. When you lift the receiver on your telephone, current is allowed to flow through the telephone and back to the line card in the telephone company switch. This signals the central office that you wish to place a call. The central office in turn sends a dial tone to you to signal that it is okay to begin dialing your phone.

When you begin dialing, the digits you dial are stored in the switch until all of the digits have been dialed. The telephone number you dialed is then analyzed by the telephone switch (according to dialing plans). If you dialed a local number, the telephone switch examines its routing tables to determine which circuit connects to the appropriate central office capable of connecting you to the caller. It may take several interim office connections to reach the final destination, especially in the case of long distance calls.

The telephone switch then generates an SS7 data message and sends it to the central office requesting that a connection be established on the selected circuit. Of course, this message must provide all of the information necessary for the remote office to make a decision, including the digits you dialed and the circuit number to which your local telephone company wants to connect the call.

The remote office then sends a response to the telephone switch, either granting permission to make the connection on the voice circuit or denying permission and providing a reason for denial. If permission is granted, there are a few more messages sent over the SS7 network before the connection is actually made and you begin your conversation.

Once your call is completed, the disconnecting office sends a message to the remote office switch to release the circuit. The circuit is now immediately available for the next call. While this explanation has been simplified significantly, you should get the basic idea of the role of the SS7 network.

It may seem that this network would actually make telephone calls painfully slow, forcing you to wait while all of these messages were sent over the SS7 network. However, the SS7 network has actually sped up the telephone network. The speed at which these signaling messages travel through the SS7 network allow telephone switches to connect circuits much faster than with analog methods. The release of these circuits is far more efficient than methods used in the past.

Remember back about 10 years ago, when you dialed the phone and waited at least a few seconds before you heard ringing? This may still happen today, but oftentimes you hear ringing immediately after dialing the last digit of the telephone number. Remember when people calling you would forget to hang up the phone, keeping your line busy for hours? Thanks to SS7, this no longer happens. As soon as you hang up your telephone, the telephone company switch disconnects the circuit you were connected to. The SS7 message is sent to all of the switches used to connect your call, dropping all of the circuits along the route.

It is the way SS7 operates that makes it far more efficient for telephone calls. Telephone facilities are set up and released much more quickly than with analog signaling methods. The SS7 network provides access to centralized databases, which saves telephone companies from having duplicate databases collocated with their switching equipment (a cost-prohibitive practice).

The real advantage of SS7 is the revenue potential for telephone companies. There are a variety of ways that telephone companies can generate revenue by giving other telephone companies access to their databases. Telephone companies can and do charge other telephone companies for this access.

5.1.3. After Signaling—Autonomous Network Control

SS7 provides network management that is not available in analog networks. The SS7 standards define a complete set of network management procedures which allow the network to automatically reroute telephone calls around switches which have failed or become congested.

The telephone companies have capitalized on this, closing down central office maintenance centers in favor of using remote maintenance centers. The remote maintenance centers can monitor the entire network from a central location and can instantly access any telephone switch within their network. Trouble can be detected and corrected much more quickly, before they affect the entire network.

When data links within the SS7 network fail, the SS7 protocol automatically starts procedures which involve rerouting traffic around the failed links and restoring the failed links to service. Since most of the problems encountered are usually timing issues, the protocol can often have traffic rerouted and the link back into service before service personnel in remote maintenance centers can even react. There are still troubles which do require human intervention of course, and SS7 provides an excellent tool for service personnel to use to determine where the problem is and what caused it.

The network management procedures used within SS7 are far more robust than those found in any other data communications network. This is both a positive and a negative. The positive side is that the network becomes far more reliable, but there is a cost associated with reliability. SS7 is not cheap to implement, and the cost of network components can be high. The offset is the number of services a telephone company can offer when they deploy SS7 in their network and the revenue potential this represents.

5.2. Intelligent Networks

The next step for SS7 networks is to provide the means for subscribers to configure their own services without talking with their local telephone company service representative. The present method of ordering new services requires a call to the business office of your local telephone company to place an order. This order must then be processed and sent through the appropriate channels within the telephone company before your service is activated.

This usually results in many unnecessary delays and drives the cost of telephone services up because of the administrative cost involved. However, if subscribers could access the telephone switch themselves and activate whatever services they wanted, the cost would plummet. No administration would be required to track a service order and get it processed because there would be no service order. Customers complete the orders themselves in real time, activating the features of the switch to configure 800 numbers, data lines, and any other services they may need.

The same type of access would allow a subscriber to deactivate services when they were through with them. For example, if you were holding a major sale and advertised your 800 number over the radio, you might expect a sudden increase in telephone calls to your 800 number. You would probably want to increase the number of lines ringing into your ordering center, but you only want the lines activated for a few days.

With the type of access we just described, you could access the telephone switch yourself and add the lines to your service. When you were finished with those lines, you could deactivate them. You only pay for the time that the 800 numbers were in use, rather than a monthly charge. You will also save because there is no telephone company administration needed to provide the service (except for billing administration).

If you were to apply the same scenario today, you would face an impossible task. Telephone companies cannot process service orders fast enough for such dynamic applications. The administrative cost is so high that they cannot afford to offer services such as 800 numbers for just a few days, so you must order them for a month or more at a time.

This is some of what an intelligent network can provide. There are of course many other services and features available in an intelligent network. The concept is to allow subscribers to configure their telephone and data services to meet their immediate needs, in real time, rather than the current method of placing orders and waiting for those orders to be implemented.

There is a cost savings to telephone companies which provide intelligent services as well. As we have seen in the scenario above, administrative costs are high. By placing the responsibility on the subscriber for ordering and implementing services themselves, the telephone companies realize savings, not to mention increased revenues from services they could not offer previously.

Intelligent networking is not just about subscriber services. Allowing telephone switches to communicate with one another and make decisions on call routing and handling is the real advantage of intelligent networking. Telephone companies have already begun utilizing the SS7 network for intelligent networking.

5.2.1. What Is Intelligence?

In the networking sense, intelligence is the ability for a network to work autonomously and for the nodes within that network to make decisions on routing and call handling without human intervention. There are a number of ways this can be done.

Databases are critical to any intelligent solution. Information on how certain calls should be routed, or how calls should be handled if a busy condition is encountered, must be stored somewhere where all of the switches in the network can access it. These databases could be located with the telephone switch itself, but the cost of maintaining multiple databases is prohibitive. Another disadvantage to distributed databases is trying to keep them all current. Any changes made to a database would have to be propagated throughout the network to all of the other databases.

SS7 allows telephone companies to use a central database for all kinds of services. Maintaining a central database is far easier and more cost effective and allows telephone companies to share their databases with other companies.

The SS7 protocol must provide the messages and procedures used to access these databases and the means for interconnecting with other databases in the telephone network. SS7 also must provide interconnectivity with other telephone switches, allowing them to communicate with one another. This allows particular features to be invoked, even while a call is in progress, by issuing a command from a local switch to a remote one.

Presently, the telephone companies are the only ones who utilize the SS7 network for this purpose. This is rapidly changing, as small and large companies discover the advantages of intelligent networking. Some trial services have already begun, allowing subscribers to send control and signaling information from their own private switches to switches in remote offices. This is done by sending that information over the SS7 network. The interface to the subscriber is called Integrated Serviced Digital Network (ISDN).

While ISDN has been around for several years, intelligent network services are just now being tapped. This is somewhat ironic since ISDN was originally developed with the concept of intelligent networking in mind. Telephone companies are just now learning how to tap the power of SS7 and utilize its potential.

5.2.2. Future Services

There are many demands being placed on telephone companies today. A good example is local number portability, which is a new government requirement. In 1995, the government mandated that subscribers with 800 numbers must be able to maintain those numbers even if they switched to another service provider. The problem was that the numbers were allocated to specific service providers, and the routing and billing information was stored in their computer databases. Therefore, service providers have to provide access to those databases to other service providers. That access is through the SS7 network. The telephone companies which provide that access charge other companies access fees every time a transaction takes place. So, rather than this being a disadvantage to service providers, it has become another avenue of revenue.

Local number portability works in much the same way. Subscribers who move from one city to another must be able to keep their telephone numbers. This is especially important to small and large companies who have purchased "vanity" numbers (numbers such as 800 COLLECT or 553-SUN1). Implementing this will not be easy, and indeed there are several thoughts about how this should be done. Whatever method or procedure is implemented by telephone companies to provide this service, SS7 will be the network to provide the intelligence to support such functionality.

This is an example of how wireline service providers are using SS7 to provide services. Cellular companies have also found that the benefits of SS7 far outweigh the implementation costs. Almost all of the cellular networks in the United States today have deployed SS7. All of the new Personal Communications Services (PCS) networks will depend on SS7 to provide the many features and services being offered.

Remember that SS7 is a data network, and it uses data messages to allow switches to communicate with databases and other switches. We have already talked about a few applications which depend on SS7. Here are a few more examples of how the SS7 network is being used to deliver services.

5.2.2.1. Intelligent Routing Many companies use call centers to answer large volumes of calls. These call centers are used for airline and hotel reservation centers, rental car agencies, and catalog sales. Calls are routed to a particular location, using conventional routing tables. What if the particular calling center location has become overloaded? How can calls be routed to another calling center based on traffic flow? The answers lie in databases which maintain routing instructions for the 800 number. This database is queried every time the number is dialed. The instructions are received by the telephone company switch and used for handling each call.

When traffic reaches a certain threshold, calls can be diverted (according to the routing instructions) to another call center. This service is now being used by many reservation centers today.

Routing can also be changed according to other parameters. The calling party number can be used to determine where the call is routed. Companies can advertise a single 800 number nationwide, but the calls can be routed to a local calling center that is determined by the first six digits (area code and office code) of the caller's telephone number. Calls can also be diverted according to the time of day and day of week. This is especially useful for customer service organizations.

5.2.2.2. Smart Custom Features Caller identification is a good example of SS7 delivered features. Caller ID is passed from one switch to another through the SS7 network and is eventually passed on to the subscriber through the subscriber interface (POTS or ISDN). When ISDN interfaces are involved, the caller ID is a part of the call setup message in ISDN that is passed to the subscriber equipment. In the case of POTS, a modem is used at the central office end to pass the caller ID over the POTS line using modulated methods.

There are, of course, many other features which can be offered only where SS7 exists. These features require databases to store special instructions regarding call handling or the ability for telephone switches to communicate with one another during the call.

Call forwarding can be customized in a way to meet the immediate needs of any business. Calls can be diverted to another location during peak hours or even forwarded to voice processing equipment. This ability provides additional revenue opportunities for telephone companies, allowing them to provide automated attendants and voice mail systems which process calls after forwarding.

Cellular providers are already using SS7 to provide additional services to their subscribers. Callers are diverted to voice mail systems, and pagers are activated when cellular customers cannot be reached. Store and forward services can also be provided, such as facsimile conversion of e-mail. These are based on special instructions stored in databases maintained by the cellular provider and accessed through the SS7 network.

5.2.2.3. Database Access—Key to Intelligence By now you should have guessed that database access is one of the key functions of any intelligent network. Routing and call handling instructions are stored in these databases and accessed through the SS7 network. Before SS7, databases were localized with the telephone switch, which meant that changes and updates were difficult to propagate to other databases.

Using the SS7 network, the databases can be more centralized and accessible by many more network nodes. The SS7 protocol provides a data transport service specifically designed for accessing these databases and transferring the data to querying switches. If the node sending the query resides outside of the providers network, the SS7 nodes are capable of trapping the address of the inquiring node. This information is then forwarded to a billing database which is used to create billing invoices to other telephone and cellular service providers.

The ability to charge for access into an SS7 network more than offsets the cost of deploying SS7. Companies charge for each transaction performed on their databases, creating millions in additional revenues from other telephone companies. Those costs eventually get passed onto the subscriber, but SS7 has helped prevent our telephone charges from becoming too expensive (a concern after the divestiture of the Bell System).

Our entire cellular network is based on databases and the ability of the cellular switches to access those databases. When a call is placed to a cellular telephone, the receiving switching office must be able to determine where the cellular customer is located, down to the antenna which is closest and most capable of handling the radio transmission to the cellular subscriber's handset.

The signal strength of the cellular handset and specifics about the cellular subscriber (such as serial number and authentication information) are stored in these databases as well. What makes these cellular databases unique is that they are dynamic. As the cellular subscriber moves around the network, the databases must be able to change information regarding the antenna serving the subscriber's telephone as well as the Mobile Switching Center (MSC) serving the subscriber.

The MSC is the central office used for cellular networks. Each MSC is capable of serving many cell sites and has access to its own location databases. The MSCs are then interconnected using the SS7 network, as well as conventional facilities for transferring calls between networks. Chapter 7 provides more information about database use in these networks.

5.2.2.4. End-to-End Subscriber Services So far, we have discussed telephone company switches and cellular switches communicating with one another through the SS7 network. ISDN is a seamless network where all telephone equipment can communicate end-to-end, including subscriber systems. For example, large corporations may use several Private Branch Exchange (PBX) systems, each one in a different location. In today's networks, these PBX systems are interconnected via voice trunks. Control and signaling information cannot be exchanged from one PBX to another, severely limiting the abilities and features that can be offered.

If the PBX could be connected to a digital data network, signaling and control information could be exchanged through the telephone network with other remote switches. The digital facility could be separate from the voice facility and could contain information regarding all of the voice circuits terminated in the PBX.

This was the original concept of the ISDN and, in fact, the purpose of the ISDN Primary Rate Interface (PRI). The D channel of the ISDN PRI is used to send signaling information for all of the voice channels on the PRI. Where multiple ISDN circuits are used, one PRI can be used for signaling only; the others are for voice and data applications only.

The D channel sends protocol messages which are compatible with SS7 protocol messages and can be transferred through the SS7 network to the remote end of a connection. Class of service information assigned to a particular extension in the PBX can be sent to the remote PBX, which can then determine which features and calling privileges the extension is allowed. There are many other possibilities when one thinks of computer telephony applications.

Some corporations are taking advantage of this service today. Many telephone companies are working on packages for their major customers which will allow their PBXs to communicate with other PBXs over the SS7 network; currently they are limited to connecting with their PBXs only through virtual connections. Eventually, as standardized features and interfaces are established for PBXs, you will be able to communicate with any PBX over the SS7 network, providing true end-to-end connectivity and feature-rich services.

5.2.3. Broadband Requirements

As telephone companies change their infrastructure from narrowband facilities over to Asynchronous Transfer Mode (ATM) facilities, the demands of the SS7 network will change. New services will be offered through the broadband network, requiring support from the SS7 protocol suite. This support will mean changes to the SS7 protocol suite.

Probably the most significant change will be found at the lower layers of the protocol. Currently, the Message Transfer Part (MTP) provides transport services for all of the SS7 applications [such as ISDN User Part (ISUP) and Transaction Capabilities Application Part (TCAP), discussed later]. However, MTP was developed for digital facilities such as DS0A and does not support broadband.

One of the changes in the transport is the ability to support ATM virtual circuits, which are provided through the Synchronous Optical Network (SONET). The MTP protocol provides procedures for digital T-span circuits and network management based on these types of data links. ATM uses a fiber optic backbone, called SONET, which does not use individual channels for transmissions. In SONET there are payload envelopes which are assigned to paths and virtual circuits. The circuits are identified by Virtial Circuit Identifiers (VCIs) and the paths that these circuits are assigned to are identified by Virtual Path Identifiers (VPIs).

The ATM protocol includes parameters used to identify these components of a circuit connection and will be used within the SS7 network to carry SS7 messages as well. The upper layers of the SS7 network will change accordingly. For example, the ISUP protocol, which is responsible for the connection establishment and release of voice circuits, will have to be modified to address the SONET equivalent of channels (or circuits).

The migration from current SS7 protocols to broadband services is not expected to happen overnight. It will most likely take several years before the telephone networks are ready to make such changes. There has been some migration already but usually as trials and not as total replacement of existing networks.

Hardware components of course will change as well. SS7 nodes will now have to support SONET interfaces. The largest challenge to manufacturers of SS7 components will be supporting the bandwidth requirements of a SONET/ATM connection, especially through the backplanes of older legacy systems. It is expected that manufacturers will introduce new platforms to support these new broadband changes.

5.3. SS7 Architecture

The SS7 network has a fairly simple design, consisting of three main functions. The endpoints originate SS7 traffic, the transfer points are responsible for routing the messages through the network, and the control points are responsible for managing traffic to databases that are resident in the network. Any of these functions can reside as a stand-alone system or as part of a switching function.

All of these functions are interconnected via data links. The data links are deployed in pairs for redundancy to ensure that there is always a path for traffic. The SS7 protocol also takes care of managing the data links and the traffic between nodes. The SS7 protocol is mostly an autonomous network, managing traffic flow and changing routes when problems in the network are detected.

5.3.1. Data Links

All of the nodes in the network are interconnected via data links. The most commonly used link in North America is 64 kbps. The links are not usually individual circuits but channels in larger facilities (such as T-3s). Telephone companies can utilize any existing channel as long as there are at least two channels (for redundancy) to any one node.

Another rule often overlooked is using channels in completely separate facilities. This means that if a T-3 channel is used between two SS7 nodes, the redundant link must be in a T-3 (or other facility) apart from the first. The cross connects, amplifiers, and channel banks must all be separate for the two links. The purpose of this level of redundancy is to guarantee the integrity of the link. If there is a failure of any of the devices used by the data link, the link could fail. If both of the redundant links are connected through the same devices, both links fail.

This seems like an obvious situation, and the solution seems just as obvious, yet in many networks this is overlooked. Either the facilities are passed onto different groups who do not understand the need for redundancy in the SS7 network, or the network administrator stops at the SS7 node, leaving the rest of the network design to a different group. The most important concepts to understand in SS7 networks are redundancy, integrity, and backup. Failures in the network can cause service outages for millions of customers and are treated with a heavy hand by the government.

Figure 5.1
SS7 Signaling Links

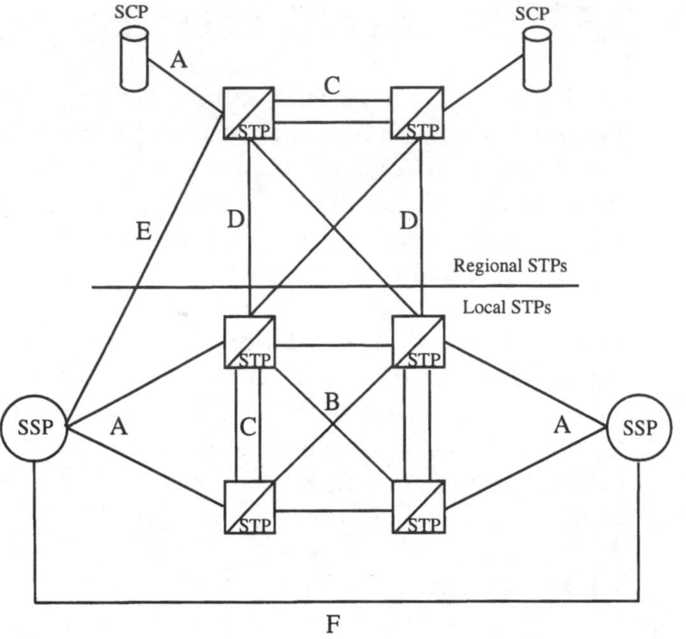

SS7 data links are labeled according to their location in the network. There is no physical difference between the various labels, but their location may determine how they are used. Network management procedures will differ on some links, all managed by the individual nodes. Almost all nodes require identification of the link by its label so that network management procedures can be invoked properly.

As shown in Fig. 5.1, there are several types of links. Links used to connect a Service Switching Point (SSP) to a Signal Transfer Point (STP), or an STP to a Service Control Point (SCP) are called access links, or A links. These links provide access into the SS7 network from an SS7 end signaling point.

Bridge, or B links, connect STPs within the same network to one another. Some networks use a hierarchical approach to STP deployment, using a pair of regional STPs as hubs. If an STP becomes congested, messages can be routed to the regional STP. These regional STPs are also used for access to network databases. The links connecting to these regional STPs are called diagonal links, or D links, and connect local STPs to regional STPs.

An SSP can also connect to an STP outside of its area. For example, there may be an STP within the same Local Access and Transport Area

(LATA) as an SSP that cannot be reached. An extended link, or E link, can be used to reach an STP outside of the LATA.

When an STP is deployed, it is always deployed in a pair. The two mated STPs interconnect through cross-links, or C links. Normally, traffic does not pass through the C links. Only network management messages are passed over the C links to the mated STP. However, in the event of an STP failure, traffic can be routed from one STP to its mate STP over the C links.

Some SSPs may have a high volume of SS7 traffic to another SSP. In this case, it would be advantageous to provide a direct connection between the two SSPs and use associated signaling rather than quasi-associated signaling. A fully associated link, or F link, provides a signaling connection between two nodes.

Deployment of these links is always in pairs. If a link is provided from an SSP to an STP, another link from the same SSP must be provided to the mated STP. The key to network integrity in SS7 is diversity and redundancy. Diversity means that links are provided on separate facilities (for instance, two different T-3s) rather than on the same facility. This way, if a cable should be cut somewhere in the network, only one link is affected and not both redundant links.

This principle applies to all equipment used to support the link. Channel banks, clocks, power supplies, and repeaters must all be deployed in such a fashion that a single point of failure does not cause an SS7 node to become isolated from the rest of the network. This is the biggest challenge in SS7 network design.

Links terminating to the same signaling point are put into groups, called linksets. A linkset is a group of links, all of which terminate to the same node. Links are deployed redundantly, which means if there are two links terminating to an STP, there will be two links terminating to its mate as well. These links can be configured into two separate linksets, or they can be configured as a combined linkset. The configuration is controlled by each signaling point through administrative procedures.

A signaling point uses routing tables to determine which linksets should be used to reach a particular point code. These routing tables define a set of routes. A route is a collection of linksets that can be used to reach a point code. There is no knowledge of intermediate point codes in the routing tables. All the signaling point needs to know is which collection of linksets will get to the designated point code.

There is usually more than one route to get to a point code. Routes can be prioritized so that the signaling point can determine which route should be selected first, which one second, et cetera. A group of routes is called a routeset. In summary, a signaling point selects a routeset, which consists of a group of routes, which consist of a group of linksets, which consist of a group of links.

5.3.1.1. 56/64-kbps Links As mentioned before, this is the most common configuration used in North America. The interface used is the DS0A, which is typically derived from a T-1 facility. The telephone office will typically demultiplex the T-1 (or T-3) using a digital cross connect. The individual DS0s are then distributed to various equipment in the central office, including the SS7 nodes. The channels used by SS7 must be dedicated channels that are only used for SS7.

One of the most critical aspects of this interface type is timing. DS0s require synchronized clock sources at both ends of the connection. There are a number of devices used in between the SS7 nodes, such as channel banks, repeaters, and digital cross connects. All of these devices must have a clock connection, which must be synchronized with all of the other devices in the network.

Such synchronization is achieved through the synchronization network. This network uses a highly accurate clock source, which is then distributed to all of the telephone offices in the network. Each office is responsible for distributing the clock signal to all of the devices in its building, using highly accurate clock distribution systems. A more thorough discussion of clock synchronization can be found in Chap. 2.

While the 64-kbps link provides more than adequate bandwidth for signaling, the full capacity of the link is protected. The links are configured to carry no more than 40 percent of their capacity. Traffic is distributed by the various SS7 nodes over all of the available links to ensure efficient use of the data links.

In the event of link failure, traffic is rerouted over the other available links. This is one of the reasons for limiting the capacity of the data link. When there is a link failure, and traffic is diverted to another link, that link can effectively carry 80 percent of its capacity. This allows enough bandwidth left over to carry network management traffic without running out of bandwidth.

5.3.1.2. 1.544-Mbps Links While the SS7 network has not yet run out of bandwidth, the type of traffic is changing, and the need for additional bandwidth is changing. At the same time, the telephone network itself is changing, and we are seeing that the present digital hierarchy is quickly becoming obsolete. What this means to SS7 network operators is a change in the type of interfaces they can use.

As the voice network is converted over first to fiber optics and then to ATM, the presence of DS0 interfaces will dissipate. SS7 networks will have to convert as well or face additional costs in maintaining separate network facilities. Bellcore issued a recommendation to SS7 network operators, converting networks in two phases.

The first phase involves changing from DS0 links to DS1 links, which provide 1.544 Mbps of bandwidth. This will mean additional capital expenditures to change the existing equipment to this new type of interface. Virtually every circuit card used to connect to an SS7 link will have to be changed. This cost can be rather prohibitive, with no real evident payback.

The second phase replaces all channelized facilities with ATM links. This is not likely to happen right away, because of the cost to support such a deployment. Most telephone companies are deploying ATM links at DS1 speeds to their SCPs first, then migrating their remaining links to ATM. By using DS1 speeds, they can continue using their existing channelized equipment (with modifications to support ATM protocols) until they are prepared to replace their outside plant with ATM switches and fiber optics.

It is not anticipated that many network operators will jump to change to DS1 links. It makes more sense to wait until ATM links (the second phase) become available and make the change then. This way, there is a one-time capital expenditure (instead of changing first to DS1 and then ATM) and an immediate payback. Telephone companies will be anxious to move all of their traffic to the new ATM network so they can decommission all of the outside plant equipment (repeaters, amplifiers, channel banks, etc.) and take advantage of facilities consolidation.

5.3.1.3. ATM Links This is the most likely direction for telephone companies. The conversion to ATM is already underway, and soon all of the telephone network will be riding on fiber optics with ATM as the transport. This means that the facilities used previously for SS7 traffic will be gone. As mentioned earlier, Bellcore issued a recommendation for conversion of the SS7 network to ATM. It is highly likely that this conversion will begin in the very near future, as manufacturers begin developing SS7 components that are ATM compatible.

SS7 does not require the bandwidth provided by ATM, but ATM does provide rate adaption. This means that signals from several sources can be multiplexed over the same cell in ATM until the bandwidth is used up. This makes ATM an efficient facility no matter what the application. In terms of SS7, this is a winning solution, providing enough bandwidth to handle all traffic mixes over a nondedicated facility.

5.3.2. Network Components

The SS7 network is a fairly flat hierarchy, consisting of only three types of components, or signaling points. When looking at the SS7 reference model, the three types of functions supported are SSP, STP, and SCP. These represent functions rather than stand-alone devices, although they certainly can be stand-alone devices. Each of these functions is explained below.

5.3.2.1. The Service Switching Point This is where SS7 messages originate. The SSP function is found most often in the end-office telephone switch. Commonly found as an adjunct processor, the SSP originates SS7 messages after determining which interoffice trunk will be used to connect a call. The SS7 message is sent to the end office on the remote end of the circuit and contains a request for connection.

The SSP also originates an SS7 message when it cannot determine which interoffice circuit to use to connect a call. For example, in the case of an 800 number, the end office cannot determine which circuit to use because it cannot determine how to route the call based on the digits dialed. The 800 number must be converted to a routing number before the switch knows which trunks to use.

The SSP originates a query to an SS7 node which will provide a connection to a database. One of the unique things about SS7 networks and databases is that the SSP does not have to know the address of the database. The message originated must only provide the digits dialed. The STP, which is responsible for routing SS7 messages through the network, can use this information to determine which database the query should be sent to.

Once the database has provided the necessary routing information, the SSP can begin connecting the necessary facilities to handle the call. Keep in mind that this process must be repeated every time a circuit is needed. It is rare that a central office will have one circuit which can handle calls to remote offices without involving other intermediate switches.

While it appears that this system would slow down call processing and result in unnecessary delays, the network operates very quickly. There are very few delays in SS7 networks, due partly to the speed of the network and partly to the implementation of the network.

5.3.2.2. The Signal Transfer Point The STP is responsible for routing traffic through the SS7 network. It is not the originator of any traffic (other than network management) and is never the final recipient of any SS7 traffic. It is an intermediate point which provides some processing and routing of SS7 messages.

The simplest traffic for the STP is circuit-related traffic. This traffic is originated by an SSP and sent to another SSP to request a connection on a particular circuit. The SS7 message is addressed to the remote SSP and does not require any processing by the STP. When the message is received at the STP, it is simply passed along without further delay. This is called ISUP traffic.

Non-circuit-related traffic will usually involve a database. We have already discussed numerous applications which use databases, such as 800 numbers. They will almost always require the processing services of the STP.

When a query is made to a database, the SSP does not typically know the address of the specific database it needs to query. This is desirable because of network management considerations. If the database were to be addressed directly, and the database were unavailable for any reason, there would be unnecessary delay in trying to determine how to best handle the query.

The STP provides Global Title Translation (GTT) to determine where a query should be routed. The STP looks at the Signaling Connection Control Part (SCCP) to determine what digits were dialed (which is part of the called party address) and makes its routing decision based on these global title digits. The STP does not have to look at all of the digits dialed, just the area and office codes.

Global title digits (usually the digits dialed in the called party address) do not have to be a telephone number. In the cellular network, global title digits are the Mobile Identification Number (MIN) used to identify a cellular terminal. GTT digits can be any kind of number.

Global title is a unique feature of SS7, and it is valuable for many reasons. In most networks, the address of an end node must be propagated throughout the network. This would mean advertising the address of all databases (or subsystems, as they are called in SS7) throughout the SS7 network. If the subsystem should fail or become congested, the address would have to be marked as unavailable. If the address is propagated throughout the network, the status information would also have to be propagated throughout the network.

Another advantage of using global title is anonymity. In some networks, the subsystem is shared with other networks. The host network does not want nodes outside of its network to be able to address the subsystem directly and will not advertise the actual subsystem address. Instead, all queries are directed to the host's gateway STP, which performs gateway screening first and then GTT to determine the best subsystem for the application.

Another feature of the STP is network security. The STP can be used in a gateway function, allowing only traffic which meets specific criteria to enter into a network. This is an important feature as telephone companies begin to share their network resources. The STP can be configured to allow messages with specific origination addresses or portions of an origination address (such as the network portion of the address or the node portion of the address).

The STP can also be configured to allow only specific types of SS7 traffic from specific addresses. For example, a screen can be configured to allow a particular network to send ISUP messages but not database queries. This function is referred to as gateway screening and is found in "edge" STPs, which connect to STPs in other networks.

The STP also manages the resources of the network. Resources in an SS7 network are primarily databases (referred to as subsystems in SS7). To connect to a database, the STP must route to a front end, the SCP. The SCP interacts with the STP to ensure that the actual subsystem is available and can process the queries being sent.

If a subsystem fails, the STP provides subsystem management procedures which will reroute queries to other subsystems, either redundant or not. This function is explained in the section below.

The STP function may change as the network evolves over to ATM. STPs will no longer be required for routing messages through the network. The new ATM network will have a flatter hierarchy, with all traffic routed by the ATM switch. The STP will still be needed for network management, subsystem management, and gateway screening.

5.3.2.3. The Service Control Point The SCP is a front end to subsystems. The SCP is not a database itself, although it can certainly be collocated with a database. The SCP function manages access to the various databases and can manage more than one subsystem.

Each of the databases is addressed through a subsystem number that identifies the application served by the database. These subsystem numbers are predefined by the network operator, and in some cases, universal subsystem numbers have been predefined to ensure interoperability.

The database itself does not have an SS7 address. Queries must be sent to the address of the SCP. The SCP then routes queries to the appropriate subsystem based on the subsystem number. This offers maximum flexibility, without having to use "locked in" address labels. If a particular subsystem should fail, there are no physical addresses to deal with. Subsystem management, which is the responsibility of both the STP and the SCP, manages routing to subsystems during failures.

There are two different configurations used for subsystem routing: solitary and replicated. When a subsystem is classified as solitary, it is the only subsystem to be used for queries. All queries for a specific application must be sent to the one subsystem. In replicated subsystems, there are two modes used, dominant and load sharing.

In dominant mode, one subsystem is marked as primary. All queries are sent to the primary subsystem first and to the secondary subsystem only when the primary subsystem cannot handle the query (because of traffic flow or failure).

In load-sharing mode, both of the subsystems share queries. Associated queries (transactions requiring more than one message) are sent to the same subsystem. The protocol used for queries (transaction capabilities application part) is responsible for managing the segmentation and reassembly queries.

5.4. SS7 Protocols

SS7 is really a family of protocols, each protocol serving a specific purpose. SS7 uses a layered hierarchy, just as Open Systems Interconnection (OSI), but because SS7 was developed before the OSI Model, there is not a one-for-one correlation. Figure 5.2 shows the protocol stack for SS7.

The packets used in SS7 are called signal units. There are three types of signal units used in SS7:

- Fill-in signal unit (FISU)
- Link status signal unit (LSSU)
- Message signal unit (MSU)

The FISU is used during idle periods, when there is nothing else to send. The idea of the FISU is to provide enough of a signal unit that

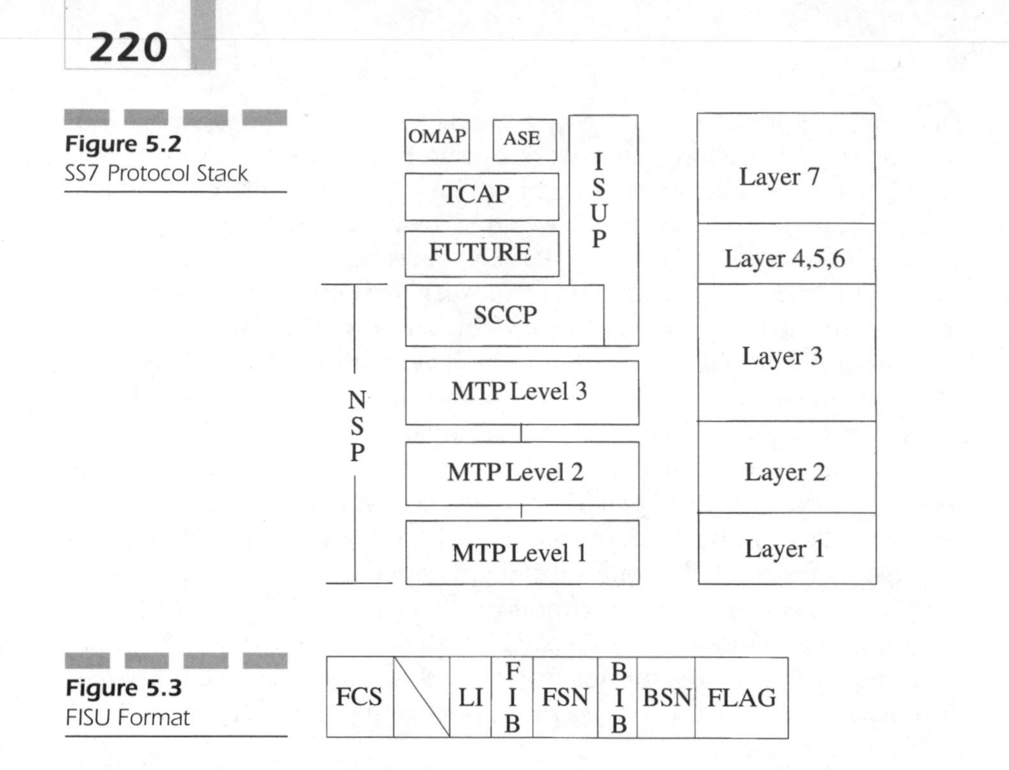

Figure 5.2 SS7 Protocol Stack

Figure 5.3 FISU Format

link integrity can be checked even when the link is idle and not carrying any real traffic.

There is no signaling information in the FISU, only a sequence number and FCS field, as seen in Fig. 5.3. The sequence numbers can be used to send acknowledgments of previously received message signal units; otherwise they carry no significance.

Also notice that there is an absence of addresses in the FISU. This is because they are of local significance only. FISUs are never routed over the network. They are sent from one node to its adjacent node only and used by the two nodes to verify that the link can still carry traffic. In the event a particular link cannot carry traffic, the protocol procedures defined for MTP are used to reinitialize the link. FISUs are once again sent to verify that the link has been reinitialized properly.

When an excessive number of FISUs have been received in error (determined by checking the Frame Check Sequence, or FCS, field), the link is reinitialized. This is monitored by the Signaling Unit Error Rate Monitor (SUERM). This is an incremental counter which counts every signal unit received in error. When 64 signal units have been received in error, the SUERM initiates link alignment procedures.

What is different about this counter is that it also decrements. When 256 signal units have been received without error, the counter decre-

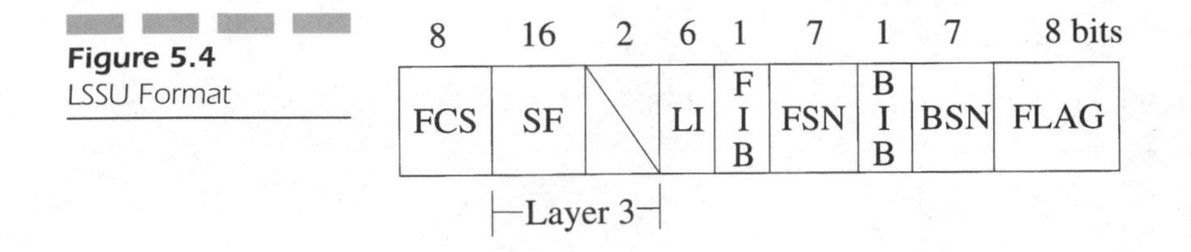

Figure 5.4
LSSU Format

ments by 1. This is sometimes referred to as the "leaky bucket" technique. Without this feature, whenever the counter reached 64, the link would be realigned, even if unnecessary. The counter is maintained by the link software and is found in all SS7 network nodes. The SUERM is also maintained on a link-by-link basis and is of local significance only. Nothing is shared between two adjacent nodes unless the link fails and is realigned. When this occurs, network management messages are exchanged.

The LSSU is actually part of the MTP protocol. It is generated by level 2 MTP and is used to send link status information to the remote end of a link. The LSSU is not routed through the network; it is only used between two adjacent nodes to manage the links interconnecting them (see Fig. 5.4).

When the link is being realigned, the LSSU is used to identify the various stages of link alignment (such as out of service, in service, etc.). This is needed so that the remote end knows when the link is capable of sending traffic again.

We will not go into a great amount of detail as to how link management works. Just remember that MTP level 2 is responsible for link management. Also understand that alignment means that the data on the link can be properly interpreted at the remote end of the link. Most errors occur because of timing problems and can be rectified by resynchronizing the link with the clock source. This is alignment of the link. For detailed information about how link alignment works and the specific sequence of events, you can refer to my book, *Signaling System #7*, which provides a comprehensive look at the SS7 network and its family of protocols.

The other type of signal unit is the MSU, which is used to carry all other types of SS7 messages. ISUP, TCAP, Telephone User Part (TUP), and network management are examples of the types of messages found in an MSU. Think of the MSU as the signal unit which always carries a payload, whereas the LSSU and FISU carry little or no information.

The MSU also has a portion used by level 4. This is the user data portion shown in Fig. 5.5. This is where you will find the application infor-

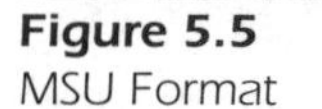

Figure 5.5
MSU Format

8	8n,n>2	8	2	6	1	7	1	7	8
FCS	SIF	SIO		LI	FIB	FSN	BIB	BSN	FLAG
	Level 3								

mation, such as the upper-layer protocols. The routing label is not part of level 4, but part of level 3, even though it is found in the user data portion of the packet.

Look again at Figs. 5.3, 5.4, and 5.5 and notice that all of the signal units contain the length indicator (LI). The purpose of the LI is to identify the type of SS7 signal unit being sent. There are only three options: FISU, LSSU, or MSU.

The length indicator does not give the total length of the entire packet, just the length of the field immediately following the length indicator. If the value is 0, the signal unit is an FISU (there is no user data portion in an FISU). A value of 1 or 2 indicates a 1- or 2-byte field, which would indicate an LSSU. Anything with a value of 3 or larger (up to 63) is an MSU.

SS7 packets can certainly be larger than 63 bytes, but again, it is not the intent of the length indicator to give the length of the entire packet, only the "user data" portion found after the LI field. It is used by level 2 software in SS7 nodes to identify the type of signal unit being received so it can determine how to handle the signal unit.

5.4.1. Message Transfer Part

The MTP is the transport protocol for SS7. The job of MTP is to move SS7 traffic through the network, providing connectionless services. MTP provides sequenced delivery of all SS7 signal units and is divided into three levels. If you look at Fig. 5.2, which shows the SS7 protocol stack, you will notice that MTP includes the first three levels of the stack.

The first level, the physical level, aligns with the OSI Model in terms of function. The physical level provides the electrical interface used on the data link and is responsible for the transmission of SS7 traffic over whatever facility is being used.

The data link level also aligns with the OSI Model and provides error detection/correction, as well as sequenced delivery of all SS7 signal units. The unusual part here is the fact that MTP provides sequenced delivery

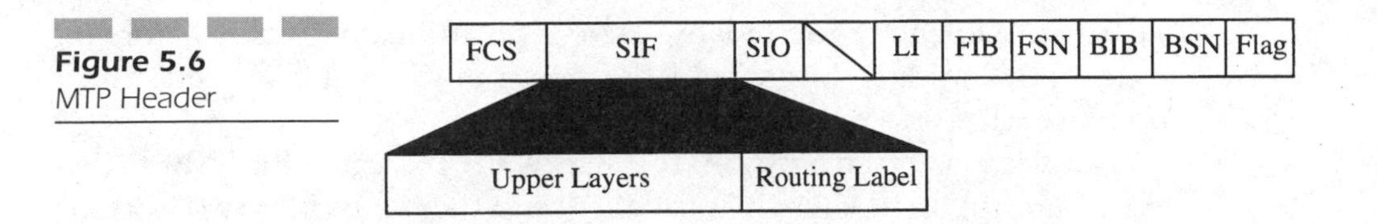

Figure 5.6
MTP Header

when it is a connectionless service. In other data protocols, connectionless services do not support sequenced delivery. In MTP, sequenced delivery is guaranteed because MTP always routes associated signal units over the same path (based on the rotation bits located in the header).

If MTP were a connection-oriented service, a connection between the two ends of the link would have to be negotiated. This would involve messages being transmitted from one node to the other, requesting a connection establishment between the two nodes. This is not the case in MTP, which makes it unique.

MTP level 3 is used for routing and network management. If one looks at level 2 MTP, there is no routing information. In fact, look back at the FISU, and you will see the components of the MTP level 2 header. MTP level 3 includes the routing label, which is contained in the user data portion of the header (see Fig. 5.6).

Sequence numbering in SS7 is used to guarantee ordered delivery of all signal units. They are used to acknowledge received signal units as well as to request retransmissions. We will look at the sequence numbering of MTP in this section. Also read the previous section to understand other aspects of MTP level 2.

Sequence numbering uses the Backward Sequence Number (BSN) field, the Forward Sequence Number (FSN) field, and the indicator bits (backward and forward). They are similar to other bit-oriented protocols which use sequence numbering, with a slight twist. The BSN is used to acknowledge the last received sequence number. In other protocols, this field would indicate the next expected sequence number.

The FSN is incremented each time a signal unit is transmitted. The only exception to this rule is in the case of the FISU. Unless the FISU is being used to acknowledge received signal units, the sequence numbers remain static, not incrementing. There is no reason to increment the sequence numbers of an FISU since they contain no data. If a signal unit is lost, it will not be retransmitted. They are used for error detection only.

If a signal unit is received in error, it is rejected by the SS7 node level 3 or level 2 (depending on the type of error), and the next signal unit sent in the opposite direction will carry a retransmission request.

Retransmission is requested by using the indicator bits. In normal mode, both indicator bits should be the same value (it really makes no difference what that value is).

To indicate a retransmission, the Backward Indicator Bit (BIB) is toggled to an opposite value (being 1 bit long, there are only two options here). When the Forward Indicator Bit (FIB) and the BIB are different values, the node interprets this as a retransmission request. The node will then look at the BSN to determine the last good signal unit received, discard all signal units up to that sequence number from its transmit buffers, and send all remaining unacknowledged signal units from its transmit buffer.

Keep in mind that sequence numbers are of local significance only. In other words, they are not carried and interpreted end to end. As a signal unit moves from node to node, the sequence number changes accordingly and is examined by each node. Several links are involved with any transmission, so this information will not only change every time a message is passed to a new node, but the sequence numbering must be processed at every link. Each link within an SS7 node must maintain its own set of sequence numbers and manage the transmission over that link (this is where distributed processing systems have a real advantage).

5.4.1.1. Network Management MTP level 3 provides network management procedures that reroute traffic around failed or congested nodes. These procedures use the MSU for sending messages to adjacent nodes, indicating the status of a node. This is different from the LSSU, which is generated by MTP level 2 to indicate the status of a link between two SS7 nodes.

There are three types of network management procedures used in SS7:

- Signaling link management
- Signaling route management
- Signaling traffic management

Signaling link management controls the activation and deactivation of signaling links. This is controlled by MTP level 3, and will result in an MTP level 2 LSSU being sent to the adjacent signaling point. The alignment procedure is invoked as a result of signaling link management.

Signaling traffic management procedures are used to divert SS7 traffic

Figure 5.7
Traffic Management Message Types

Traffic Management Messages	
0001 0001	Changeover Order
0001 0101	Changeback

Figure 5.8
Route Management Message Types

Route Management Messages	
0100 0001	Traffic Prohibited
0100 0010	Transfer Cluster Prohibited
0100 0011	Transfer Restricted
0100 0100	Transfer Cluster Restricted
0100 0101	Transfer Allowed
0100 0110	Transfer Cluster Allowed

from a failed link to another link in the same linkset (see Fig. 5.7). They are also used for flow control procedures over a data link. Messages are sent using the MSU signal unit.

Signaling route management procedures are used to identify failed or congested SS7 signaling points. MTP level 3 responds to these messages by rerouting traffic around the indicated point codes. The messages used in these procedures also use the MSU signal unit (see Fig. 5.8).

Level 3 software is not concerned with the ordered delivery of signal units but is responsible for the routing of a signal unit. There are three functions within each link's software that are responsible for routing signal units.

The procedures used in network management often interact with one another. For example, if a link or series of links fails, an LSSU (or multiple LSSUs) will be sent over the link to indicate the status to the adjacent node. The LSSU can only be sent if the link will support MTP level 2. If several links fail, the signaling point itself could become isolated or congested, resulting in traffic management procedures and possibly even route management procedures. Each of the procedures works at varying levels, beginning with the signaling link management (which is responsible for individual links).

Next is signaling traffic management, which works with an adjacent signaling point to manage the diversion of traffic from a failed link to a good link. The worst case is route management, where a node is no

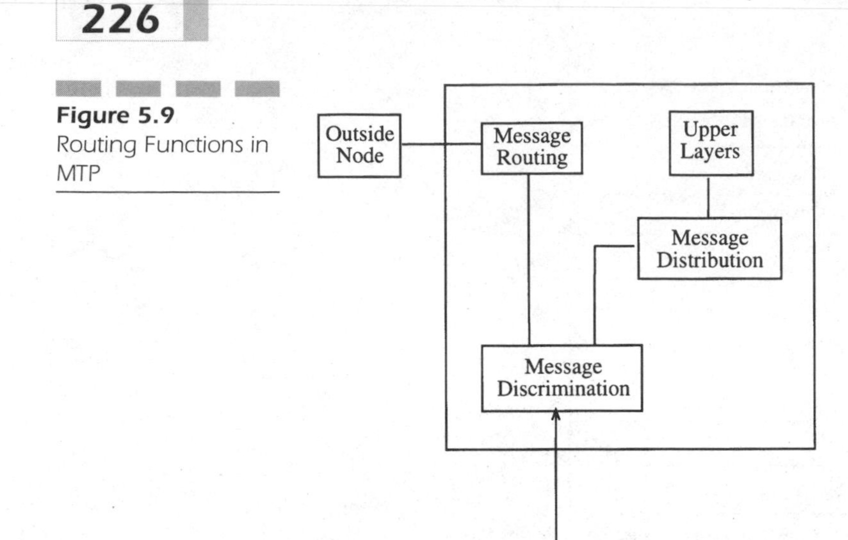

Figure 5.9
Routing Functions in MTP

longer capable of handling SS7 traffic or has become congested to the point where traffic must be diverted away from the entire node. This results in SS7 traffic being rerouted to other signaling points.

As Fig. 5.9 shows, a message is received and examined by the message discrimination function. From here it is determined whether or not the signal unit is addressed to the receiving node or another remote node. If the signal unit is addressed to the receiving node, it is passed to message distribution, which is then responsible for processing the header information and passing the user data on to the next level.

If the signal unit is addressed to a remote node, the signal unit is passed to the message routing function. Here is where the new sequence number and a new BSN (based on the link to be used) are assigned. Remember that the BSN is acknowledging the last signal unit received on that link, not by that node. This is of some importance to understand since a device such as an STP could have over 500 links terminating to it.

Figure 5.10 shows the routing label and its location. The routing label

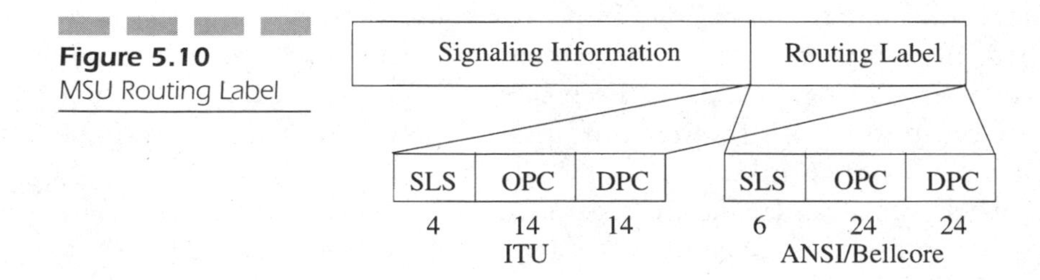

Figure 5.10
MSU Routing Label

is found in the user data portion of the signal unit. It contains the actual address of both the source and the destination. The addresses used in SS7 are called point codes and consist of different parts, depending on whether the point code is a national or international point code.

The national point code format is different for every country. In the United States there are three parts to the point code, the network identifier, the cluster identifier, and the member identifier. Each of these address portions is three digits long (or 3 bytes long in the binary code). The point codes must be defined by a central body with authority over all point codes in the United States. Currently, Bellcore is the authorized body responsible for the allocation of point codes to network operators.

The international point code format consists of zone, area, and member identifiers. The zone identifier is a 3-bit field used to identify a country or a group of countries. The area identifier is an 8-bit field that identifies the network. The member identifier is a 3-bit field that identifies the member within a network.

There are two planes, or levels of networking: the international and the national planes. The international plane uses the International Telecommunications Union (ITU) standards and allows all countries to interconnect their networks using a common interface. An international gateway STP is used to connect at the international plane, providing protocol conversion from the national protocol (such as the American National Standards Institute, or ANSI) to ITU (see Fig. 5.11).

The national plane allows every country to use its own flavor of SS7, without compatibility problems with the rest of the SS7 network. As long as a country can connect to the international plane, it can still interconnect with other networks.

Once access has been gained at the international gateway, any SS7 traffic can be converted by a protocol converter and routed into any international network, regardless of the country. This is important to all SS7 network operators, who must be able to connect to networks anywhere in the world. It is especially important for cellular and future communications networks, where transparent transfer of service is critical.

While MTP provides the procedures necessary for routing all SS7 traffic through the network, it is not sufficient for sending queries to a subsystem. Additional routing information is needed for this level of routing. MTP addresses SS7 nodes only and does not have the means for addressing subsystems, nor does it provide the procedures for managing those subsystems. This is handled by the SCCP.

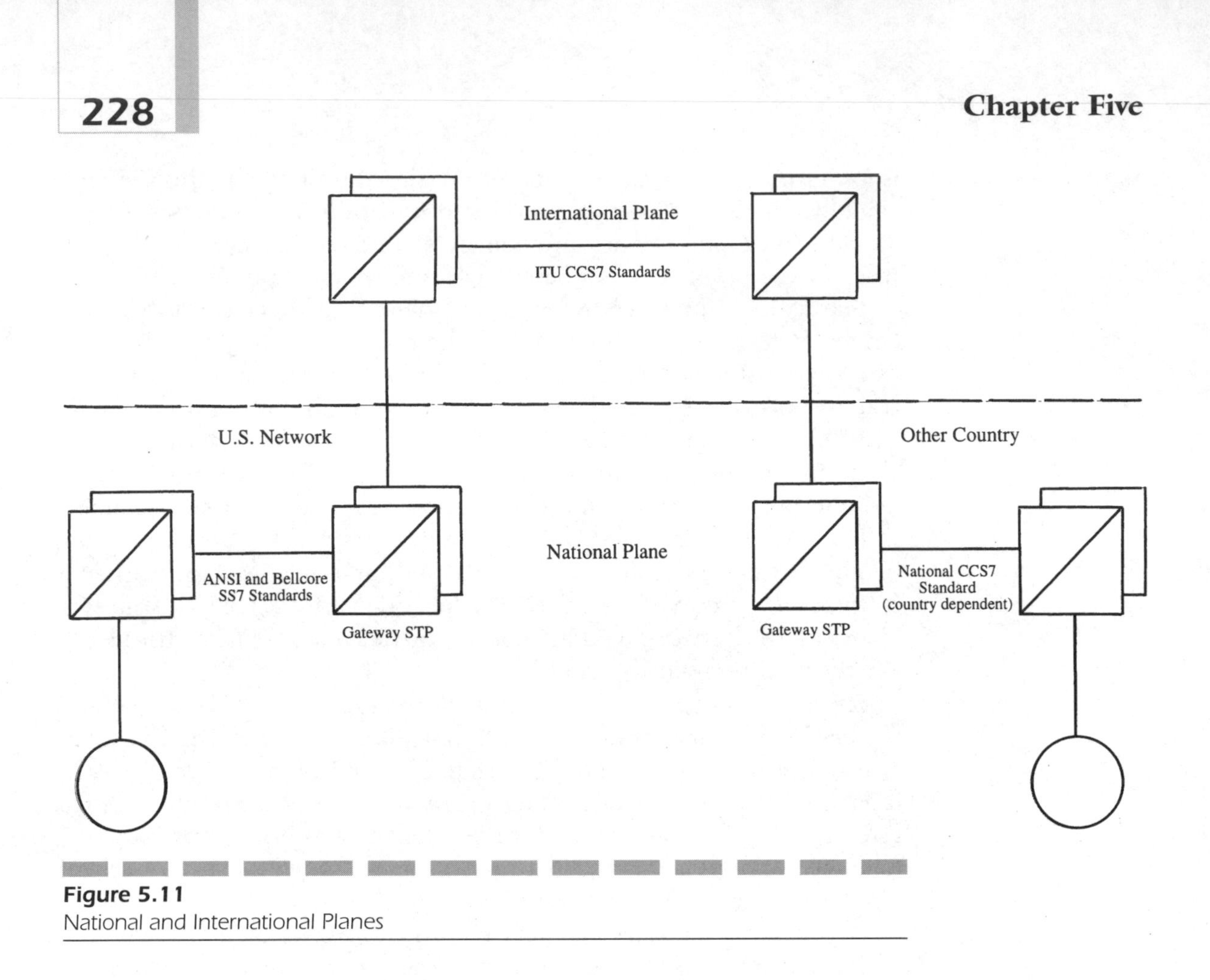

Figure 5.11
National and International Planes

5.4.2. Signaling Connection Control Part

SCCP can be found between the MTP layer and the application layers (such as TCAP and ISUP, in some cases). In Fig. 5.2 SCCP was shown as providing services to ISUP, as well as TCAP. In reality, there are no current uses for SCCP and ISUP. The standards do not address this other than mentioning the possibilities for future growth in this area.

So we will not discuss uses for SCCP and ISUP but will concentrate instead on SCCP and TCAP applications. As we mentioned in the section on MTP, the MTP protocol does not provide the procedures needed for accessing subsystems. In fact, there is only one address in the MTP header, and that is the SS7 point code used to identify the end node for an SS7 connection (end node being an SS7 device). Databases used in SS7 do not have point codes and are identified by a more generic subsystem number, meaning that the database does not have a unique address assigned specifically to that logical entity. Instead, a subsystem number is assigned which identifies the application type instead of the entity. For example, to identify an 800 database application, the subsystem number 256 would be used. This same subsystem number would be used for all 800 number databases in the network.

To route to a subsystem, the STP determines which database is best suited for the query received (based on the called party address in the SCCP header) and creates an MTP header addressed to the SCP that is acting as the front end to the subsystem. The SCP, which may provide front end services to several subsystems, looks at the subsystem number in the SCCP header and routes the data (minus the SS7 information) to the database.

When the query has been processed by the subsystem, the response is sent to the SCP, which is responsible for placing the data into an SS7 envelope and transmitting the response back to the originating SSP.

The response is addressed directly to the originator of the query, not the STP that provided the GTT. The originating point code is found in the calling party address field of the MSU when it is received by the SCP. This portion of the calling party address field is then used to create the header for the response.

SCCP provides several types of services for accessing subsystems. They are divided into connectionless and connection-oriented services. The differences between the services is mostly in the level of service (such as sequencing and segmentation/reassembly). The following are the various classes of service provided by SCCP

- Class 0—Basic connectionless
- Class 1—Sequenced connectionless
- Class 2—Basic connection-oriented
- Class 3—Flow control connection-oriented
- Class 4—Error recovery and flow control connection-oriented

Connection-oriented SCCP is not used in the United States but is used in some international networks. In the United States, connectionless services are supported, with some connection-oriented features.

Class 0 services are commonly used in North American networks. It provides delivery of messages, such as database queries, without sequencing. This is similar to other connectionless services found in other protocols. As with all connectionless services, there is no guaranteed delivery.

Class 1 services are also connectionless, with the addition of sequence numbering. MTP sequencing is separate from SCCP sequencing.

Remember that MTP is used to deliver SS7 messages from one node to the next. SCCP provides delivery of messages from end to end. Sequence numbers found in SCCP are used when there are multiple packets being sent for one transaction. An example of this would be a switch invoking a feature in another switch, where TCAP would send more than one message to the remote switch. Sequenced delivery is required to ensure proper operation.

Even though sequenced delivery is guaranteed with Class 1, it is still considered connectionless because SCCP does not request a logical connection with the remote node. This is somewhat unique to SS7.

Class 2 services provide connection-oriented services, which require a logical connection between the two communicating nodes. A reference number in SCCP is used to identify the logical connection since there are usually multiple transactions to the same node. This is similar to X.25 logical connections.

Class 3 services add flow control and expedited data. Flow control controls the flow of data to a remote node, while expedited data identifies messages with a higher priority. This allows SCCP to identify data which must be processed immediately.

Class 4 services (not currently defined in the Bellcore standards but defined in the ANSI standards) provides for error recovery. When an error is detected, SCCP can request a retransmission. This is apart from the MTP error recovery procedure.

SCCP provides end-to-end delivery of signaling data in the SS7 network. While the standards identify SCCP as providing services to TCAP and ISUP, SCCP can work with any other type of protocol (in theory) and was designed after the X.25 transport protocol.

5.4.3. Transaction Capabilities Application Part

TCAP is used for non-circuit-related signaling. Non-circuit-related means the signaling does not pertain to any one circuit or connection. Examples of non-circuit-related signaling include database queries for 800 services. TCAP is used whenever a database is involved or when invoking a feature in another switch.

TCAP is used in cellular networks as well. Cellular networks use application protocols which require the services of both SCCP and TCAP. These application protocols include IS-41 and the Mobile Application

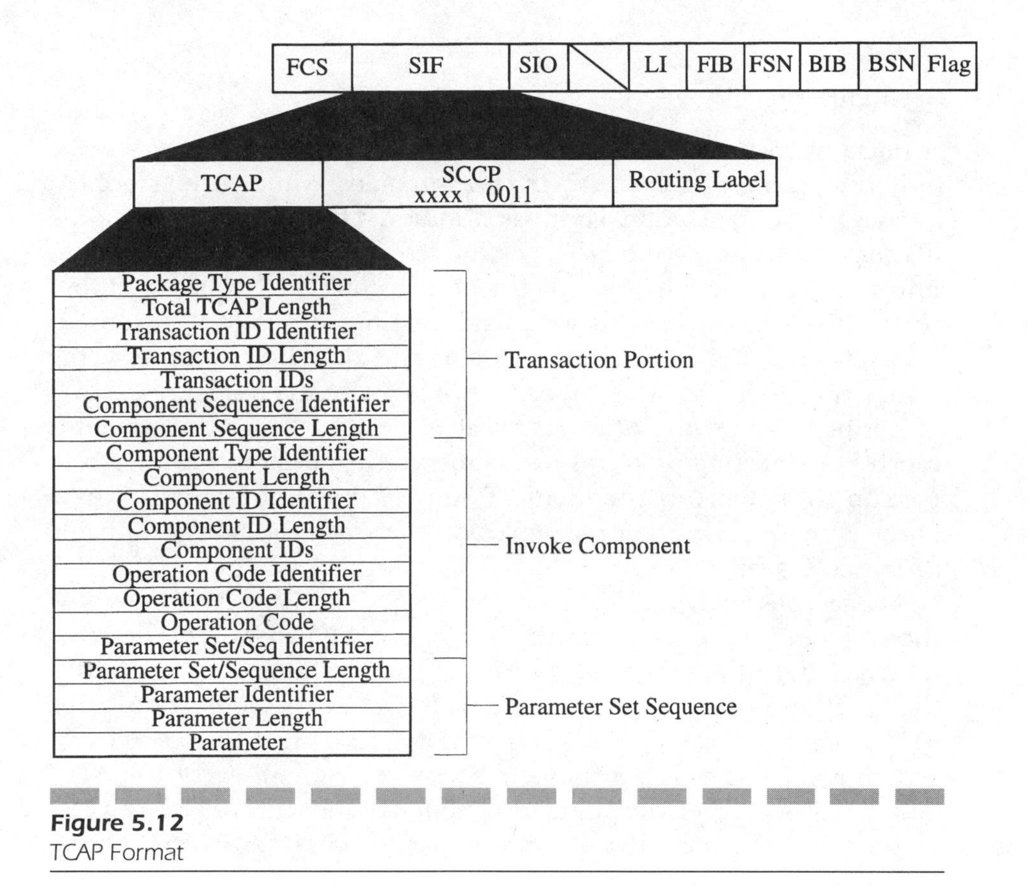

Figure 5.12
TCAP Format

Part (MAP). What TCAP provides is management of queries and support for multiple transactions in one transmission. TCAP also provides error detection/correction, segmentation, and reassembly.

Figure 5.12 shows that TCAP is divided into three different sections. The transaction portion identifies whether or not the following component portion carries a single transaction or multiple ones. It also indicates the type of transaction by providing a package type identifier.

The package type identifier consists of the following values:

- Query with permission
- Query without permission
- Response
- Conversation with permission
- Conversation without permission

- Abort
- Unidirectional

A query with permission is a database query. Typically, there will be only one query to a database, but there could be multiple associated queries in one transaction. Permission indicates that either end, the originating node (sending the query) or the remote node (the SCP), can terminate the transaction before it has been completed. A query without permission indicates only the origination point can terminate a transaction. This forces the SCP connecting to the database to complete the transaction no matter what happens at the remote end.

It is probably worth noting here that SCCP subsystem management is capable of detecting problems with subsystems and rerouting queries to backup subsystems in the event of subsystem failure. This is how a query without permission is completed, even when the subsystem being addressed has failed.

The response package type identifies a TCAP message that is carrying the response to a query. This of course is what you would see being returned from the subsystem to the originating node. Remember that the MTP destination address for queries is usually the point code of the STP adjacent to the SCP. The STP provides GTT so that the correct subsystem address can be determined. The response will carry the MTP destination address of the originating node, and not the STP that provided the GTT (remember that the calling party address in the SCCP header also contains the origination point code.

The package type conversation is used when a dialog between two nodes is needed. If a subsystem receives a query, it may be determined that additional information is needed to complete the transaction. The subsystem then has the option of returning a package type of conversation, opening up a two-way dialog between the subsystem and the querying node. This can also be used when two switches need to exchange signaling information while a call is in progress.

The abort package type allows either node (depending on whether or not permission was granted) to terminate a transaction. A reason code is included in the abort message to indicate why the transaction was aborted.

The unidirectional package type is used when a message is sent without a response required. This is the only package type which does not require a transaction identifier.

The transaction identifier is used to track transactions sent to a node. This identifier is of local significance only; that is, it is used only by the originating node so that it can correlate responses with queries. When a response (or a conversation) package type is returned, it will contain the same transaction identifier so that the originating node can associate the response with the proper query.

The second portion of the TCAP packet is the invoke component. There may be more than one component in this section. This allows a single transaction to invoke multiple features or send multiple queries (if associated) to the same subsystem.

Each component carries its own unique identifier so that the originating node can keep track of responses to various components. The correlation identifier is found in the component identifier and may be accompanied by an invoke identifier. The correlation identifier is also of local significance only.

The operation code identifies the type of operation to take place. There is one operation code for each component of a transaction. If there are multiple components in a transaction, there will be multiple operation codes. Operation codes are divided into families, with various values for each family. Think of an operation code family as a class of operations. The following are the operation families:

- Parameter
- Charging
- Provide instructions
- Connection control
- Caller interaction
- Send notification
- Network management
- Procedural
- Operation control
- Report event
- Miscellaneous

The operation codes do not provide enough information by themselves and must be followed by the parameters which provide the data needed to complete the operation. Parameters are found in the next section of TCAP, the parameter set sequence. There can of course be multiple parameters for an operation.

We will not go into detail about how each of these operation codes are used since that is beyond the scope of this book. For a thorough overview of TCAP and all of its operations, refer to my book *Signaling System #7*.

As mentioned above, each of the operation codes requires additional parameters. The last section of the TCAP packet consists of those parameters. The organization represented in Fig. 5.12 suggests that all of the fields are organized in linear fashion. This is not the case. The transaction portion identifies how many components are to be found in the component portion. Each component identifies an operation code, followed by its respective parameters in the parameter set sequence section. The component section is then repeated for the next component, followed by its respective parameter set sequence.

Now that we have examined the types of transactions identified in the TCAP protocol and have a basic understanding of the values provided in this protocol, let us look at some applications which use TCAP. We have already talked a little bit about one application, 800 services. Let us take a look at some other applications.

A relatively new service is the 900 service. This is similar to 800 services except that the calling party is charged a set amount determined by the length of the call (usually by the minute, with a minimum duration). When a caller dials a 900 number, the SSP function in the local switch first initiates a TCAP query to a subsystem to obtain routing instructions. In addition to returning routing instructions, the subsystem may also return billing information.

The billing information includes the amount (or rate) to be applied to the caller, as well as the name and billing data for the subscriber (owner of the 900 number). Once this information is known, the SSP can initiate an ISUP message to begin the call setup procedures. Once the call is released, another TCAP message is generated to a billing database to create a billing record for the call. The duration of the call and any other billing data are sent along with the TCAP message.

This is an example of database usage. TCAP can also be used to invoke features. For example, one feature offered in some networks is automatic callback (it may be known by a different name in some networks). When a caller dials a number and receives a busy signal, he or she can enter a code and hang up. When the number becomes available, the caller is called back, and when he or she answers, the number is redialed. This feature requires interaction between the two end office switches.

When the calling party receives a busy signal and enters in the callback code, the local SSP function sends a TCAP message to the remote SSP, invoking the callback feature. When the called party hangs up, the remote SSP sends a TCAP message back to the local SSP, alerting it that the number is now available. The remote switch temporarily blocks the number from any other incoming calls until the calling party can establish a connection.

The local SSP function then rings the calling party telephone, usually with a distinct ring. When the calling party answers, normal ISUP call setup procedures are initiated to the remote SSP. The called party's phone then rings, and the connection is established. There are many other similar features which utilize the TCAP protocol to invoke features in telephone switches. Theoretically, the same function could be provided to owners of PBX equipment.

TCAP is not limited to wireline services. It is also widely used in cellular networks. Cellular networks rely on database access for virtually every call placed over the wireless network. These databases contain location information for every active mobile telephone in the network. Before a call can be connected, the location of the mobile subscriber must be determined. This is accomplished through the use of TCAP and cellular protocols (such as IS-41 and MAP).

Cellular switches also communicate using TCAP, providing update information regarding the location of mobile subscribers. The various cell sites in the wireless network send location updates to the mobile switching center in their area, which sends the data to a Home Location Register (HLR) and Visitor Location Register (VLR), which are usually collocated. See Chap. 7 for more information about how the various databases are used.

TCAP usage was not frequent until telephone companies began providing new services. Its original use was 800 services, but now it has become widespread to include many different wireline and wireless services. As the telephone network continues to evolve, TCAP usage will become more widespread and will account for the majority of SS7 traffic.

5.4.4. Telephone User Part

TUP is an obsolete circuit-related protocol, found primarily in European networks. TUP does not support ISDN circuits, prompting the development of a newer protocol which would support the requirements of ISDN. When we talk about the requirements of ISDN, we are speaking specifically of identifying the channels of an ISDN circuit.

TUP was designed specifically to handle voice-type calls on voice circuits. When ISDN was developed, the signaling requirements changed since ISDN is capable of carrying both voice and data. The signaling protocol must be able to address both voice and data signaling. If you look at the ISDN protocol defined in Q.931, you will see many similarities between ISDN and SS7 ISUP protocols.

To handle data, the Data User Part (DUP) was used. This meant having two separate protocols for handling voice and data. The DUP protocol does not support the parameters of ISDN, however, and is not sufficient for this purpose. DUP is also an obsolete protocol, found primarily in European networks.

Since TUP is obsolete, and many of the networks currently using TUP are migrating to the ISUP protocol, we will not spend any more time on this protocol.

5.4.5. ISDN User Part

ISUP is the protocol used throughout North America and many parts of Europe for circuit-related signaling. Circuit-related signaling applies to all voice and data circuits. We have already learned of two obsolete protocols, TUP and DUP, which have been replaced by ISUP.

The ISUP protocol supports both physical circuits and logical channels. This is something its predecessor, TUP, is not capable of supporting. However, ISUP cannot support the virtual circuits and virtual paths used in ATM networks. To support broadband services, a newer version of ISUP, Broadband ISUP (BISUP), is being developed.

ISUP provides two types of services, basic and supplementary. Basic services provide support for establishing connections for voice and data circuits in the Pubic Switched Telephone Network (PSTN). Supplementary services are used for the exchange of signaling data that is related to a call already in progress.

There are two methods defined for passing circuit-related information for supplementary services. The signaling information being sent is used by the originating and destination nodes and is ignored by any intermediate nodes along the signaling path. To understand how this works, we should first examine how a call is set up.

As shown in Fig. 5.13, the originating SSP function sends a setup message to an adjacent SSP. This is typically not the terminating SSP and may be a tandem switch. The setup message is routed through an STP. The adjacent SSP then must initiate a setup message to another adjacent SSP (which may or may not be the destination SSP) to establish an end-to-end connection. The use of an STP as the signaling hub is referred to

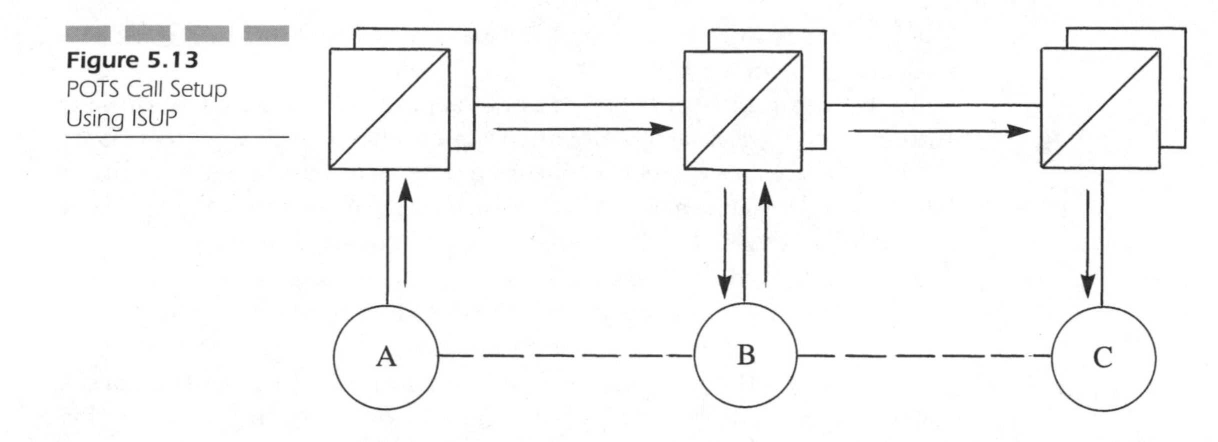

Figure 5.13 POTS Call Setup Using ISUP

as quasi-associated signaling because the STP performs an intermediate function. When STP is used and the two SSP functions are directly connected, the mode is referred to as associated signaling. This is found when fully associated links (F links) are used.

It should be noted that the MTP destination address in each case is the point code of the adjacent SSPs and not the final destination. The called party address in the ISUP message is used to determine the destination point code for the next hop. The setup procedure must be repeated for each hop used to establish the call connection.

Once a call connection has been established, additional circuit-related information can be sent to the remote SSP. There are two methods used with supplementary services to provide end-to-end signaling once a connection is established and a call is in progress. The pass-along method uses the services of MTP to send ISUP supplementary information, while the SCCP method uses the services of SCCP to send ISUP supplementary signaling. Only pass along is currently used in North America.

When pass along is used, the ISUP message is passed along the same path as the call setup messages. This means the message must be sent through all of the intermediate nodes along the connection path. This will of course introduce additional delay and is not the best method of transport, but it is commonly used today.

The SCCP method allows the ISUP message to be sent from the originating SSP to the remote SSP, bypassing all of the intermediate SSPs along the signaling path. This is possible because SCCP can route end to end, using STP services to determine the shortest path. MTP services can only support routing from one SSP to another adjacent SSP (using an STP as an intermediate step in quasi-associated signaling). The use of

SCCP may become more commonplace as new intelligent network services are implemented.

There are a number of ISUP message types used for the establishment and tear down (release) of voice and data circuits. Figure 5.14 shows how a call from an ISDN telephone (labeled as a terminal) interacts with an ISDN switch (or end office switch), which in turn generates SS7 messages for the SS7 network. The ISDN messages are explained in Chap. 6.

The initial address message (IAM) is used to request a connection between two SS7 signaling points. The IAM contains both the calling party and the called party telephone numbers. This information is passed through the network without alteration, allowing the final destination SSP to identify the telephone number of the calling party. If this SSP can access a name database, a query can be made to the database, providing the name of the calling party as well. This information is then passed on to the called party telephone display.

If the remote end of the connection is not using an ISDN circuit, but an analog circuit (POTS service), a special modem is used at the end office switch to pass the calling party number and name. A special display at the subscriber's location receives data from the modem between ring cycles and displays the information for the called party prior to the call being answered. This is mentioned so you might understand more about how signaling works. Without SS7, calling name display would not be possible.

When an SSP receives an IAM message, it must determine whether or not it has the resources to maintain a connection (resources meaning processing capacity, as well as circuit availability). The circuit to be used is identified in the IAM message as well. However, just because the circuit has been identified by the originating SSP as available does not mean that the same circuit is available at the other end. There are many circumstances which may render the circuit unavailable at one end of the connection.

If the receiving SSP determines that the connection can be made on the designated circuit, it returns an address complete message (ACM). This indicates that the IAM was received correctly and the destination SSP can support the connection. The destination SSP then must determine the circuit to use for the next hop on the connection. Once the circuit has been identified (determined by examining the called party address in the IAM message), the SSP generates an IAM message to the next SSP on the connection path.

Once the circuits have been established end to end, and the called party is alerted (by ringing the telephone), an answer message (ANM) is

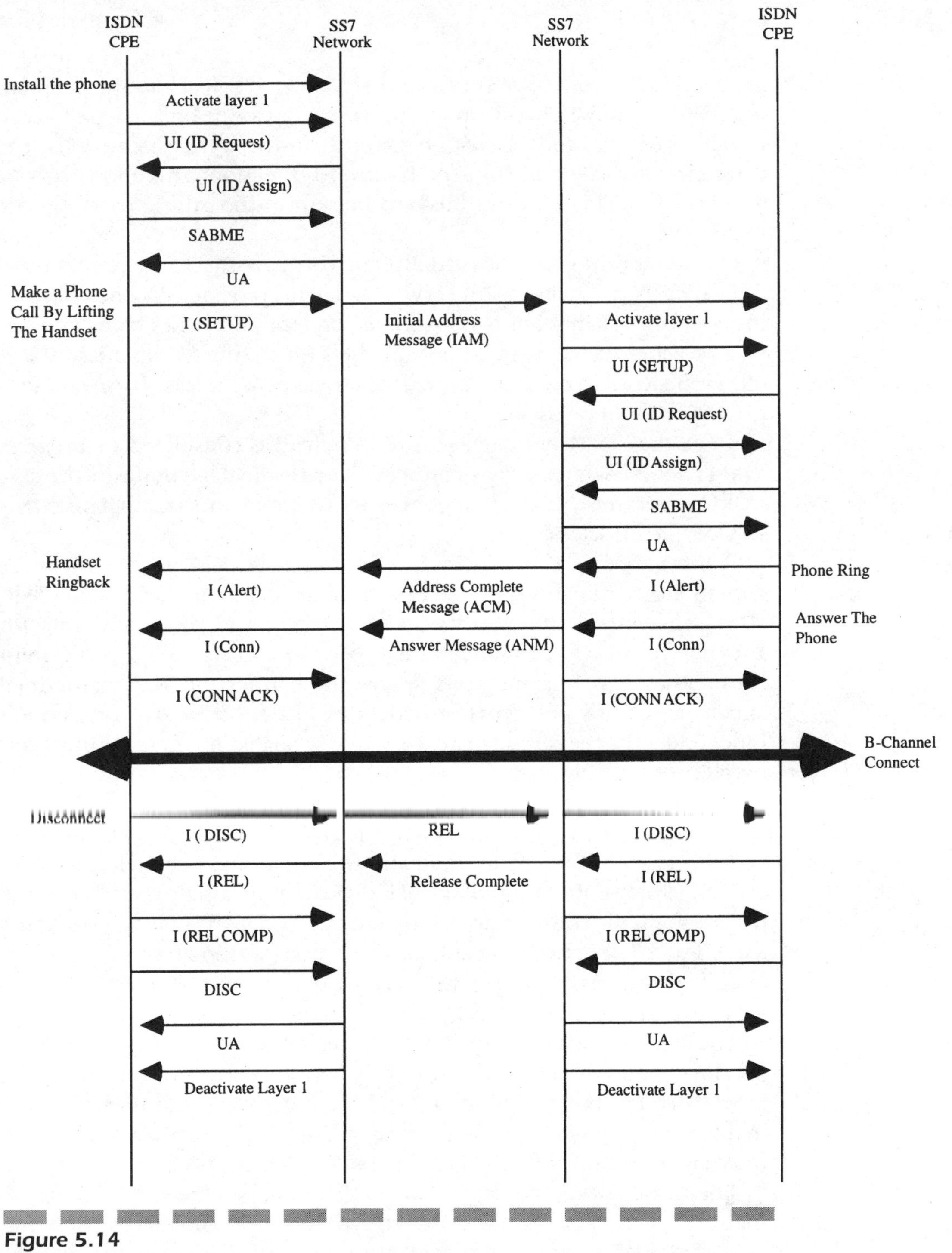

Figure 5.14
ISDN Call Setup Using ISUP

generated and sent backward from each of the SSPs (back to the originating SSP). The voice circuit up to this point has been only partially connected. Some networks initiate backward cut-through, which means the voice circuit is connected in the backward direction. This allows service tones (such as ringback and busy) to be sent to the calling party by the remote SSP.

Other networks do not actually cut through the voice circuit until the called party answers. All service tones are provided by the originating SSP. Implementation is network-dependent and varies from one service provider to the next. Almost all Bell Operating Companies (BOCs) follow Bellcore recommendations, which vaguely indicate backward cut-through should be used.

Once the ANM has been received, the call is considered in progress. The ISDN B channel is now connected (or the analog circuit in the case of POTS service). No other messages will be sent unless supplementary services are initiated.

When either party hangs up, a release message (REL) is sent. It can be sent in either direction. The release message is sent to each of the SSPs, allowing each of the circuits to be released and made available for another connection independently of one another. This is different from other signaling methods, where the connection was maintained end to end until both parties had hung up. By releasing each circuit independently, the circuits can be made available for new connections much more quickly.

When an SSP receives the release message (which will also identify the reason for releasing), a release complete (RLC) message is sent as confirmation that the REL was received and the circuit has been released.

Now that we understand the ISUP procedures used to establish and release a circuit connection, let us look at the ISUP message structure itself. The ISUP message consists of three parts, as shown in Fig. 5.15. The circuit identification code (CIC) is also part of the ISUP message and is located after the routing label.

The CIC is not found in BISUP since it cannot support broadband services. The CIC can only identify narrowband circuits used for interoffice trunking. The identification is a simple number, which is maintained by each node. The circuit identification for any one circuit must have the same identity at both ends of the connection.

The mandatory fixed part contains the message type and a pointer to the next portion of the ISUP message. Mandatory implies that this portion of the ISUP message must always exist, and fixed indicates that the

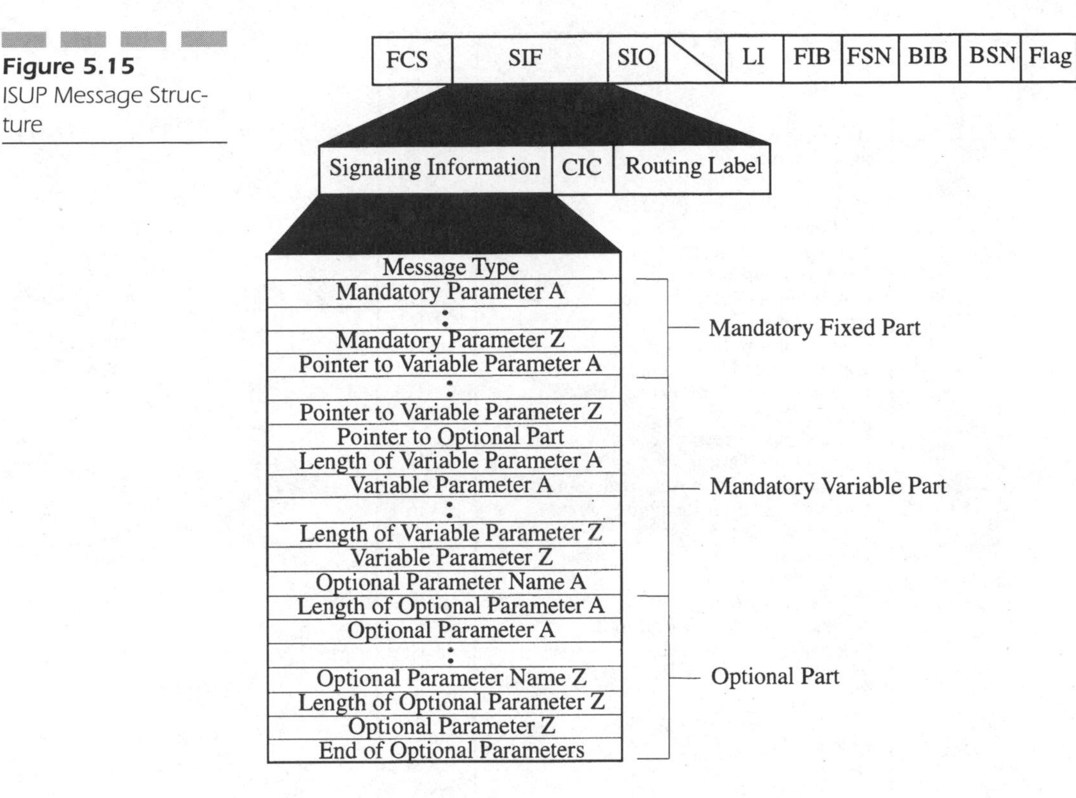

Figure 5.15
ISUP Message Structure

fields in this section are of a fixed length. The pointer is used to identify where the next section, the mandatory variable part, begins. It provides the offset, which is the number of octets to be counted from the message type to the beginning of the mandatory variable part.

There are a number of message types supported in ISUP and BISUP. Figure 5.16 identifies the message types used in ISUP and BISUP and whether the message types are used in ITU or Bellcore networks. Describing the use of each of these message types is beyond the scope of this book. For a more thorough description of each of these message types and their parameters, refer to my book *Signaling System #7* (another commercial, but it really is a good book for learning SS7 in detail).

The next section of the ISUP message is the mandatory variable part. This section is not usually mandatory, depending on the message type. It is possible to have an ISUP message with only the mandatory fixed part and an optional part. It is also possible to have an ISUP message with only a mandatory fixed part. There can be any number of variables in this section, depending on the message type. Again, we will not

Figure 5.16 ISUP and BISUP Message Types

Address Complete	ACM
Answer	ANM
Blocking	BLO
Blocking Acknowledgment	BLA
#Call Modification Completed	CMC
#Call Modification Reject	CMRJ
#Call Modification Request	CMR
Call Progress	CPG
#Charge Information	CRG
Circuit Group Blocking	CGB
Circuit Group Blocking Ack	CGBA
Circuit Group Reset	GRS
Circuit Group Reset Ack	GRA
Circuit Group Unblocking	CGU
Circuit Group Unblocking Ack	CGUA
Circuit Query	CQM
Circuit Query Response	CQR
Circuit Reservation	CRM
Circuit Reservation Acknowledgment	CRA
Circuit Validation Response	CVR
Circuit Validation Test	CVT
Confusion	CFN
#Connect	CON
*Consistency Check End	CCE
*Consistency Check End Ack	CCEA
*Consistency Check Request	CCR
*Consistency Check Request Ack	CCRA
Continuity	COT
Continuity Check Request	CCR
#Delayed Release	DRS
Exit	EXM
#Facility Accepted	FAA
#Facility Reject	FRJ
#Facility Request	FAR
Forward Transfer	FOT
*IAM Acknowledgment	IAA
*IAM Reject	IAR
Information Request	INR
Initial Address Message	IAM
Loopback Acknowledgment	LPA
*Network Resource Management	NRM
#Overload	OLM
Pass Along Message	PAM
Release	REL
Release Complete	RLC
*Reset	RST
Reset Circuit	RSC
*Reset Acknowledgment	RAM
Resume	RES
*Segmentation	SGM
#Subsequent Address	SAM
Suspend	SUS
Unblocking	UBL
Unblocking Acknowledgment	UBA
*User Part Available	UPA
*User Part Test	UPT
#User-to-user Information	USIS

go into details of these parameters since that would require a book of its own. I will mention that it is in this mandatory fixed part that you will find the called party address in the IAM message.

The section following the mandatory variable part is the optional part. This is the most frequently found section in ISUP messages; that is, the majority of ISUP messages use the mandatory fixed part and the optional part. The calling party address is in the optional part.

There are many ISUP parameters which may use additional parameters, which are found in the optional part. If we look at the overall structure, assuming all sections of ISUP are used, we will find the message type (such as IAM) followed by a parameter or series of parameters. Any one of the parameters could have additional parameters found in the optional part.

ISUP was the first SS7 protocol to be implemented in North American networks. As the SS7 network grew, TCAP services became more prevalent. Future networks will see more TCAP traffic than ISUP traffic as databases become more and more important in the intelligent network.

This is just an overview of the SS7 protocols. SS7 is a very complex network, and the procedures used in the network take up volumes of standards. As the network grows, SS7 becomes more and more complex, providing newer services and features to wireline and wireless network operators. There is no doubt that SS7 will be the future of the telecommunications network.

5.5. Chapter Test

1. Out-of-band signaling utilizes the bandwidth within the voice band (300 to 3800 Hz) to send signaling tones.
   a. True
   b. False
2. The TCAP protocol allows an end office switch to:
   a. Connect to another end office switch to invoke a feature
   b. Connect to network databases
   c. Send non-circuit-related signaling through the network
   d. All of the above
3. What configuration allows an SSP to connect directly to another SSP and send signaling messages without going through an STP?
   a. Associated signaling

b. Quasi-associated signaling

4. What type of an SS7 data link connects two mated STPs to one another?
   a. B link
   b. D link
   c. F link
   d. C link
5. What is the first message type sent when establishing a circuit connection between two offices?
   a. Release
   b. Address complete
   c. Initial address
   d. Query with permission
6. What message type is used in the TCAP protocol to request a data lookup in a network database?
   a. Query
   b. Invoke
   c. Abort
   d. Conversation
7. The SCCP protocol provides several classes of service. Which ones are used in North America?
   a. Classes 0 and 1
   b. Classes 1 through 4
   c. Classes 2 through 4
   d. All classes
8. A group of routes is called what?
   a. Linkset
   b. Route
   c. Routeset
   d. Combined linkset
9. What function provides a front-end service to various subsystems in the SS7 network?
   a. Service switching point (SSP)
   b. Signal transfer point (STP)
   c. Service control point (SCP)
10. The transfer prohibited (TFP) message is used by which network management function?
   a. Link management
   b. Traffic management
   c. Route management

11. Which SS7 protocol uses the Circuit Identification Code (CIC)?
    a. SCCP
    b. MTP
    c. ISUP
    d. TCAP
12. The Message Transfer Part (MTP) provides end-to-end transport services for SS7.
    a. True
    b. False
13. Which protocol provides end-to-end sequenced delivery of ISUP and TCAP messages?
    a. MTP
    b. SCCP
    c. Both of the above
14. Which network management function is responsible for diverting SS7 traffic away from a failed link to another link in a linkset?
    a. Link management
    b. Traffic management
    c. Route management
15. What is the name for a group of links that are connected to the same signaling point?
    a. Linkset
    b. Route
    c. Routeset
16. Which network management function is responsible for diverting SS7 traffic away from a failed or congested signaling point?
    a. Link management
    b. Traffic management
    c. Route management
17. Connection establishment for a broadband virtual circuit is supported by which SS7 protocol?
    a. ISUP
    b. TCAP
    c. BISUP
    d. All of the above
18. What type of signal unit is used by MTP level 2 to indicate the status of a link?
    a. FISU
    b. LSSU
    c. MSU

19. What does the length indicator found in the header of MTP indicate?
    a. The type of signal unit being sent
    b. The entire length of the SS7 packet
    c. The length of just the MTP portion of the signal unit
20. What are a group of links called?
    a. Route
    b. Linkset
    c. Routeset
    d. Combined linkset

CHAPTER 6

ISDN and Broadband ISDN

6.1. ISDN—An Overview of Its Capability

The telephone companies have been revising their networks for a long time. The first objective was to automate long distance dialing, eliminating the need for an operator to place the call. This was accomplished in the 1950s through Direct Distance Dialing (DDD). The next objective was to provide a network which would provide end-to-end connectivity, access and service integration and standard interfaces and would allow customers to have more control over the services they needed. The Integrated Services Digital Network (ISDN) addresses these objectives.

ISDN provides a standard interface for all types of digital transmission. Digitized voice, low- and high-speed data, video, facsimile, and imaging can be sent over ISDN facilities, using a common pair of copper wire, coaxial cable, or fiber optics. ISDN provides end-to-end connectivity and digital transmission over the local loop.

ISDN is as much a concept as a technical solution. Talks about an integrated network began in the 1960s, as telephone companies began to research ways to consolidate their facilities. It had already become apparent that circuits riding on copper cable were in high demand, and the telephone companies needed a plan to consolidate their facilities before they ran out of copper.

ISDN as a concept implies that the whole public network is integrated. Integrated means that all transmissions, voice, data, and video, can use the same facility instead of using separate specialized circuits. Digital network means that all transmissions must be converted into a digital format so they can be transmitted over the ISDN.

Before ISDN could be successful, one missing element was needed. The signaling used to set up and release circuits had been transmitted over the same circuit as the transmission itself. This was no longer an economical means for managing connections. ISDN calls for common channel signaling, where all signaling is placed on one common channel for an ISDN interface.

The signaling through the public network had to be resolved. Common channel signaling for the public network uses a packet switching network called Signaling System #7 (SS7). You will find that the signaling used in ISDN is compatible with the signaling messages used in SS7. However, SS7 provides much more than signaling to support ISDN. It also provides access to network databases and intelligent service to the network.

The original concept of ISDN was to extend the services of the signaling portion out to the subscriber. Telephone companies were concerned that providing this level of access to their network resources would jeopardize the security of the telephone network, and they asked for a subscriber interface different from that of SS7. The result was what we know today as ISDN.

ISDN is really the user interface to the integrated network. ISDN standards define the interfaces to the subscriber and the means by which voice, data, and video are transmitted over the user interface. There are other protocols which are used within the public network, such as SS7.

Before discussing the interfaces and procedures used in ISDN to connect calls through the network, let us look first at what ISDN provides. Then we will look at the ISDN protocols, procedures, and architecture of the network. We will also look at the future of ISDN, referred to as Broadband ISDN (BISDN).

6.1.1. ISDN Standards

There are many standards defining ISDN, both international and national. The International Telecommunications Union (ITU) is responsible for defining the international standards, while the American National Standards Institute (ANSI) is responsible for defining national standards used in the United States. Bellcore defines requirements for network components used in Regional Bell Operating Company (RBOC) networks. The ITU Recommendations, listed below, define all of the aspects of ISDN from an international perspective:

- I.100 Series—General Concepts
- I.200 Series—Service Aspects
- I.300 Series—Network Aspects
- I.400 Series—User-Network Interface Aspects
- I.500 Series—Internetwork Interface
- I.600 Series—Maintenance Principles

General concepts describe the objectives of ISDN, the interfaces to be provided, and some of the principles to be defined in other standards. Services aspects describe the various services to be delivered by ISDN and how those services are to be delivered. Network aspects describe the functions to be provided in the ISDN network. User-network aspects define the user interface to the ISDN network, while the internetwork interface standard defines the means by which ISDN networks are connected to one another. Maintenance principles define the procedures to be provided by the protocols for maintaining an ISDN network.

6.1.2. ISDN Features

ISDN was originally defined as the integration of the entire Public Switched Telephone Network (PSTN), supporting circuit- and packet-switched services, voice and data transmission, and nonswitched services. Many now refer to ISDN as the user interface to the integrated PSTN.

ISDN indeed supports all of the services mentioned above. Unfortunately, ISDN was deployed before its time. The ISDN protocols defined in the ITU and ANSI standards only define the interface to the user network. The public network itself uses a different set of standards, explained a little later. For ISDN to support end-to-end services as it was designed requires the deployment of SS7 in the public network.

SS7 is the network suite of protocols used to establish connections between telephone switches. It also gives end office switches access to intelligent databases used to support the many services offered by ISDN. It appears that just as we are about to realize widespread deployment of true ISDN services, the network will begin deployment of newer broadband services, requiring a new flavor of ISDN solutions.

6.1.3. Services and Applications

ISDN supports a number of services from the subscriber equipment to the public telephone network. The concept is to give subscribers one facility and one interface for all of their communications needs. This means ISDN must be able to support not only digital voice transmissions (in the form of telephone calls) but also data transmissions, video (in the form of video conferencing), and packet switching.

Circuit-switching services are those typically related to voice transmissions. The telephone network is a good example of a circuit-switched network. Connections are made by request through the network using switching equipment until a connection has been established from end to end.

Switches are different from data routers because they establish a connection from end to end before actually transmitting anything. Once the circuits have been connected, transmission can begin. When transmission is complete (or in the case of the telephone network, conversation has ended and either party hangs up), the connection is released. Data routers do not establish a connection over circuits but instead route data from multiple sources over the same circuits.

Packet-switched services encapsulate data into envelopes, adding headers containing routing information (as well as other overhead), and transfer the packets through the network using any circuits available. Connections can be established (as is the case with connection-oriented services), but these connections are virtual. A virtual connection means a message is sent through the network to the destination, requesting that resources be allocated for a transmission.

Resources in this sense are processor and software resources rather than circuits. One circuit may be used to transmit packets to a variety of different destinations. Another unique feature of packet-switched services is that transmissions from one source may follow different routes in the network. With circuit-switched networks, all transmissions from one source always follow the same route, using the same circuits.

One of the biggest advantages of ISDN is the fact that both voice and data can use the same facility. Instead of using a dedicated circuit for all data transmissions (which are routed over a separate network from voice transmissions), data can be sent over ISDN, along with voice transmissions. ISDN is digital and uses time division multiplexing to assign transmissions to channels on a digital circuit. These channels can be assigned voice or data at any time, on demand.

Nonswitched services are typically permanent connections established between two endpoints. These connections are used for the transmission of audio (different from voice, audio is high fidelity, such as that used for radio broadcasts), video, or data. A permanent connection is usually established at the time the ISDN interface is installed at the customer premise.

One of the objectives of ISDN is to support all of these services using a limited number standardized interfaces. Two interface types have been defined, Basic Rate Interface (BRI) and Primary Rate Interface (PRI). Both of these are explained in greater detail later on. The principal difference between the two interfaces is the amount of bandwidth made available.

ISDN and its services are defined in a number of standards. The ITU (formerly the CCITT) I series of recommendations defines the various aspects of ISDN, including the protocols used to deliver ISDN services. ITU Recommendation I.120 (1988) defines the concepts of ISDN and was the first recommendation to be completed by the ITU.

ANSI has also defined a set of U.S. standards for ISDN. ANSI adopts ITU standards and then modifies them for use in the United States. There are many similarities between the two standards, but Europe has different needs than the United States, and their digital transmission facilities are different from ours.

The telephone companies follow yet another set of standards and requirements, based on the ANSI versions. Bellcore defines how ISDN services and interfaces will be implemented in Bell Operating Company (BOC) networks. These are really voluntary requirements, designed to influence manufacturers into following one method of implementation. Unfortunately, earlier attempts at standardization of ISDN deployment failed, with many vendors' equipment being incompatible with one another.

ITU standards define digital facilities as the means for end-to-end connectivity, providing 64 kbps per connection. This was based on existing capabilities within the network. Digital facilities were already in place that could support individual 64-kbps channels, without replacing a substantial amount of network infrastructure. It is recognized in the standards that later implementations of ISDN would support higher and lower data rates.

Another concept of ISDN was to use layered services. The OSI Model introduced this principle in 1984, but the OSI Model was not designed for more modern networks. Nevertheless, the concept of layered protocols was already a standard practice.

Layered protocols can use any type of protocol (or service) below and above it. The idea is that one layer is transparent to another; in other words, you can use ISDN on a Frame Relay circuit or even on an X.25 network because the ISDN protocols are transparent to the X.25 protocols (in theory).

Another reason for using this layered approach was so that existing protocols could be used wherever possible. For example, Link Access Protocol—B (LAPB) used at layer 2 in X.25 networks was used to develop LAPD. This is the protocol used at layer 2 in ISDN and is defined in ITU Q.921.

Layering also allows a telephone company to implement various functions and features when they are ready for them, instead of implementing ISDN all at once. For example, a telephone company may choose to offer BRI services to their residential customers, giving them high-speed access to the Internet, but may want to wait before offering end-to-end signaling services for their commercial subscribers.

One important objective of ISDN is that ISDN services must be transparent. This means any protocol or application can be used over an ISDN interface, and the ISDN protocol does not affect the transmission in any way. For Internet access, the Transmission Control Protocol/Internet Protocol (TCP/IP) protocol suite is required to communicate with other nodes in TCP/IP networks. If ISDN is used to access these networks, the TCP/IP protocols must be encapsulated into ISDN packets before they can be transmitted over the ISDN interface. The ISDN protocols must not alter or modify any portion of the original TCP/IP packet since this will render it undeliverable in the TCP/IP network.

Now that we have an understanding of the types of applications supported by ISDN, as well as the protocols used in supporting those applications, we can look at specific services. There are three services defined in ITU Recommendations I.200:

- Bearer services
- Teleservices
- Supplementary services

Bearer services include digitized voice, data, and other forms of user data. Bearer services also support transfer of user data using packet-switched services (using X.25). Connection-oriented and connectionless services are supported as well. The ITU Recommendations define bearer service in three parts: the definition of bearer services (I.230), circuit-mode bearer service categories (I.231), and packet-mode bearer service categories (I.232).

Teleservices are defined in the I.200 series as well. Recommendation I.240 defines teleservices, while I.241 describes teleservices supported by an ISDN. Teleservices usually support computer-to-computer types of applications. This includes file transfers, terminal access to remote databases, and other dial-up-type applications.

Teleservices also include teletex (which is ASCII-based text communications via terminal), telefax (facsimile), videotex (enhanced with mailbox functions and graphics), and telex (interactive text communication via a terminal). These are information processing services that are often provided by the telephone company for a fee.

Supplementary services are defined in a number of recommendations. Following is a list of ITU Recommendations which define supplementary services:

- I.250—Definition of Supplementary Services
- I.251—Number Identification Supplementary Services

- I.252—Call Offering Supplementary Services
- I.253—Call Completion Supplementary Services
- I.254—Multiparty Supplementary Services
- I.255—Community of Interest Supplementary Services
- I.256—Charging Supplementary Services

Supplementary services are best defined as enhancements to bearer services. Users of Private Branch Exchange (PBX) equipment probably recognize many of these services as inherent features of their PBX. Telephone companies are not able to offer many of these services using existing equipment either because the equipment is not capable or they cannot obtain the proper tariffs to offer these services. With ISDN, they can offer these services and much more.

Number identification includes calling party identification, both by telephone number and name (calling name delivery). Call offering includes services such as hunting, forwarding, and call transfer. Call completion includes services such as call hold, call waiting, and call busy. Multiparty services are teleconferencing and three-party services. Community of interest is also referred to as a closed user group, which is analogous to extensions from a PBX. Charging includes credit card calls, reverse charging, and usage-based charging.

The ITU Recommendations define how these services are to be delivered and the protocol procedures to be used to deliver them. Like other protocols we have discussed so far, ISDN protocols are message-based. This means that each packet of information carries a specific meaning, or message, which is to be acted upon by the recipient. The messages used in ISDN protocols provide the instructions and parameters necessary to deliver services to the subscriber.

6.2. Subscriber Interface to SS7

ISDN extends the signaling network already in use by the telephone companies out to the subscriber. This means subscribers can send control information from their PBXs to other PBXs at a remote office, using the ISDN. The ISDN messages are translated at the local telephone company office into SS7 messages and are routed over the SS7 network to the destination exchange, where they are then converted back into ISDN messages and forwarded to the destination PBX.

There are many other advantages to having access to signaling information. Consider this scenario. A large corporation sells its products through a mail order catalog. They advertise an 800 number, which routes to their ordering center. When a call is received, the ISDN D channel provides the calling party's telephone number. This is then sent to a database residing in an adjunct processor.

The database looks up the telephone number and finds a record providing the calling party's name, address, and past ordering history (if they have called in before). The operator then receives the call, along with the database information (displayed on the terminal screen). The operator can now provide more personal service and needs to only verify the address and credit card information since that information is already provided by the database.

The same company can also use ISDN to dynamically route callers based on operator availability. Callers can be routed to customized recordings, voice response systems, and other automated systems, reducing the number of calls requiring a live operator. If the caller has placed an order before, the entire ordering process can be automated by callers entering in the code for the products they wish to order and verifying their address and credit card information via a voice response system.

Signaling aside, there are other reasons for deploying ISDN. Many RBOCs are offering Internet access, bundled with ISDN service. As the Internet matures and the World Wide Web sites start using interactive software and video, the need for ISDN and other high-speed access lines will be paramount.

Many Internet Service Providers (ISPs) are offering ISDN service to support the new multimedia sites appearing on the Internet. With new standards from the Internet Engineering Task Force (IETF) supporting multimedia and conferencing (voice and video), as well as programming languages such as Sun's Java, ISDN access to the Internet is becoming unavoidable.

6.2.1. End-to-End Signaling with DSS1

End-to-end signaling allows companies with private networks to send signaling information from their switching equipment to switches in other remote locations. The signaling information is sent from the ISDN interface (using the D channel) through the public network's signaling network, SS7. Not all telephone companies provide this service, but this was one of the original reasons for providing ISDN.

End-to-end signaling allows a PBX to send the same information used in routing, call offering, call completion, multiparty, community of interest, charging, and other call information to their other switches transparently through the public network. This is of value to large corporations with multiple locations.

Digital Subscriber Signaling System No. 1 (DSS1) is a collection of protocols that are used to support end-to-end signaling. Q.930, Q.931, and Q.932 make up this signaling system. DSS1 is the signaling used from the user to the network, where it is transferred to the SS7 network for transport to the remote ISDN. Once received at the remote end, DSS1 is again used to deliver the signaling information to the remote PBX.

6.2.2. Private Intelligent Networks

Private intelligent networks can be of benefit to large corporations because they allow them to more efficiently handle calls to their networks. Before ISDN, corporations used tie lines to connect to their remote switches. The switches often had a star configuration, with one PBX acting as the hub.

The hub was used to route internal calls to other locations, without the users having to dial through outside lines. This let office personnel call other offices by simply dialing an extension number, even when the extension was in another PBX. Complex routing tables and numbering plans had to be programmed into the switches to allow them to provide this capability.

Outgoing calls could be routed through the hub, which would then use routing tables to determine which PBX in the private network should be used for the outgoing call. Corporations could realize big savings because what would normally be a long distance call could be routed to another switch over tie lines, where it would be treated as a local call. Unfortunately, the only information that could be provided to the remote switch was the dialed digits, which were transmitted over the tie lines using tones (such as Dual-Tone Multifrequency, or DTMF).

With ISDN, tie lines are no longer necessary. All of the switches are interconnected to each other using the public network. The private network becomes a virtual private network because there are no dedicated circuits interconnecting the switches. With end-to-end signaling, information regarding call handling can be sent over the signaling channel of the ISDN interface and then transported through the SS7 network to the remote switches.

With digital protocols, there are many more possibilities in messaging. Sending information using tones is very limited because of the possible combinations. In digital information, an unlimited amount of information can be sent regarding virtually anything.

Think of reservation centers that receive millions of calls per day. They may have several locations located nationwide receiving those calls (such as hotel and rental car agencies). When one particular reservation center becomes congested from too many calls, signaling information can be sent to the main switch to indicate that calls should be routed to another location until congestion subsides. This is achieved using end-to-end signaling.

Telephone companies are now offering intelligent services to their larger customers, in place of ISDN solutions. For example, telephone companies use complex databases which reside in their own signaling network to store routing information for a company. Callers may dial an 800 number, and the telephone company makes decisions about where the call is routed. The routing is based on the call volume for each of the company's reservation centers, or it can be based on the zip code of the calling party.

Such services exist today and show the power of the telephone network when digital technologies such as SS7 and ISDN are provided. There are many more possibilities, which we will talk about throughout this chapter (Chap. 5 discussed the many services provided by telephone companies).

6.3. Early ISDN Issues

Many problems have plagued ISDN. Cost, ease of installation, configuration, and availability have been ongoing problems. One of the first problems (and still an ongoing issue) has been ISDN availability. One would think that digital services would be simpler to deploy for telephone companies, which should already have plenty of copper wire to send digital services through. However, telephone companies some time ago installed repeater and multiplexing equipment in their outside plant to make more efficient use of their infrastructure.

Multiplexers use analog techniques to multiplex multiple conversations over a copper pair. Many of these devices are not compatible with ISDN and interfere with the digital transmission of ISDN. Telephone companies have to replace this equipment in their outside plant before they can offer ISDN services to their customers. This is one of the biggest costs associated with deploying ISDN.

The telephone switches used today are ISDN ready. New line cards and software upgrades are required, but the switch architecture is in place. Had it not been for the outside plant problems, ISDN would be readily available throughout the United States.

Another issue in the United States has been the cost of ISDN services. For residential users, ISDN is an expensive service. At anywhere from $50 to $100 a month, ISDN can be hard to justify. Many are installing ISDN for Internet access because ISDN can provide Local Area Network-(LAN) like speeds to the Internet. Yet ISDN still lacks the killer application that will make it as common as cable television.

In Europe, a different marketing strategy was used to sell ISDN services. European telephone companies realized the expense of deploying and maintaining analog circuits, so they priced ISDN well below analog lines. This encouraged people to buy ISDN because it was cheaper, instead of trying to justify ISDN for its features.

The European telephone companies have saved millions of dollars because digital transmission allows them to send many conversations over the same copper pair. If they can transmit 24 conversations over one copper wire instead of 1, they can collect that much more revenues on that single copper pair than they would with analog services. This is also referred to as pair gain and can be a significant advantage to telephone companies.

In Japan, public telephones include an ISDN interface. This allows workers to connect their laptop computers to the public telephone and access their corporate networks from anywhere. Japan has a large mobile workforce, providing a good model for mobile communications. ISDN is widespread throughout Japan, in an effort to support their mobile workers.

6.3.1. The Cart Before the Horse—Premature Offering

Another issue that hurt the popularity of ISDN was its premature offering. When ISDN was first deployed, the only real marketing tool was calling party identification. Many PBX vendors capitalized on this feature, adding the capability to their switching equipment so that the calling party information could be passed right down to the desktop.

Many data applications used in calling centers were based on this information as well. With calling party information made available to the PBX, the calling party telephone number could be passed to a separate database, which would then look up the number and provide the caller's name, address, and any other information available.

Special interfaces were provided on the PBX to pass this information from the PBX to adjunct computers. IBM formed a niche market for its AS400 computer by providing telecommunications applications which relied on the ISDN interface to provide call information from the telephone companies.

This is all fine and good, as long as calls originate from the same exchange that the PBX is connected to. If the call originates from another exchange or is carried by a long distance carrier, the calling party information is not available. That is because the telephone companies have not yet deployed SS7 throughout their networks.

Even after widespread deployment of SS7 began, telephone companies had not yet negotiated interconnect agreements with other carriers to pass SS7 messages through one another's networks. SS7 remained a private network within each phone company, with the exception of call setup.

It was not until the late 1980s and early 1990s that SS7 access agreements were drafted between carriers, and access to valuable databases (which contain all the caller information) was provided. ISDN suffered because the full capability of ISDN could not be offered until SS7 was deployed throughout the public telephone network. Telephone companies put the cart in front of the horse.

Today, SS7 is deployed not only in the public telephone network but in the cellular networks as well. Access agreements continue to be negotiated between carriers as more and more SS7-based services are being offered. The Federal Communications Commission (FCC) has issued a number of mandates in the last year as part of the Telecommunications Act of 1996, which will require more of these access agreements and will increase the use of SS7 networks. This can only help the ISDN cause since then more features can be supported.

6.3.2. Interoperability—Where Did the Standards Go?

When ISDN was first deployed, it was riddled with problems. Vendors were following the standards, but there were many implementation issues not defined in them. Vendors began choosing their own methods of implementation, making equipment incompatible. Companies had to buy all of their ISDN equipment from one vendor and hope that the telephone company switch in their city would interconnect with the vendor's equipment.

While the standards defined the services and interfaces to be used with ISDN, they did not define the central office switch and how it would support ISDN features. This in itself has caused many implementation nightmares.

6.3.3. Configuration—The Consumer Nightmare

ISDN is not easy to get. Indeed there are many consumers who work from their homes or have home-based businesses who would love to get ISDN services but cannot for a variety of reasons. One of the biggest problems in ordering ISDN has been knowing what to order and weeding through the mountains of specifications and configuration rules, all of which are vendor-specific.

To compound the problem, the telephone company personnel know very little about digital services, especially ISDN. In many circles ISDN became known as "I Still Don't kNow." This is a problem we still have today. Getting an ISDN connection is difficult, and once the order has been placed and the lines installed, making them work can be frustrating.

When you buy ISDN equipment for your home or office, you must configure your connection according to the type of central office switch that is providing you with service. A Service Profile Identifier (SPID) is a string of numbers that identifies the type of central office equipment you are connecting to as well as other service parameters necessary to make the switch and your ISDN equipment talk to one another.

Needless to say, not many users are going to be savvy enough to know what questions to ask when they purchase equipment and, worse, what to tell the telephone company when they order their ISDN service. When the telephone companies have trouble understanding the technology, getting your connections to work can be frustrating and time consuming.

Some ISPs have come to the rescue, using their expertise in digital communications and networking to install and configure ISDN equipment for their customers. This has helped in many cases, but the lack of digital telephony knowledge in the telephone companies will continue to hurt the success of ISDN.

The North American ISDN Users Forum is establishing new standards for ordering ISDN. These new standards will simplify the configuration of ISDN equipment and the ordering process. Already they have begun implementing these standards in areas where ISDN is heavily deployed (Pacific Bell in California has a large ISDN base already).

The ISDN Forum (a different group) has developed standards where the ISDN device is configured automatically. When the device is connected to the ISDN interface, it sends messages to the central office switch and obtains the SPID automatically. Users do not have to get this information on their own.

6.4. ISDN Network Architecture

I mentioned earlier that ISDN is really a concept of a fully integrated network. The standards define the user interface to this network. There are two interfaces defined for user access to the ISDN: BRI and PRI.

6.4.1. Basic Rate Interface

BRI is a digital circuit, providing two "bearer" channels (both at 64 kbps) and one signaling channel (at 16 kbps). A bearer channel is one in which user data can be transmitted. Only two wires are needed from the central office to the subscriber equipment to support these three channels.

The BRI is a bidirectional interface. In a two-wire circuit, an encoding scheme called 2B1Q is used to transmit over one wire and receive over the other. Most BRI today use 2B1Q because it allows ISDN to be delivered over existing twisted pair. The connection used at the interface is an RJ-45, where pins 3 and 6 are used for transmission and pins 5 and 4 are used to receive.

The BRI supports point-to-multipoint configurations, which is consistent with the intent of the BRI: to support small businesses and residential services. Up to eight devices can be supported by the Terminal Equipment (TE) on a passive bus up to 200 m from the Network Termination 1 (NT1).

6.4.2. Primary Rate Interface

PRI provides 23 bearer channels (or B channels) and one signaling channel (the D channel), all at 64 kbps, for a total of 1.544 Mbps in U.S. networks and 2.048 Mbps in European networks. The signaling channel supports signaling for all 23 channels.

PRIs are used in commercial applications, usually where a digital PBX exists. If the digital PBX supports direct ISDN connections, the PRI can be terminated to the PBX line card. If the PBX does not support ISDN, a channel bank is required.

The channel bank is responsible for receiving the ISDN messages and converting them to analog circuits to be connected to the PBX. This means that the channel bank must be an ISDN-compatible device.

PRI only supports point-to-point configurations, which has been the objective for PRI. The original intention of this interface was to support switching devices such as PBXs on the ISDN.

The physical connection to ISDN PRI is a 4-, 6-, or 8-pin modular connector, similar to those found in your home. In the United States, the connector is an 8-pin RJ-45, providing one pair for transmit, one pair for receive, and two pair for power.

ISDN telephones receive power either from the network or from an external source. For PRI applications, the power is provided from the telephone company network. In BRI applications, power is usually provided by an external source (such as a power adapter). The phones operate on −48 VDC.

6.4.3. Channel Usage

The B channel (also referred to as the bearer channel) is used to transmit digital voice (using 64-kbps Pulse Code Modulation, or PCM). In Chap. 1, we discussed digitizing voice using PCM. The B channel is also used for high-speed data (either circuit or packet switched), facsimile, and slow-scan video. There is not quite enough bandwidth to support broadcast (full motion) video.

The D channel (signaling channel) is used for signaling (basic and enhanced) for all of the B channels on the ISDN interface. The D channel can also be used for low-speed data (such as videotex or terminal communications) and telemetry. Telemetry is used for emergency services (such as alarm monitoring) and allows utility companies to read electrical and water meters. This application has not been widely accepted, especially when cellular services can provide the same capability cheaper than ISDN.

The BRI channel supports multiple devices over the 2 B channels, which means there must be arbitration to determine which device can use the D channel first. This arbitration is handled by the layer 1 protocol. The BRI is not activated permanently, so there is also a need for activation/deactivation bits. The BRI is only active when a device wishes to transmit.

The PRI D channel supports multiple channels (24 in North America and 32 in Europe). In North America, the D channel is always channel 24. In Europe, the D channel is always time slot 15. Time slot 0 is used in European interfaces for framing. The interface is activated permanently, so there is no need for activation/deactivation bits. The PRI only supports point-to-point configurations, so there is no need for arbitration on the D channel.

Individual channels can be combined to provide additional bandwidth for applications such as video and high-speed access. H channels are available in multiple configurations. An H0 channel supports data at up to 384 kbps. Two H1 channels, H11 and H12, provide 1.536 and 1.92 Mbps, respectively. These are used for high-speed data, near-broadcast-quality video, high-fidelity audio, and voice applications.

Some applications do not require even the 64 kbps provided in a B channel. Terminal applications may only require a 9600-baud connection. When this is the case, rate adaption is used. With rate adaption, only the portion of the channel needed is used, while the rest of the channel is filled with binary 1s. Transmissions can be multiplexed (interleaved) with other data over the same channel, allowing the full bandwidth to be used.

6.4.4. The Nodes and the Reference Points

ISDN standards identify the functions to be provided at various points in the ISDN. The point between each of these functions is a reference point. There is really no physical entity associated with reference points or ISDN functions. A device can provide one function or several functions and can bridge more than one reference point.

Figure 6.1 shows the ISDN Reference Model. Each of the functions is identified in the boxes, while the lines between the boxes represent reference points.

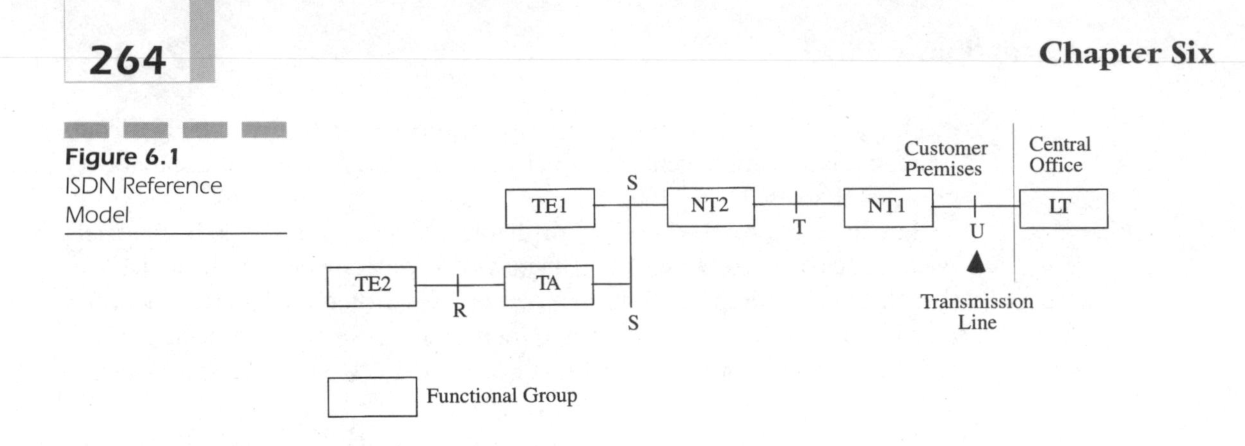

Figure 6.1 ISDN Reference Model

6.4.4.1. ISDN Functions NT1 isolates the user equipment from the local loop. This allows the telephone company to use digital loop techniques (such as 2B1Q encoding) without extending their encoding methods into the customer premise. Customer Premises Equipment (CPE) connects to a standard connector, usually at the NT1. The NT1 provides the functions of the physical layer, such as 2B1Q. From the NT1 to the NT2, a four-wire circuit is used (2B1Q is a two-wire circuit).

Network Termination 2 (NT2) can support up to layer 3 if required. Switching functions and line concentration can be supported if needed. Switching functions are used when connecting ISDN to a PBX. Line concentration is used when connecting multiple ISDN telephones or terminals to one ISDN connection (which is the case with BRI and residential services).

Terminal Equipment 1 (TE1) represents functions provided by ISDN telephones and terminals. These are digital telephones, modems, and other devices that can support the ISDN protocol and connect to the ISDN interface at the NT1. Non-ISDN devices must use a different interface which provides ISDN support as well as analog to digital conversion.

Terminal Equipment 2 (TE2) is equipment that is not ISDN compatible. The functions provided here support analog telephones, modems, and terminals using RS-232 serial connections. Devices used to connect to X.25 networks are supported by TE2 as well.

The Terminal Adapter (TA) supports the connection of TE2 equipment to ISDN circuits. The TA is usually a function provided by devices supporting NT2 functions, such as an ISDN telephone. Many of these devices have a connection for interfacing with non-ISDN equipment.

6.4.4.2. ISDN Reference Points The User (U) reference point separates the network side of the ISDN from the subscriber. This is a full-duplex subscriber line. The U interface is not well defined in ITU Recommendations but is defined in the ANSI standard T1.601. This standard defines the U interface using two binary one quartenary (2B1Q) encoding, supporting full-duplex operation on one pair of twisted pair cable.

2B1Q uses four voltage levels, which permits 2 bits of information to be conveyed by each "pulse." Each signal has four possible values, which is how 2 bits can be represented per pulse.

The Terminal (T) reference point separates the network termination equipment from the user's termination equipment. For example, with BRI the subscriber loop is a two-wire circuit with 2B1Q encoding. The NT1 on the user side supports a four-wire circuit, with separate transmit and receive pairs.

The System (S) reference point separates the user terminal equipment from network functions. The user TE does not need to be concerned with network functions; they are only concerned with applications. This reference point represents that separation. The Rate (R) reference point separates non-ISDN equipment from adapter equipment.

Again, the purpose of establishing a separate interface for connecting to ISDN (instead of connecting CPE directly to the ISDN circuit) was to provide a means by which the network could continue to evolve without affecting the CPE. The NT1 isolates the CPE from changes made in the subscriber loop.

6.4.5. Protocols of ISDN

ISDN uses several protocols to support various functions within the network. In this section, we will examine these protocols and the functions that they provide. As with the rest of this book, this section will provide an overview of these protocols rather than a detailed description of each.

Protocols used in ISDN differ according to the layer of services and the type of bearer services. ISDN layers do not go beyond the network layer since this is a transport technology for user information. The protocols and the ITU Recommendations that define them are listed below:

- Physical Layer—I.430 (BRI) and I.431 (PRI)
- Data Link Layer—I.441/Q.921 (LAPD) on the D channel
  I.465/V.120 for circuit switched and semipermanent on B channel
  LAPB (X.25) for packet switching on B channel

- Network Layer—I.451/Q.931 Call control (signaling) on D channel
  X.25 packet level for packet switching on D channel
  X.25 packet level for packet switching on B channel

6.4.5.1. Link Access Procedure for the D Channel Layer 2 is supported by the link access procedure for the D channel (LAPD) protocol. LAPD in a High-Level Data Link Control (HDLC) protocol similar to LAPB used in X.25. In fact, LAPD was derived from the LAPB protocol used in X.25 packet-switching networks. LAPD provides error detection/correction, flow control, and addressing. The addressing in LAPD is different from the addressing used in X.25, for obvious reasons. ISDN has different address requirements than X.25 packet switching.

LAPD provides unacknowledged information transfer and acknowledged services. Unacknowledged is the same as datagram services, or connectionless. Acknowledged is the same as connection-oriented. Both acknowledged and unacknowledged can be supported simultaneously over the same D channel since they are sent in correspondence to specific B channels (one B channel can use unacknowledged while others are sending acknowledged).

LAPD addressing supports multiple data link connections. This is different from LAPB, which only supports single data link connections. This is important to ISDN because LAPD must be able to address many different connections at the same time and maintain those connections.

The LAPD frame is shown in Fig. 6.2. The flag is a keep-alive signal, used to maintain synchronization on the link. Like other HDLC protocols, the flag is a pattern of a 0, six consecutive 1s, and a 0 (01111110). When the link is idle, LAPD transmits nothing but flags to maintain timing on the link.

To prevent data from duplicating this pattern, bit stuffing is used. The transmitting node (at layer 2) generates the flag and then begins transmitting a frame. If it detects a pattern of five consecutive 1s in the bit stream, layer 2 inserts a 0 after the fifth binary 1.

The receiving node detects the flag and then reverses the process. Whenever a pattern of five consecutive 1s is detected, layer 2 at the receiving node discards the 0 bit that follows. If the receiver detects that

Figure 6.2
LAPD Frame Format

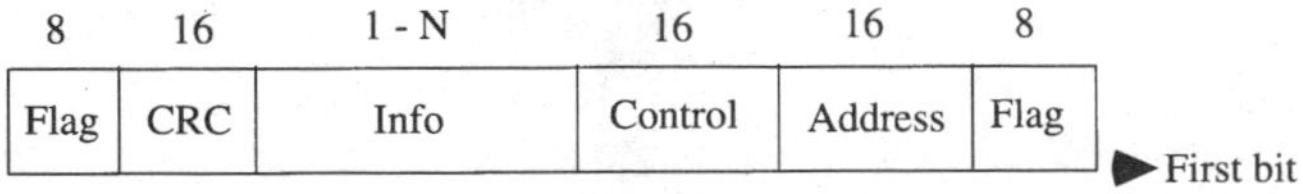

8	16	1 - N	16	16	8
Flag	CRC	Info	Control	Address	Flag

▶ First bit

the sixth bit is a 1 and the seventh bit is a 0, it is accepted as a flag. If both the sixth and the seventh bits are a 1, the frame is considered an abort frame.

Two levels of addressing are provided to identify the end node as well as the service. For example, several terminals may be sharing the same ISDN line, and each terminal may be sending a combination of packet-switched traffic and signaling (both on the D channel). The address used in LAPD identifies both the terminal endpoint and the service to be accessed. This is provided in the LAPD address field.

The address field is a combination of two elements: the Service Access Point Identifier (SAPI) and the TEI. The combination of these two elements is called the Data Link Control Identifier (DLCI). This is the same as in Frame Relay.

The SAPI identifies a layer 3 user of LAPD. For example, if a frame is carrying X.25 data, the SAPI value is 16. This SAPI value indicates to the receiver that the payload is X.25 data and will require the services of the X.25 resources available at the receiving node. The SAPI values presently defined are:

- SAPI 0—Signaling and call control procedures
- SAPI 1—Packet-mode switching using Q.931 procedures
- SAPI 16—X.25 packet-mode services
- SAPI 32-62—Frame Relay services
- SAPI 63—Layer 2 management procedures
- All others—Future standardization

SAPI 1 is recent, supporting Q.931 signaling with packet-mode services for applications such as user-to-user signaling (end-to-end signaling). SAPIs are unique within a terminal (the same SAPI cannot be used twice with the same TEI on different connections).

The TEI identifies a connection endpoint within a SAPI. The following TEIs are defined in Q.921:

- 0–63—Nonautomatic TEI assignment
- 64–126—Automatic TEI assignment
- 1–27—Group TEI

Nonautomatic TEI assignment is used when the user configures the TEI address into the equipment at implementation. This means when the equipment is initially connected to the network, the TEI must be assigned through administration procedures in the equipment (which

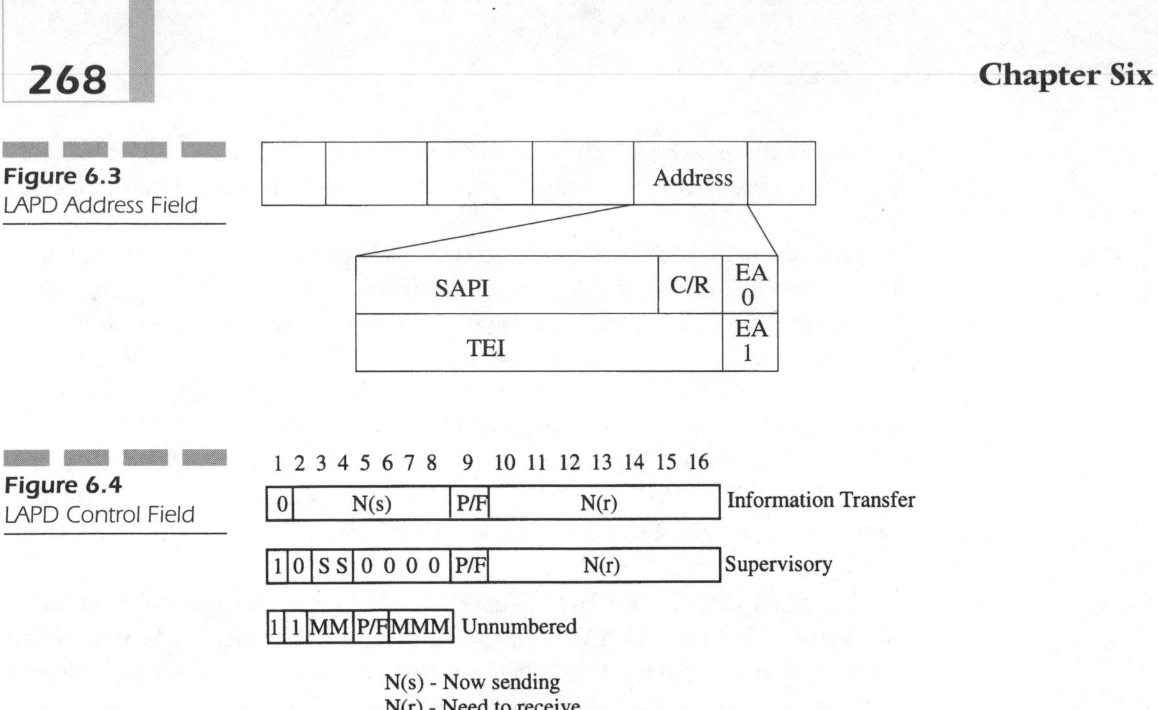

Figure 6.3 LAPD Address Field

Figure 6.4 LAPD Control Field

may be through a terminal or switch settings). A manufacturer can also assign the TEI to the equipment when it is manufactured.

Automatic TEI assignment allows the network to assign an available TEI at the time a connection is requested. Each time the connection is released, the TEI is assigned to another device that is requesting a connection. A device can have more than one TEI, depending on what type of device it is. A terminal concentrator would have multiple TEIs. Figure 6.3 shows the LAPD address field.

The control field carries a control command. These are similar to X.25 control commands. As seen in Fig. 6.4, the control commands are divided into three groups: information, supervisory, and unnumbered. Each group uses its own format.

Information control frames are used for sending user data. They may also carry an acknowledgment of previously received information frames. Information frames are only sent when there is user data to be transferred across the ISDN.

Supervisory frames are used for flow control and error detection. There are three possible commands used in supervisory frames, receiver ready (RR), receiver not ready (RNR), and reject (REJ). When there is no user data to send, a supervisory frame is used to acknowledge received information frames. Supervisory frames are also sent to maintain a connection.

The receiver ready (RR) is used to acknowledge received frames. Receiver not ready (RNR) sends an acknowledgment but also requests the suspension of transmission for flow control purposes. Reject (REJ) indicates that the last frame received was rejected and needs to be retransmitted. The N(s) field provides an acknowledgment of the last good frame received. All sequences sent after the sequence indicated in the N(s) field are retransmitted (more on sequence numbering later).

Unnumbered frames are used for connectionless services. These frames do not use sequence numbers (which is why they are referred to as unnumbered). There are several commands associated with unnumbered frames:

- Set Asynchronous Balanced Mode Extended (SABME)
- Disconnect Mode (DM)
- Disconnect (DISC)
- Unnumbered Information (UI)
- Unnumbered Acknowledgment (UA)
- Frame Reject (FRMR)
- Exchange Identification (XID)

The SABME command is used to establish a connection with an endpoint. This command is only used with acknowledged (connection-oriented) services. Connection establishment begins with a request for a connection. The SABME sets sequence numbering to 7 bits (0 127) and establishes a logical connection between two peer entities. This is different than in most data communications, where there is no peer relationship but a master/slave relationship (Data Terminal Equipment/Data Circuit Terminating Equipment, or DTE/DCE).

One important note about sequence numbering is that it begins at 0 for a connection and is incremented sequentially with every frame sent. This continues for the life of the connection. The numbers are significant to the connection only. When a connection is established, sequence numbering is reset to 0.

The Disconnect Mode (DM) command is used to reject a connection request. A connection can be denied for any number of reasons. The DM command identifies the cause for denying the connection (although the cause codes are not very specific).

The Disconnect (DISC) command is used to indicate a released connection. It can be sent by either the originator of the connection or the other endpoint. The DISC will release all resources dedicated to the connection.

Unnumbered Information (UI) commands are used by TEI management. When a TEI number is requested by an endpoint, a UI with the assigned TEI value is sent in response. TEI management will be discussed more fully later.

An Unnumbered Acknowledgment (UA) is sent as an acknowledgment without any sequence numbering. The UA is sent in response to a SABME. The UA indicates that a connection has been established and transmission of information frames can begin.

A Frame Reject (FRMR) is sent when a frame is received that violates protocol. There are several reasons for an FRMR:

- Undefined or nonimplemented control field (according to the standards)
- Incorrect length for supervisory or unnumbered frames
- Invalid N(r) (must be between last acknowledged and last sent)
- Information field in I frame exceeds maximum established length

The FRMR will also return the control field of the frame in question as a reference for the receiver.

The Exchange Identification (XID) command is used to exchange DLCI addresses with another endpoint. The XID can be sent as a request or a response. In other words, if an endpoint requests the ID of another endpoint, it sends an XID. The endpoint receiving the XID will return its ID in another XID.

The last field in the LAPD frame is the Frame Check Sequence (FCS) field. This is similar to other error detection schemes, where an algorithm is used on the contents of the header or the entire frame and the results placed in the FCS field. When received, the same algorithm is used and the results compared to the FCS field received. If there is a match, the frame is considered good. If there is no match, the frame is discarded.

6.4.5.1.1. LAPD Sequencing Before continuing with LAPD, we need to discuss sequence numbering. Sequence numbers are used for connection-oriented services. The sequence numbers are started when a connection is established. They always begin at sequence number 0.

Looking at the format of the Q.921 control field, the sequence numbers can be found in two forms: the N(s) and the N(r). The N(s) can be considered as "now sending." In other words, the sequence number is assigned to the frame in which it resides.

The N(r) can be considered as "need to receive." This is an acknowledgment of previously received frames. For example, if you send me frames with sequence numbers 1, 2, 3, 4, and 5, I will acknowledge with a N(r) value of 6 (assuming of course I received all five frames).

Since ISDN is asynchronous in nature, it is not uncommon to receive acknowledgments for some frames and not for others sent afterward. There is usually a delay in the acknowledgments, which is why timers are used in all network nodes. The timers are set when the frame is transmitted, and if the timer expires before an acknowledgment is received, the frame is retransmitted. Remember that all frames transmitted are kept in the transmit buffer until they are acknowledged.

As I mentioned before, sequence numbers are always reset to zero when a connection is established. This ensures that both endpoints of the connection know where they are starting. LAPD differs from LAPB because it uses a modulo of 128. This means that sequence numbers run from 0 to 127 (compared to LAPB, which uses modulo 8, or sequences 0 to 7).

6.4.5.1.2. Management There are two types of management at layer 2: TEI management and parameter negotiation. When a user obtains an automatic TEI assignment, a UI frame with a SAPI of 63 and an TEI of 127 is sent. The information field contains a message type of identity request and a reference number.

In response, the network returns a UI with a message type of the identity assigned. The reference number is the same as the one sent in the request. The assigned TEI value is returned in the information portion of the frame.

Some LAPD parameters can be negotiated between entities. An XID frame is used to change these parameters. The parameters set the timers and counters used for connection establishment, flow control, and error detection/correction.

6.4.5.1.3. LAPD Connection Establishment Before layer 3 can establish a connection end to end, a data link connection must first be established between two adjacent nodes. This must be repeated at every connection point along the path. LAPD uses the SABME message to request a layer 2 connection to an adjacent node, which replies with a UA frame. Once the UA has been received, LAPD can begin transmitting information frames, which will carry the layer 3 messages and user data (once connection establishment has been achieved at layer 3).

So there are really two connection establishment procedures: one for the data link layer between each connection point and one at layer 3 providing the end-to-end connection through the network.

6.4.5.1.4. LAPD Flow Control and Error Detection/Corretion When an entity becomes busy, it will send a Receive Not Ready (RNR) to an adjacent node. The entity receiving the RNR will then periodically send a Ready to Receive (RR) with the poll bit set to 1. The busy entity must then respond by sending either an RR or RNR (still busy).

When a node sends an RNR, it means that the link is busy, not the entire node. Remember we are still at layer 2 here, and layer 2 only deals with data link procedures. If the entire node were congested, layer 3 management would be responsible for advising all nodes in the network of the busy status.

When a link becomes busy, it means that either the buffer is full or the link cannot process the data fast enough. Think of a printer connected through a serial interface to a computer. The leads on the serial interface are used to either stop the transmission of data to the printer or start the transmission again. This is flow control. The same is true for the data link layer in ISDN. The LAPD protocol provides flow control procedures using the supervisory frame and its associated commands at the link level only.

Timers are used for error detection/correction. During the connection phase, timers and counters are reset. For example, the T-200 timer is started by the initiating entity (the one requesting a connection) after layer 2 transmits the SABME. If the timer expires before a UA is received, another SABME is transmitted. This is repeated until a UA is received.

A counter is also used to prevent layer 2 from constantly sending SABMEs with no success. The counter keeps track of the number of attempts made by layer 2 to establish a data link connection, and after *n* number of tries, the connection phase is aborted. Network management is made aware of the unsuccessful attempts, and network management procedures are then used to determine and report the fault.

Another use for timers is error recovery. If an entity detects an error in an information frame, it discards the frame. It cannot send a Reject (REJ) because it does not know where the information frame came from (the DLCI cannot be determined). There is no possible way to alert the sender that the frame they sent was found in error.

The sender of the information frame uses a timer to determine when information frames should be retransmitted. Timer T200 is used for retransmission. When an information frame is transmitted, T200 is set to zero. If an acknowledgment has not been received before T200 times out, it sends an RR to determine the status of the adjacent node.

The receiver of the bad information frame will send an RR or an RNR, depending on the state of the link, with an acknowledgment of the last good frames received. The originator then discards all of the acknowledged frames from its transmit buffer and retransmits all unacknowledged frames.

When sending unnumbered frames, LAPD does not provide any error correction or flow control. Error detection is provided, but frames found in error are simply discarded without retransmission requests. LAPD relies on the upper layers to manage retransmission of unnumbered frames.

6.4.5.2. B Channel Data Link Protocol ITU Recommendation I.465 defines a data link protocol used on the B channel. The LAPD protocol is used over the D channel only. The protocol for the B channel is also known as V.120. The V.120 protocol allows TE2s to communicate with TE1s, or two TE1s with one another. This means the protocol is used at reference points R, S, and T.

The V.120 frame was derived from the LAPD protocol. When you look at the frame in Fig. 6.5, you will see that they are almost identical. The address field is different from the LAPD address field. Instead of using the TEI and SAPI, V.120 uses a Logical Link Identifier (LLI). The address format is shown in Fig. 6.6.

There is also an information field, which is used to carry the user data (see Fig. 6.7). The information field also has a header, which contains

Figure 6.5
V.120 Frame Format

8	16	8or16	8or16	16	8bits
Flag	Address	Control	Information	FCS	Flag

Figure 6.6
V.120 Address Field

7 6 5 4 3 2	1	0
Logical Link Identifier (LLI0)	C/R	EA0
Logical Link Identifier (LLI 1)		EA1

Figure 6.7
V.120 Information Field

8bits	8	Variable
H	CS	V.120 Information

1	1	1	1	1	1	1	1bit
E	BR	Res	Res	C2	C1	B	F

Figure 6.8
V.120 Terminal Adaption Header/CS Header

control information necessary for the processing of the contained data. The first part of this header is the TA header. This is an 8-bit optional field that is used when terminal communications is being used. Terminal communications supports applications such as telex, where ASCII characters are sent to a dumb terminal.

There is also an optional header extension for control state information. The control state (CS) header is used for flow control. The values are shown in Fig. 6.8.

The address field contains the LLI, which is the data link address used over the B channel. The values for LLI are shown below:

- 0—In-channel signaling
- 1 to 255—Reserved
- 256—Default LLI
- 257 to 2047—For LLI assignment
- 2048 to 2190—Reserved
- 8191—In-channel layer management

As seen in Figs. 6.7 and 6.8, only LLIs 257 to 2047 are available for actual assignment. LLI 256 is used for connectionless services. For connection-oriented service, a logical connection must first be established. This can be done over the B channel, but if SS7 is available end to end (and user-to-user signaling services are supported), the D channel is commonly used.

The bits in the control state header are used to communicate lead transitions on modems. For example, most modems are connected using an RS-232C or V.35 interface. The bits in the control state header communicate the lead transitions from these interfaces over the ISDN interface. Not all of the interface leads are represented.

6.4.5.2.1. V.120 Connection Establishment If connection establishment messages are to be sent over the B channel, the messages are encapsulated in the V.120 frame and sent with an LLI of 0. If they are sent over the D channel, the messages are encapsulated in an LAPD frame and sent with an SAPI of 1. The messages sent over the B channel are similar to those used at layer 3 over the D channel: CONNect, SETUP, RELease, and RELease COMPlete. These messages are not part of V.120 since V.120 is the layer 2 protocol.

LLI assignment can be made by the calling or the called party. If the calling party is assigning the LLI, an available LLI within the range of 257 to 2047 is included in the SETUP message. If the called party is assigning the LLI, the address is provided in the CONNect message, which is the response to a SETUP message. Release procedures are the same as LAPD, using the RELease and REL COMPlete messages at layer 3.

The V.120 protocol allows multiple logical connections to be established on one circuit between two end users (but not multiple users). The protocol is used for non-ISDN terminal devices (TE2s) connecting to a TA, which in turn connects to a TE1 or NT1. The protocol is not used over the public network, only the user network and only between the devices described above.

6.4.5.3. ISDN Layer 3 So far we have only covered layer 2 protocols and procedures for ISDN. Layer 3 is responsible for the end-to-end connection establishment and transfer of user data. There are also management procedures provided by layer 3, which are covered in this next section.

ITU Recommendation I.451/Q.931 defines the layer 3 protocol used for signaling and control. The protocols defined in this and other related recommendations make up the DSS1 specification. The Q.931 protocol was developed for use over the D channel to support both functional terminals (intelligent devices) and stimulus terminals (digital telephones). Functional terminals are capable of utilizing the full range of Q.931 message types and their parameters, while stimulus terminals only utilize a subset of these parameters.

The message structure of Q.931 can be seen in Fig. 6.9. It consists of a protocol discriminator, a call reference value, and the message type. The protocol discriminator identifies the frame as containing a Q.931 format-

Figure 6.9
Q.931 Message Format

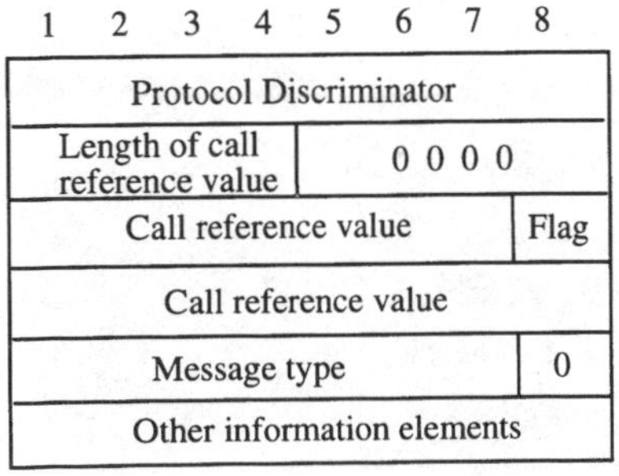

1 2 3 4	5 6 7 8	
Protocol Discriminator		
Length of call reference value	0 0 0 0	
Call reference value		Flag
Call reference value		
Message type		0
Other information elements		

ted message (versus a X.25 or V.120 message). The call reference value identifies the B channel the message is associated with. If this is a BRI interface, only 1 octet is needed (since there are only two B channels). If this is used over an PRI, 2 octets are needed (to support 24 or 32 channels).

The message type identifies what control message is being sent. A complete list of message types is provided in Table 6.1. The remainder of

TABLE 6.1

Q.931 Message Types

0000 0001 Alerting	ALERT
0000 0010 Call Proceeding	CALL PROC
0000 0011 Progress	PROG
0000 0101 Setup	SETUP
0000 0111 Connect	CONN
0000 1101 Setup Acknowledge	SETUP ACK
0000 1111 Connect Acknowledge	CONN ACK
0010 0100 Hold	HOLD
0010 1000 Hold Acknowledge	HOLD ACK
0011 0000 Hold Reject	HOLD REJ
0011 0001 Retrieve	RET
0011 0011 Retrieve Acknowledge	RET ACK
0011 0111 Retrieve Reject	RET REJ
0100 0101 Disconnect	DISC
0100 1101 Release	REL
0101 1010 Release Complete	REL COMP
0110 1110 Notify	NOTIFY
0111 0101 Status Enquiry	STAT ENQ
0111 1011 Information	INFO
0111 1101 Status	STAT
1111 1011 Key Hold	KEY HOLD
1111 1100 Key Release	KEY REL
1111 1101 Key Setup	KEY SETUP
1111 1110 Key Setup Acknowledge	KEY SETUP ACK

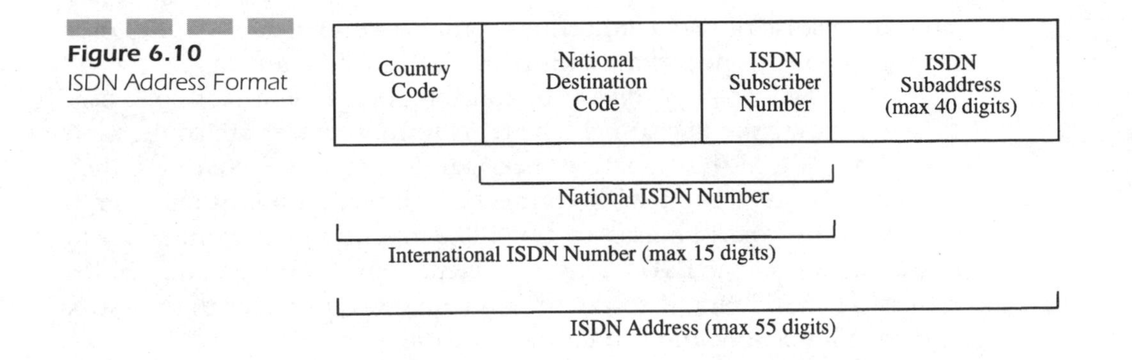

Figure 6.10
ISDN Address Format

the frame depends on the message type, since each message type provides particular parameters.

ISDN addressing consists of an ISDN number and an ISDN address. The ISDN number identifies the subscriber's connection to the network (at the T reference point). The ISDN address identifies an ISDN terminal (at the S reference point). This is layer 3 addressing, which is used end to end (unlike LAPD addressing, which is used link to link). There are many methods for addressing, depending on the network configuration. The ISDN address structure is shown in Fig. 6.10.

The country code identifies the country in which the message originated or a geographical area. A variable number of digits (1 to 3) defined in ITU Recommendation I.163 are used. The national destination code is also a variable-length field. It can be used to reach a network within the country code or to route to a region within a network.

The ISDN subscriber number is a variable-length field containing the number used to reach the subscriber in the same local network or numbering area. The ISDN subaddress can be a maximum of 40 digits. This is not part of the ISDN numbering plan but provides additional addressing information.

6.4.5.3.1. Q.931 Message Applications There are four applications supported by Q.931 procedures. They are circuit-mode connection control, packet-mode access connection control, user-to-user signaling, and global call reference.

Circuit-mode connection control involves the set up, supervision, and subsequent release of B channel connections and their resources for call control. This is mostly related to voice and data calls over a circuit-switched connection.

Packet-mode access connection control involves the set up, supervision, and release of data connections using packet-switched services over circuit-switched connections. This is an ISDN-specific feature.

User-to-user signaling consists of control messages sent from one user to another over the D channel. The users in this context are connection endpoints. While the signaling messages do not use B channels, they may well be associated with connections already established over B channels. This could be used by two PBXs to send signaling messages to one another over the PSTN. This is different from Q.931 signaling, which relates to ISDN circuits. User-to-user signaling is transparent to the ISDN and the SS7 network and is used solely by the user.

Global call reference is used during both call setup and call clearing. It can be used to clear one or more B channels simultaneously. We will talk more about the applications for global call reference a little later.

There are four classifications for Q.931 messages: call establishment, call information phase, call clearing, and miscellaneous messages. We will look at each of these messages in the order in which they would appear during their respective phases.

These messages are not used in the public network. The D channel is not extended from telephone office to telephone office; it is only used from the user network to the public network. Once the D channel message has been received by the local exchange, it is forwarded to a function in the switch called the Service Switching Point (SSP), which is an SS7 function. The ISDN signaling message is then changed to an SS7 message, which is then forwarded through the public network. We will discuss this SS7 relationship a little further later on.

6.4.5.3.2. Call Establishment Messages Table 6.2 lists the call establishment messages.

The SETUP message is used to initiate a call connection. It can be

TABLE 6.2

Call Establishment Messages

Call Establishment Messages
Alerting
Call Proceeding
Progress
Setup
Connect
Setup Acknowledge
Connect Acknowledge

sent in either direction (network to user or user to network). If the network is sending the SETUP message, it means that a user is requesting a connection, and the network is forwarding the SETUP message to the destination user (network to user). If the user is sending the SETUP message, obviously the user is requesting a connection to be established with another user.

A SETUP ACKnowledge is sent in response to a SETUP message. It indicates call establishment has begun but more information is needed. If more information is not needed, this message is not sent.

CALL PROCeeding indicates that call establishment has been initiated. It can be sent in either direction and is of local significance only.

ALERTing indicates that the called party is being alerted (the ISDN phone is ringing). This message can also be sent in either direction (user to network or network to user).

CONNect means the called terminal has accepted the connection request (the called party has answered). The B channel should now be connected, and conversation can begin once the message is acknowledged. This message is also sent in both directions.

CONN ACKnowledgment indicates the call has been awarded to the user and is sent in response to a CONNect message. When this message is received by the user sending the CONNect message, conversation begins.

PROGress is used to report the progress of a call establishment. It can be sent in either direction during any part of the connection establishment phase (but only after a SETUP message has been sent).

6.4.5.3.3. Call Information Phase Messages The call information phase messages used are listed in Table 6.3. These messages are sent after connection has been established. They are sent in relation to connections on B channels where conversations are in progress. The RESume message

TABLE 6.3

Call Information Phase Messages

Call Information Phase Messages
Hold
Hold Retrieve
Hold Reject
Retrieve
Retrieve Acknowledge
Retrieve Reject

is used to resume a call that has been previously suspended. It is only sent in the user-to-network direction. Calls can be suspended temporarily, allowing the calling or called party to initiate a connection to another entity. We will discuss suspension later on.

RESume ACKnowledge indicates the resume request has been granted, and the call is reestablished. This message is only sent in the network-to-user direction.

RESume REJect is sent if the call cannot be resumed. For example, a B channel connection may not be available to resume the call, resulting in the reject message. This is only sent in the network-to-user direction.

SUSpend allows a specified call to be temporarily suspended. The B channel is made available for other calls but not through Q.931 release procedures. The difference is that a RELease message flushes all buffers associated with the specified connection, and the network no longer maintains called/calling party identities.

With the SUSpend message, the identity of the called and calling parties is maintained, and resources dedicated to the connection are maintained. This makes reestablishment of the connection much quicker (the call establishment phase is not used). The SUSpend message is only sent by the user to the network and is of local significance only.

SUSpend ACKnowledge is returned in response to a SUSpend message and indicates that the B channel has been released. Charging is also stopped at this point. This message is sent by the network back to the user.

SUSpend REJect is sent by the network to the user, indicating that the request to SUSpend a call has been rejected.

User information can be sent in either direction, but it is always initiated by a user (not the network). User information is sent by one user to the user on the other end of a connection. The ISDN standards do not define these messages, although it is assumed that this would be call control information sent by two devices at user networks.

6.4.5.3.4. Call Clearing Messages Table 6.4 shows the call clearing messages.

DISConnect is sent in either direction and is always initiated by

TABLE 6.4

Call Clearing Messages

Messages
Disconnect
Release
Release Complete

either user. It is used to begin release procedures for a channel and all associated circuits. When received by the network, the network begins releasing circuits through the PSTN used for the connection.

RELease is sent in response to a DISConnect and indicates the channel(s) to be released. All associated circuits and resources reserved for the connection are released at this point.

RELease COMPlete is sent in response to a RELease message, indicating that all resources and associated circuits have been released. It should be noted here that this does not necessarily mean that circuits connected through the PSTN have also been released. Only the channel(s) between the user and the network are indicated in this message. SS7 messages then proceed through the PSTN to release circuits through the PSTN.

6.4.5.3.5. Miscellanous Messages Table 6.5 shows the miscellaneous messages used.

Congestion control can be sent in either direction and is used to begin or end flow control procedures on messages associated with user information (end-to-end signaling messages).

FACility is used to request a supplementary service. It can be sent in either direction but is always initiated by a user.

INFOrmation provides additional signaling information during any of the call phases we have just discussed. It is initiated by either user and is sent in both directions (user to network or network to user).

NOTIFY is used to send information which pertains to an established call.

STATUS is used to send error information as well as the status of a call in progress. It is sent in response to a STATUS enquiry message. The STATUS enquiry message is sent to initiate a STATUS message (request status of a call in progress).

6.4.5.3.6. Q.931 Message Parameters There are a number of parameters used with Q.931 bearer capability messages. These parameters are described in this section.

TABLE 6.5

Miscellaneous Messages

Miscellaneous Messages
Notify
Status Enquiry
Information
Status

Bearer capability parameters provide detailed information needed to establish the desired services over a B channel. They are defined in I.231 and I.232. The protocol options at each layer are defined. The information is divided into four groupings: bearer services, access attributes, information transfer, and general attributes.

Bearer services define the access methods to network functions or facilities, how information is to be transferred over the network, and general attributes such as quality of service and supplementary services.

Access attributes define the type of channel to be used as well as the user data rate. The signaling and information access protocols for layers 1, 2, and 3 are also identified.

Information transfer attributes define the bit rate for circuit-switched connections; whether circuit-switched, packet-switched, or Frame Relay services are to be provided for the call; and the call configuration (point-to-point, point-to-multipoint, or broadcast).

General attributes are used to identify supplementary services, quality of service, and interworking parameters. Supplementary services are defined as:

- Number identification
- Call offering
- Call completion
- Multiparty
- Charging
- Community of interest

The following is a description of the parameters provided for bearer services.

Call Identity. Used with the SUSpend message to identify a suspended call. It is used as a reference so that when a RESume message is sent, it can be correlated with a call. The call identity is assigned by the initiator of a SUSpend and is assigned at the start of suspension (when the SUSpend message is sent).

Call State. Identifies the state of a call. The state may be indicated as active, detached, or disconnect request.

Called/Calling Party Number. The subnetwork of both the called and the calling party. It is the subnetwork portion of the ISDN address. There is a separate field for the called party and one for the calling party.

Called/Calling Party Subaddress. The subaddress of the called and calling party. This is also part of the ISDN address. There is a separate field for the called party and one for the calling party.

Cause. Provides diagnostic information in a number of messages. For example, in the RELease message, the cause parameter identifies the reason for call clearing. Cause codes are not necessarily specific. In the case of the RELease message, one possible value is "normal call clearing," which indicates someone hung up.

Channel Identification. Identifies the B channel for call establishment messages as well as call information phase messages.

Congestion Level. Originally intended to provide congestion levels, this parameter currently supports two values: receiver ready and receiver not ready. Eventually, this parameter may be expanded to provide individual levels of congestion, but as of yet this has not been defined.

Display. Provides information in ASCII format to be displayed on an ISDN terminal (such as the display found on ISDN telephones).

Facility. Indicates the invocation of supplementary services.

Feature Activation. Used in the call establishment phase to activate specific features.

Feature Indication. Provides status information to the user of supplementary services.

High-Layer Compatibility. Used in the SETUP message during the call establishment phase to identify the terminal type connected at the S/T reference point. This information is passed transparently through the network as user-to-user information.

Keypad. ASCII representation of characters entered on an ISDN terminal's keypad.

Low-Layer Compatibility. Provides information transfer capability and transfer rate and identifies protocols at layers 1 through 3. Used to check compatibility of lower layers in the network.

More Data. Indicates that a USER INFORMATION message has been segmented and additional information is being sent in another USER INFORMATION message.

Network-Specific Facilities. Allows the specification of certain facilities that are unique to a network.

Notification Indicator. Currently the only values supported are user suspended, user resumed, and bearer service charge. This information is sent in the NOTIFY message to provide information pertaining to a call but not the status of a call.

Progress Indicator. Used in both call establishment and call clearing messages to indicate an event that took place while the call was in progress.

Repeat Indicator. Indicates that information elements have been repeated, and only one possibility should be selected.

Restart Indicator. Initiates a restart on a channel. A restart will flush all of the buffers and reset all associated counters and timers to zero.

Segmented Message. Indicates that the message received has been segmented and the rest of the message will be sent in a subsequent message.

Sending Complete. Used in the call establishment phase to indicate completion of the called party number. This is an optional parameter used in the SETUP message.

Signal. Used during the call establishment and call clearing phases. Stimulus terminals (ISDN telephones) interpret this parameter and generate tones (such as ringback, busy, and dialtone).

Switchhook. Tells the network the status of a stimulus terminals switchhook. The status can be either on hook or off hook.

Transit Network Selection. Identifies a network or sequence of networks that should be used to complete a call. If a sequence of networks is identified, the parameter is repeated within a message.

User-to-User Information. Has no network significance and is transparent to the network nodes. Used to send user information end to end (between two private switches).

I have described all of these parameters not as a reference, but so you can see the type of information provided by various ISDN Q.931 control messages. This should help in your understanding of what ISDN does and how call setups are initiated. Figure 6.11 shows the sequence of events and the message exchange that takes place during the call establishment, call information, and the call clearing phases.

The figure also shows the relationship between ISDN messages and SS7 messages. Remember that ISDN messages are not sent through the PSTN but are "converted" to SS7 messages and passed through the network as SS7 signaling messages. User-to-user information is passed transparently through SS7, using the pass-along or end-to-end method (Chap. 7 defines these methods).

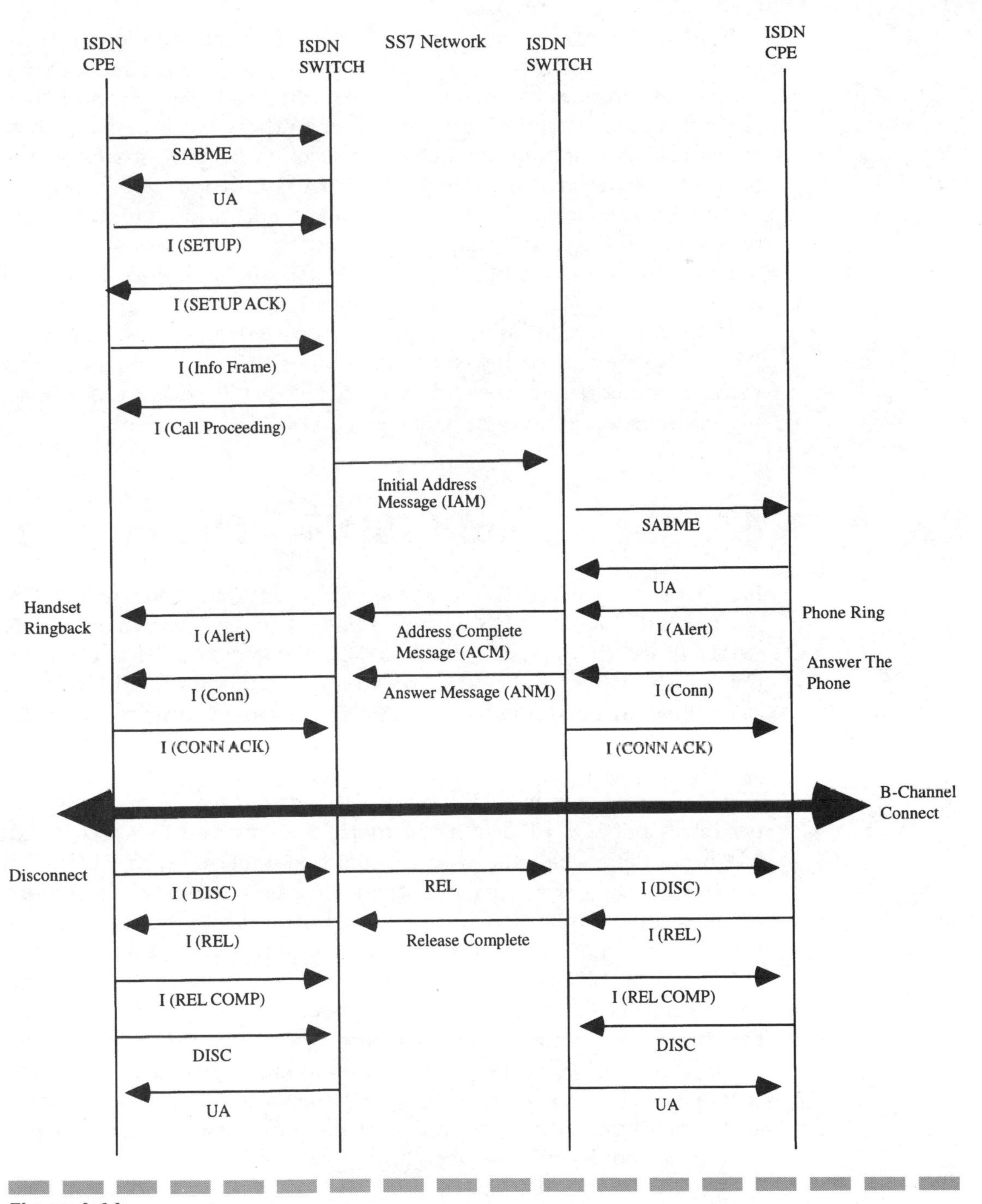

Figure 6.11
ISDN and SS7 Call Establishment/Call Clearing

Notice that the SS7 messages must be sent independently from exchange to exchange. In other words, the IAM message is sent from one exchange to another but not any further. The two exchanges must then establish a connection between each other. In the meantime, the farthest exchange begins setting up another connection between itself and the next adjacent exchange by sending an IAM message. This process is repeated until a route is established end to end, and connections are reserved until the called party answers.

Packet-mode call establishment procedures are similar to those used for circuit mode. A subset of the same messages are used.

ISDN does not address the needs of future networks. As we have seen, ISDN is sufficient for low-speed data, digitized voice, and slow-scan video. To support high-speed data, broadcast-quality video, and interactive multimedia applications, a new version of ISDN is needed.

6.5. Broadband ISDN—The Future

Broadband has been defined by the ITU as any rate faster than ISDN (paraphrased a bit). ISDN is now referred to as narrowband ISDN (NISDN), and the new solution is BISDN. BISDN was developed to support much higher bandwidths than the 1.544 Mbps now available. Up to 600 Mbps can be supported in BISDN, using Asynchronous Transfer Mode (ATM) as the transport and Synchronous Optical Network (SONET) as the transmission medium (physical layer).

ATM can be transmitted over coaxial and shielded twisted pair copper cables, but SONET delivers far more reliability and increased bandwidth not obtainable otherwise. It should be understood that BISDN is not like NISDN. All new technologies are required to deliver this service. There is some duplication in messages and protocol structure, but the entire architecture changes dramatically with the advent of ATM.

Reports of BISDN touted switch availability as early as 1994. This never occurred because the industry could not agree on a standard. Even today, the standards have not been completed, and work may not be finished until 1998. The ATM Forum thought they could provide the answer to fast deployment of ATM, but they found that when you collect several hundred vendors into a room and vote on an implementation, agreement is difficult to reach.

BISDN is designed to carry two types of traffic: Constant Bit Rate (CBR) and Variable Bit Rate (VBR). CBR is negotiated at call setup time.

Changes to the bit rate may be negotiated while the call is in progress, making this service different from circuit-switched services.

VBR is negotiated at call setup as well, based on the following parameters:

- Minimum capacity
- Maximum capacity

Examples of VBR are high-speed data, some forms of video, multimedia, and facsimile. CBR includes voice, video, and audio applications. CBR means there is a constant stream of bits sent at a constant rate, whereas VBR is more bursty in nature.

6.5.1. Overview of BISDN Advantages

There seems to be some confusion over what BISDN is and what its role will be in future networks. The ATM Forum says that ATM to the desktop is highly likely, delivering 600 Mbps right to the CPE. They proclaim that ATM throughout the network will replace many older technologies such as TCP/IP and even SS7. This is not true, which will be evident in this section.

ATM and BISDN concepts began in the laboratories of AT&T (Bell Laboratories) back in the late 1960s. The objective was to find a new switching method that would eliminate the need for channelized services and make more efficient use of the available bandwidth. There were many computer models built during this time, and a lot of research was done, but it was not until the late 1980s that ATM and BISDN development made any real progress.

The first published standards for BISDN were released in 1988 by the CCITT (now known as the ITU-T). The standards are known as the "Blue Book." The standards released by the ITU every 4 years were color coded to indicate which release they were.

The ATM Forum was started to help expedite the standards process and get ATM development started as soon as possible. The ATM Forum was started primarily by data communications companies interested in building ATM equipment for use in data networks. In the meantime, the telephone companies and Bellcore were actively defining the role of ATM and BISDN for use in their networks. There are a number of requirements documents published by Bellcore available on BISDN and ATM equipment.

The original intent of ATM was to provide a single transmission facility through which all traffic within the telephone network could be routed. This would alleviate the need for different circuits and specialized equipment to maintain data, voice, signaling, and video networks. All of these networks could be merged into one big integrated network, which brings us back to the ISDN concept, a fully integrated digital services network.

The BISDN concept is the same as in NISDN: extend the capabilities of the telephone network to the subscriber and give subscribers one interface on which all of their communications traffic can be routed. Today, it appears the pipe dreams of the ATM Forum to deliver ATM right to the desktop have been shot down since many companies are ignoring ATM and opting instead for more cost-effective technologies such as Fast Ethernet, and the Fiber Distributed Data Interface (FDDI).

In this section, we will look at the services delivered by BISDN and the applications it supports. We will also look at what is happening with other related technologies, such as SS7, which are expected to support BISDN in the public network.

There are two classifications for broadband services: interactive and distribution services. Interactive services include conversational, messaging, and retrieval services.

Distribution services include services with and without user presentation control. Distribution services are primarily one-way, from the network to the user.

Conversational services include video and audio. The transfer of sound, moving pictures, and scanned images and documents is considered broadband video telephony. Sound is different from voice because voice is considered as low fidelity. Sound and audio are high fidelity in most cases, but they do not have to be.

Videoconferencing includes the transmission of voice, video, scanned images, and documents in a point-to-multipoint configuration. The originator of the videoconference must be able to broadcast to several other points. The applications for videoconferencing go beyond the boardroom. Many companies are already using videoconferencing to reach remote locations and for meetings, presentations, and training sessions.

Universities are now offering courses using videoconferencing centers to broadcast the course sessions to other campuses around the world. This broadens their market area, allowing them to offer courses in areas once too far away to consider. Distance learning has become a major attraction for busy professionals who want to pursue graduate degrees but have no time to attend classes. They can now attend through video-conferencing centers near their workplace or even by dialing from their home computer (a possibility when BISDN reaches the residential market).

Video surveillance is another possibility. It is the transmission of sound and video from a building security system. Remote monitoring centers can monitor the activity within a building, listen to perpetrators, and report eyewitness reports to authorities.

The same systems can be used for traffic monitoring. Those of us who have lived in Los Angeles and enjoyed the commute in and around its suburbs know how this works. The transportation authorities use cameras mounted in strategic areas on their highways to monitor traffic flow. They can then send messages to billboards to report traffic conditions ahead and even provide alternate routes. In Los Angeles, the same traffic information is available over the Internet, making is possible to determine your route home before leaving the office. There is a rumor that in the near future, the cameras will be connected to the Internet as well, allowing commuters to see what traffic is like before leaving their desks.

High-speed data support means that all forms of data can be transported over the public network, regardless of size. File transfers of very large files, imaging applications (like those used in the medical profession), and high-volume transfers (where there are many files to be transferred) can be supported by BISDN. This includes documentation transfer and high resolution digital images. We can also connect to remote interactive games and game networks from our homes. Medical imaging can allow doctors to reach experts in other areas for opinions, sending them scanned x-rays in a matter of seconds. All of these fall under the category of conversational services.

Messaging services are more like e-mail and paging services. A unique twist to e-mail is video mail. A mailbox with sound and video would allow you to see the messengers as well as hear them. Calls to grandma's house would be extra exciting if she could see the kids as well as talk to them.

Retrieval services involve the retrieval of different files, such as video, sound, and high-resolution images. These files could be stored in data warehouses on corporate networks and retrieved over the BISDN interface for downloading on a home computer. This is a perfect application for telecommuters.

Distribution services have generated a lot of interest in the last few years. We have all heard about the many mergers and joint ventures between telephone and cable television companies. They offer distribution services, sending broadcast television signals to many homes in their networks.

Broadcast-quality video in different formats must be supported. Video comes in PAL, SECAM, and NTSC formats for television broadcasts and TV program distribution. Electronic newspaper distribution, electronic publishing, and remote education (distance learning) applications, as well as the new High Definition Television (HDTV), are all supported by BISDN distribution services.

6.5.1.1. BISDN Architecture There are two types of interfaces, or access points, defined in the BISDN standards: the network-to-network interface (NNI) and the user-to-network interface (UNI). The NNI allows multiple networks to be interconnected, forming one large cohesive global network. The UNI allows subscribers to interconnect equipment from any vendor to the BISDN. This is not unlike the basis for ISDN.

The physical layer for BISDN must rely on fiber optics. Broadcast video requires high sustained data rates, which existing copper infrastructure cannot support. Deployment of SONET is a requirement to support data rates up to 600 Mbps.

ATM is the transport for BISDN. It uses cells (a cell is the equivalent of a packet, with some significant differences). One of the most significant differences between cell relay and packet switching lies in the header and the fixed cell size (compared to packet switching, which uses large headers and trailers and variable-sized packets). Packets are routed according to the destination address in the packet header, while cells are routed based on a connection, identified in a small header.

ATM is independent from the physical layer. This allows ATM services to be offered over SONET or existing copper-based services. While twisted pair copper cable cannot support the speeds of some ATM services, lower speeds can be supported. ATM is really a multiplexing method which uses nondedicated slots for transmission of cells, making it incompatible with channelized (T-1) services.

Another unique feature of ATM is that it can support variable transfer rates for the same connection, whereas channelized services must provide the same transfer rate throughout the duration of a connection. In other words, with ATM the transfer rate can be changed while a connection and data transfer is in progress, something channelized services cannot support.

Figure 6.12 shows the BISDN network. This model was derived from the NISDN model, using the same functions found in NISDN. The

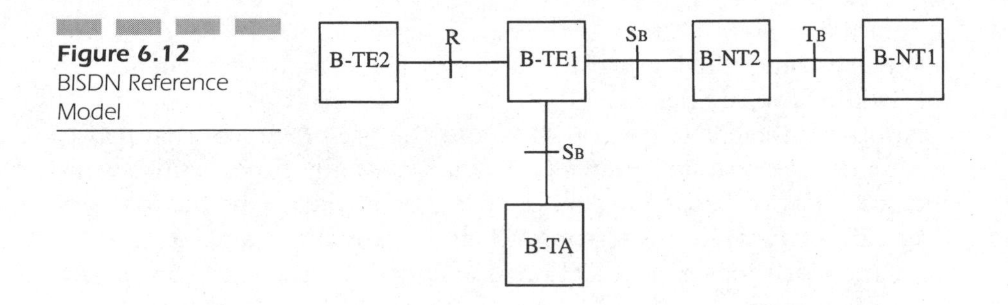

Figure 6.12 BISDN Reference Model

functions change somewhat, due to the nature of broadband transmission.

For more information about ATM and its functions, see Chap. 9.

6.5.2. BISDN and ATM—What Do They Have to Offer One Another?

The basis of BISDN is to provide a more efficient mode of transport for all forms of data. As was the case with NISDN, BISDN provides an interface from the public network to the user network. The public network will use ATM in a very different way and will obviously require much more bandwidth than the user interface. BISDN services are really designed to meet the needs of the subscriber rather than those of the network.

The public telephone network will use the same components that exist today. That is, voice switches will still exist as will the devices used in the signaling network. The major change for the public network comes in the form of support devices, such as repeaters, multiplexers, cross connect equipment, and some switching equipment. To understand some of these changes, we need to compare the existing infrastructure with ATM.

There are two ways to allocate channels in communications circuits for a transmission. Synchronous Transfer Mode (STM) circuits use dedicated time slots for transmission of data, while ATM circuits use labels in the header to designate the connection assignment. The ATM method uses virtual rather than "fixed" channels. A virtual channel does not require a dedicated time slot. Instead, in ATM there is a guaranteed number of cells offered during a specified time period. During that time interval, anyone can send data in available cells. In channelized STM facilities, a time slot is dedicated to a device for transmission to the endpoint of the connection. If the device has nothing to send, the channel remains idle. Obviously, ATM provides much more efficient use of bandwidth and facilities.

Another inherent feature of ATM and the use of virtual channels is its support of dynamic bandwidth allocation. At any time during a connection, a device can request more or less bandwidth. Channelized services cannot provide this feature. When a channel is assigned in STM, the bandwidth for the connection is assigned for the duration of the connection.

ATM will also support point-to-multipoint configurations, which are needed for certain distribution applications. For example, the cable television provider broadcasts video signals over the network to thousands of receivers (multipoints). This capability is not supported in NISDN.

BISDN provides the signaling procedures needed to establish these multipoint connections, and ATM provides the functions to make the connections and maintain them. This part of the ATM standards is still evolving, and ATM switches designed for "multicast" are just now entering field trials.

The answer to the question What do they offer one another? is Everything. BISDN provides the signaling procedures and functions needed to establish the various services provided by ATM, while ATM provides the services needed to support any communications application over one facility.

6.6. Frame Relay

I cannot talk about ISDN without describing Frame Relay. This technology was developed for bearer services over ISDN circuits. The intent was to provide a streamlined transport protocol for data transfer, eliminating many of the procedures found in packet-switching networks such as X.25.

The Frame Relay standards were developed by the ITU, ANSI, and now the Frame Relay Forum. The ITU continues development of international Frame Relay standards, while the ANSI continues its development of Frame Relay standards for use in the United States. Frame Relay was developed from the X.25 standards and uses protocol procedures that are similar to those found in X.25 and ISDN.

Frame Relay is based on packet switching. Time slots are used whenever data needs to be sent but are not dedicated to any one connection (as is the case in circuit-switched services). Frame Relay is different from cell relay in at least one aspect: Frame Relay supports variable-sized frames, while cell relay uses fixed-sized cells (they are the same as a packet, just different terminology).

The Frame Relay standards are documented in ITU I Series Recommendations and are broken up into different descriptions. The ITU I.233 (and the ANSI T1.606) define the services provided by Frame Relay, as well as the overall functions in Frame Relay protocols. ITU I.370 defines the procedures used in Frame Relay to manage network congestion. The functions of access signaling and data link control are defined in ITU Q922, "Core Aspects." ITU Q933 defines the protocol for signaling information used to establish and maintain virtual connections over the Frame Relay network. ITU Q922 defines the protocol used for end-to-end delivery of data as an option. ANSI has not defined this option in any of their standards.

There are two different data rates defined in Frame Relay, the Committed Information Rate (CIR) and the access rate. CIR is the rate the network has agreed to carry data at and is different from the access rate. The access rate is the data rate provided at the physical connection at either end of a Frame Relay virtual connection.

The CIR is enforced through rate enforcement. Frames which exceed the CIR will be carried over the network only if there is available bandwidth. If sufficient bandwidth is not available, the frame is discarded. Any frames with the discard eligibility bit automatically set, and found to exceed the CIR, are automatically discarded regardless of the available bandwidth.

Another parameter that must be negotiated at connection establishment is the committed burst size. This is the maximum number of bits per second the network will transfer during any measurement interval.

Frame Relay was developed for data transfer and was never intended for voice transmission. However, due to the slow deployment of ATM, many companies have found Frame Relay suitable for voice transfer. In fact, the Frame Relay Forum is actively defining standards to support voice over Frame Relay networks. The savings can be 25 to 35 percent over conventional T-carrier facilities.

As with T-carrier (T-1, T-3, etc.), the voice is converted first into digital form (using PCM). It is then compressed using Digital Speech Interpolation (which deletes any pauses or hesitations in the audio) and sent over the Frame Relay network with as few hops as possible (decreasing delay). Currently, voice over Frame Relay is limited to one network. Voice cannot be sent between two Frame Relay networks under today's Frame Relay standards.

Another limitation is that Frame Relay cannot support prioritized traffic. Voice traffic is treated with the same priority as data traffic. When the two are mixed, voice does not take precedence over data, which could cause additional delays if the traffic mix got heavy. Since Frame Relay allows for variable packet sizes, data packets can be up to 100 megabytes long, causing delays in the delivery of smaller voice packets. This may change as new standards evolve.

A Frame Relay Access Device (FRAD) can be used to segment large data packets into smaller packets, reducing the potential for delays. A FRAD is a device that sits on the edge of the network, acting as an interface between the voice switch and the Frame Relay network. FRADs can be used for a variety of applications, such as accepting traffic from non-Frame Relay networks and encapsulating the data into Frame Relay packets for transmission over the Frame Relay network.

One such device even allows SS7 traffic to be transferred over a telephone company's Frame Relay network, reducing the cost of SS7 facilities. In the case of this product (made by Tekelec), the FRAD actually discards extraneous Fill-In Signal Units (FISUs), reducing the amount of traffic sent over the Frame Relay network. This is where the cost savings come in, since it reduces the amount of traffic required between SS7 nodes. The SS7-FRAD (patent pending) performs other functions as well, but this gives you the principal idea behind them. They provide interworking between different network types and Frame Relay networks.

A FRAD can also prioritize traffic before sending it over the Frame Relay network. The FRAD does not actually add priority information, but it receives traffic from other protocols, uses the procedures of those protocols to apply prioritization, and buffers lower-priority traffic while sending high-priority traffic. Some FRADs also provide voice compression, reducing the bandwidth requirements from 64 to 4 kbps.

FRADs are currently available to support ATM, TCP/IP, SS7, and other protocols. They are point-to-point devices, requiring a FRAD at both ends of the Frame Relay connection. This is analogous to a modem, which requires another modem at the other end of the connection.

The advantage of using Frame Relay is its speed and low overhead. Most protocols require a lot of processing at each of the network nodes. Frame Relay requires very little network node processing, relying instead on the upper protocols and applications to provide management procedures within the end devices. This of course reduces the delay time in the network and allows packets to be sent much more quickly (which means more throughput).

Figure 6.13 shows a Frame Relay packet and its fields. As you can see,

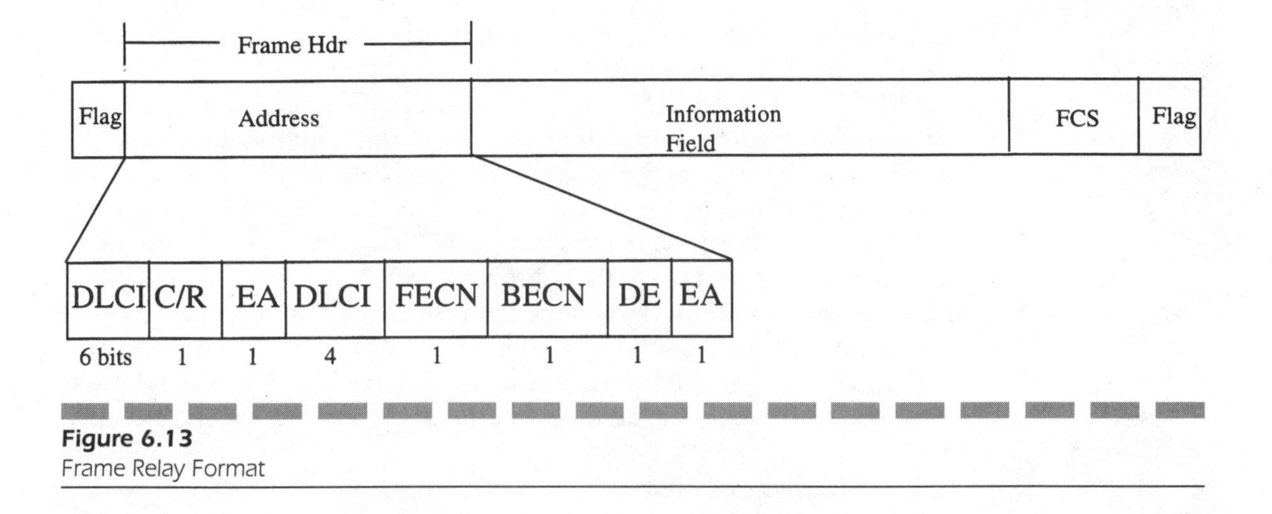

Figure 6.13
Frame Relay Format

there is very little to Frame Relay. The addressing used by Frame Relay is the same concept used in ISDN, using a DLCI to identify the connection.

The Forward Explicit Congestion Notification (FECN) and Backward Explicit Congestion Notification (BECN) fields are used by Frame Relay nodes in the network to notify other network nodes of congestion. When congestion conditions are encountered, Frame Relay makes no attempt to manage frames that are lost. The Frame Relay nodes simply discard the frames and rely on the upper layers to keep track of lost frames.

Congestion notification cannot be sent to the source of the congestion. Only the nodes that are adjacent to the congested node act on the FECN and BECN, slowing transmission to the congested node until the congestion subsides. Most carriers do not implement congestion control using the FECN and BECN because of this limitation.

Instead, many use Consolidated Link Layer Management (CLLM) at the end devices because this protocol enables end nodes to send congestion and status information to other nodes in the network. CLLM is a management protocol, sent in the payload portion of Frame Relay frames.

Nodes in the network do not provide layer 3 processing and provide very little layer 2 processing (such as error correction and flow control). Instead, Frame Relay follows a very simple rule: If the data is not in error, it is routed to the destination. If the data is found to be in error (or the frame is in error), the frame is discarded. The end nodes are then responsible for retransmission and end-to-end flow control. This enables network nodes to process frames much more quickly and more efficiently.

Originally, Frame Relay was designed for data transfers with large throughput requirements. Applications sending small-sized frames in bursty form are not good candidates for Frame Relay (it will work, but it is not as cost effective).

Frame Relay is not dependent on ISDN. There are many Frame Relay networks in use today that are running independently of ISDN. The Frame Relay protocol was developed to provide packet-switching services for ISDN subscribers without having to use X.25. The X.25 protocol was designed for use on circuits that were not as reliable as today's, and it has a lot of overhead and procedures requiring processing by the network nodes.

The idea behind Frame Relay was to eliminate all of the processing requirements in the network and to move the recovery procedures out to the end nodes. This is a good idea, as long as the circuits are reliable and there is not a lot of noise on the line. If the line is noisy and data is corrupted as a result, Frame Relay is no longer an efficient means of data transport because the upper layers will require retransmission too often, flooding the network with retransmitted frames.

6.7. Chapter Test

1. What does rate adaption provide?
   a. Supplementary services
   b. Data from multiple sources transmitted over one common B channel
   c. Data sent from one source transmitted over multiple B channels
   d. Changes to the transfer rate of data after a connection has been established
2. What is the SS7 protocol that supports ISDN Q.931 signaling called?
   a. Message Transfer Part (MTP)
   b. Transaction Capabilities Application Part (TCAP)
   c. Service Connection Control Part (SCCP)
   d. ISDN User Part (ISUP)
3. Applications such as digital telephony, X.25 packet-switched data, and Frame Relay data are supported by which ISDN services?

a. Bearer services
b. Supplementary services
c. Teleservices
d. All of the above

4. E-mail, videotex, and facsimile are supported by which ISDN services?
   a. Bearer services
   b. Supplementary services
   c. Teleservices
   d. All of the above
5. How many devices can an NT1 support?
   a. 1
   b. 2
   c. 8
   d. 12
6. What is a Service Profile Identifier (SPID)?
   a. String of numbers identifying the type of central office switch in the local exchange
   b. The address of the subscriber's ISDN equipment
   c. The identification assigned to a BISDN connection
   d. Channel assignment identifier found in the header of the ISDN packet
7. What is the bandwidth of H channels?
   a. 384 kbps
   b. 1.536 Mbps
   c. 1.92 Mbps
   d. All of the above
8. What is the maximum length of the ISDN address?
   a. 40 digits
   b. 15 digits
   c. 12 digits
   d. 55 digits
9. The Primary Rate Interface (PRI) supports point-to-multipoint configurations.
   a. True
   b. False
10. What is the bandwidth of a PRI in the United States?
    a. 1.544 Mbps
    b. 192 Mbps
    c. 2.048 Mbps
    d. 64 kbps

11. What is the bandwidth of the Basic Rate Interface (BRI)?
    a. 1.544 Mbps
    b. 192 Mbps
    c. 2.048 Mbps
    d. 64 kbps
12. What is the layer 2 protocol used over the D channel in ISDN?
    a. LAPD
    b. LAPB
    c. X.25
    d. Q.931
13. What type of line coding is used over the U interface on a BRI?
    a. ZBTSI
    b. HDB3
    c. 2B1Q
    d. AMI
14. What does a Service Access Point Identifier (SAPI) identify?
    a. The ISDN device on a connection
    b. Services to layer 3
    c. A connection endpoint within a service
    d. The DLCI
15. What does a Terminal Endpoint Identifier (TEI) identify?
    a. The ISDN device on a connection
    b. Services to layer 3
    c. A connection endpoint within a service
    d. The DLCI
16. What SAPI is used for layer 2 management?
    a. SAPI 63
    b. SAPI 16
    c. SAPI 0
    d. SAPI 42
17. What is the combination of the SAPI and the TEI called?
    a. Network Termination 1
    b. Service Access Point Identifier
    c. Data Link Control Identifier (DLCI)
    d. None of the above
18. What does the call reference value identify?
    a. The endpoint terminal
    b. A particular connection at the local network
    c. A particular connection at the remote network
    d. The type of message being sent

19. What Q.931 message is sent over the D channel to request a connection?
    a. REL
    b. CONNect
    c. SETUP
    d. SABME
20. In LAPD, the sequence number can be any value from 0 to what?
    a. 127
    b. 128
    c. 8
    d. 7

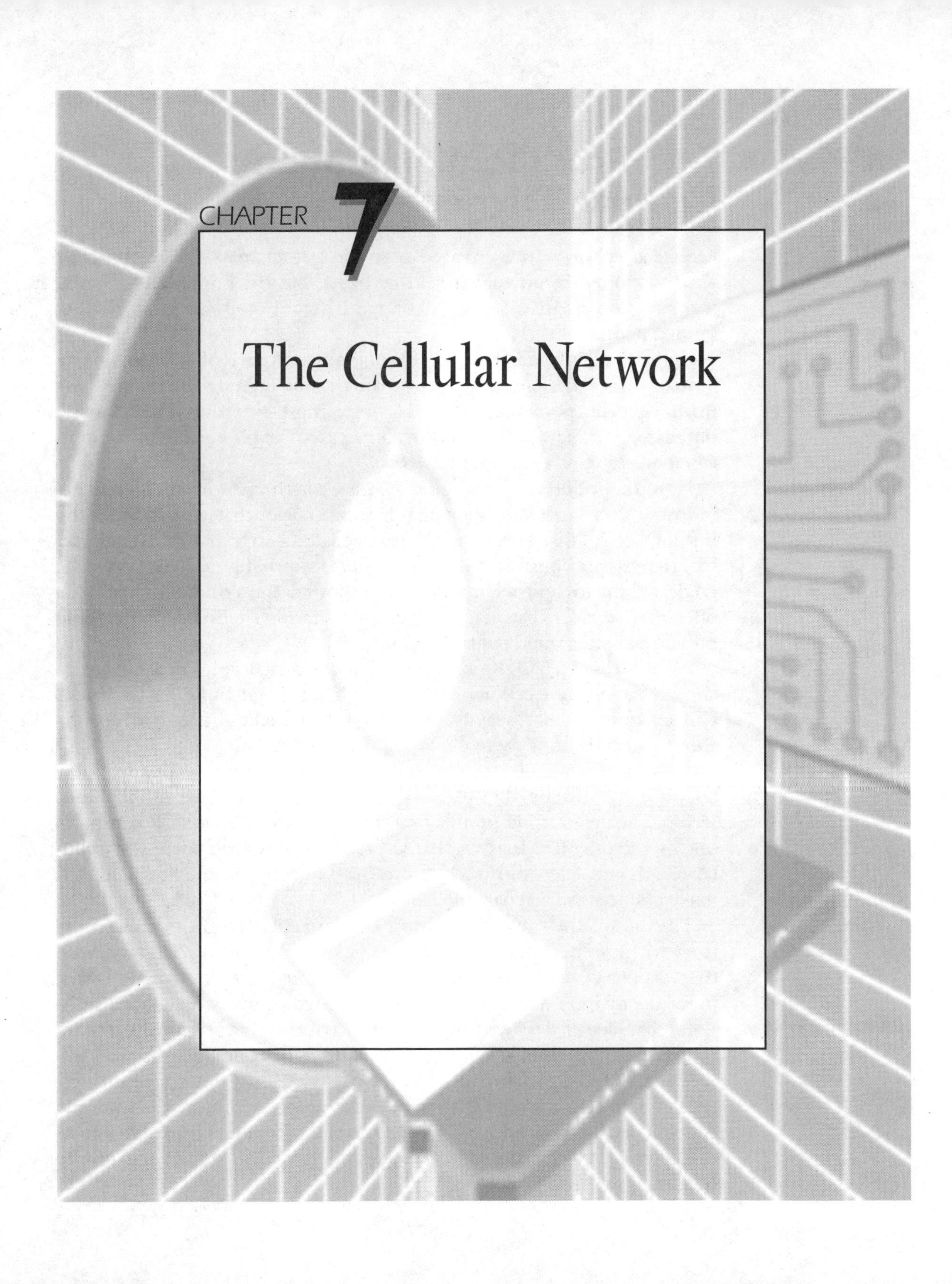

CHAPTER 7

The Cellular Network

7.1. From Radiotelephone to Cellular Telephones

Mobile telephones have changed over the last 20 years. Originally they were supported by an analog radio network, but the limitations of radio transmission quality forced a change in technology to what is now known as cellular service.

Cellular telephones have become a normal part of our lives. The industry has experienced explosive growth, and aggressive marketing has made cellular telephones affordable for almost everyone. They can be purchased for a few dollars, and in some cases, the phones are given away when a user signs a contract for 1 year.

Now, the industry is changing again. Growth has forced the cellular industry to begin looking for newer technologies that will support the millions of cellular subscribers. The need for better fraud control and security has also pushed cellular providers into changing their networks. Digital transmission techniques have allowed the cellular industry to offer new services that may compete with local wireline services, while providing better fraud control and security.

All this change has also brought a level of confusion. There are many choices facing new cellular service providers when building networks, and existing cellular service providers must make difficult decisions about which technology to deploy in their networks.

Making these decisions especially hard is interoperability. Cellular telephones must be able to work in any network since mobile subscribers "roam" outside of their service provider's network. Roaming is one of the premiere features that led to the widespread success of cellular telephones. Users can freely move from one city to the next and still make and retrieve cellular calls.

This means their telephones must be compatible with the transceivers used in those networks. New technologies make this difficult because they do not work the same way. Time Division Multiple Access (TDMA) uses time division multiplexing, a method used in many digital transmission technologies deployed in wireline networks. Voice transmissions are multiplexed into time slots over radio channels in the cellular network.

Coded Division Multiple Access (CDMA) encodes the digitized speech with special codes. These codes are used to delineate the multiple transmissions broadcast over the air interface. Receivers must be able to detect each of these individual codes and ignore all of the rest of the transmissions being received.

The fundamental difference between the two technologies lies in the air interface. TDMA telephones send a signal requesting service from a cell site. The cell site in turn sends a signal to the cellular telephone, telling it which frequency to transmit on, and which time slot is assigned to that transmission. These signals are sent over common control channels.

CDMA cellular telephones share the same frequency as other transmissions. Rather than assigning a time slot to the transmission, a special code is assigned. CDMA cellular telephones (or handsets, as I will refer to them from here on) receive all transmissions on a channel and must be able to find the transmission with the right code and ignore the rest of the transmissions.

The type of air interface is only one decision. Within the network, various technologies are used to allow cellular network nodes to communicate with one another. All sites must communicate with a central switching center (called the mobile switching center in most networks). Different transport technologies are used to interconnect these nodes, and different protocols are used to send control and signaling information.

Within the United States, X.25 has been the most commonly used transport technology. However, many cellular networks have been converting to Signaling System #7 (SS7) because of the services it can support. The higher layers that support cellular-specific applications differ as well. In most U.S. networks, Interim Standard-41 (IS-41) provides the signaling for the cellular network (using the services of either X.25 or SS7). However, in Europe, the Global System for Mobile Communications (GSM) provides the signaling for European cellular networks. Some U.S. cellular providers are looking at using GSM technology for new Personal Communications Services (PCS) networks, partly because GSM provides a standardized solution for the entire network, not just the air interface.

There are differences in the frequency ranges as well. U.S. networks operate within the 900-MHz range, while European networks operate in a higher-frequency range. The new PCS networks will operate in the 2-GHz range. This means that those who purchase PCS handsets will not be able to use their cellular handsets in existing cellular networks and that PCS providers must be able to offer the same capability that cellular offers today.

Depending on how quickly PCS operators can get their networks built, PCS users may find themselves restricted to metropolitan areas when using their telephones. It may take years before they can enjoy the same mobility that exists in cellular networks today. This represents a huge challenge to PCS network providers, who do not have the same amount of time to build their networks that cellular providers have had. Cellular providers have been building their networks for many years, but PCS providers must be able to roll out networks with full coverage when they open for business or risk competition from cellular networks.

To understand more about cellular networks, it helps to understand where the network started and how it evolved. In the next section, we will look at the predecessor to cellular phone, the mobile telephone.

7.1.1. Overview of Radiotelephone Networks

Older radiotelephone networks were designed after radio dispatch systems. In a radio dispatch system, a central dispatcher transmitted to specific mobile radios. The radio network was designed to cover a large geographic area with the entire radio spectrum allocated for the application.

Interference was avoided by placing special encoder/decoder chips in the radios so that only those radios with the same encoder/decoders could hear one another. This works even today for radio systems but is far too limiting for mobile telephones.

First, the large coverage area means that frequencies cannot be reused in the same area. This limits the number of simultaneous conversations the network can support. This was evident during the 1960s when mobile telephones began to become popular. Many users were forced to wait before they could place a call because the system was already busy. In some areas, only 300 simultaneous connections could be supported.

With the number of subscribers limited, the cost of mobile telephones became steep. Obviously, if you have a lot of subscribers sharing the cost of the infrastructure, you can make your service affordable. If only a limited number of subscribers can be supported, the cost will be much higher.

The mobile telephone was also an analog system. Interference was a problem in most networks, and there were lots of spots which caused dropouts. This was because the antennas were placed on hill tops in an effort to cover as much of an area as possible. Valleys and areas behind other hills caused problems because additional antennas could not be installed to reach those areas.

Cellular phones solved the problems mentioned above. We will discuss how these problems were solved later. But even cellular service has its share of problems. When the first cellular networks were deployed, they could only cover a limited geographical area. There was no roaming between networks, and the cost was still fairly prohibitive. It took cellular providers nearly 10 years to build their networks to what they are today.

As new networks are deployed, they will not have the same luxuries cellular networks did. New networks using new technologies will be forced to provide the same areas of coverage as cellular networks when they are first deployed, and they will need to provide the same level of services. Otherwise, it will be impossible to compete. After all, would you trade in your cellular telephone for a newer model you knew would only work in your home area?

7.1.2. The Cellular Solution—Architecture and Distribution

There is a limited number of radio channels available for cellular use. These channels can only support a limited number of subscribers. If the coverage area is made smaller, and the frequencies divided into small geographic areas, more subscribers can be supported. The available frequencies can be reused in other cells, as long as the cells are not adjacent to one another. This allows networks to use the existing spectrum, reusing the same frequencies within their network.

This concept is known as frequency reuse. If the available frequency spectrum is divided into blocks of frequencies, those blocks can then be allocated into cells. Each cell would then use a different range of frequencies. If allocated properly, the same frequency would never be assigned to two adjacent cells, and there would be sufficient distance between cells with the same frequencies that they would not interfere with one another.

If the cells covered a small geographic area, the transceivers used would not require as much power. The less power used to transmit, the less interference with neighboring cells. In today's cellular networks, frequencies are divided into seven blocks and are assigned to cells accordingly. Figure 7.1 illustrates this structure.

A cell can cover up to a 10-mile radius in any kind of pattern. Antenna technology allows antenna arrays to be arranged to provide any type of coverage needed for a geographical area. Directional antennas and new smart antennas have enhanced the control given to cell sites. This provides better-quality audio as well as more efficient use of the frequency spectrum.

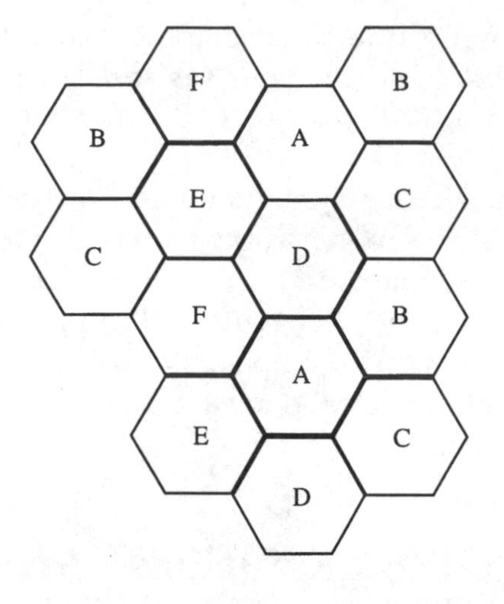

Figure 7.1 Cell Structure and Frequency Reuse

This is the basic architecture of modern-day cellular networks. How can we possibly improve? If the cells could be made even smaller, the transceivers would require less power. If transceivers required less power, cellular phones would require less power as well. Their batteries would last longer and could weigh less, meaning that the phones could be made smaller. The frequency reuse pattern could be repeated many more times, providing support for many more subscribers. This is the concept being used today for new PCS networks.

PCS is not a technology but a level of services. There is not much difference in the operation of a PCS network and a modern day cellular network, apart from the higher frequencies allocated for PCS networks. Both are digital, both use the same types of entities, and both rely on intelligent networking to access databases and communicate with other networks. It's the type of services provided, such as one number access for both your cellular and home phone, voice mail, and pager. There are many other services being offered by PCS providers, which we will discuss later.

7.2. Cellular Network Architecture and Protocols

There are many standards defining wireless communications. Cellular systems, cordless telephones, wireless Private Branch Exchanges (PBXs), and local area networks are all defined under the classification of wireless communications. In this book, we will focus on the cellular market and data services offered through cellular providers, but we need to first understand the standards used in the cellular industry and the concept of "mobility." Cellular networks address three issues:

- Terminal mobility
- Personal mobility
- Service portability

Terminal mobility allows a cellular subscriber to move from one network to another network, or from one cell to another, without losing the ability to send and receive cellular calls. This is often referred to as roaming. In early cellular networks, subscribers had to obtain a roaming number and arrange for calls to be forwarded to that number. In today's networks, roaming is seamless, allowing subscribers to move around without special arrangements with the cellular providers.

The problem with terminal mobility is that it only addresses the mobility of the cellular terminal, and not the subscriber. Personal mobility allows the subscriber to move around without being tied to a particular handset or terminal. This is easiest achieved with wireless PBX, where an extension number is assigned to a user, and a special transponder worn by the user gives the PBX the location. It still requires some sort of device worn by the user to keep the network informed about the user's location. An alternative to the transponder is a Personal Identification Number (PIN) entered by the user at any phone. This tells the PBX where calls for that user should be routed.

Personal mobility is an issue under study and an objective of the newer PCS networks. Service portability is now under investigation. Part of service mobility is maintaining the same services you have at home in other networks. Number portability is part of service mobility and now is mandated by the Federal Communications Commission (FCC).

In July of 1996, the FCC issued an order that all telephone companies in the top 100 MSAs must provide local number portability (LNP) by the year 1997. Cellular providers actually have until 1999 to complete the implementation of this order. Local number portability allows subscribers to keep their telephone numbers even when they change carriers. The first phase of LNP will allow subscribers to keep their numbers if they change telephone companies or service type but only within their rate center. The next phase of LNP is location portability, which will allow subscribers to keep their telephone numbers (including area code) when they move outside their rate center (a rate center is the local calling area subscribers reside in). This is far more complex and is not part of the FCC mandate.

As discussed in Chap. 2, telephone numbers for both wireline and wireless services are assigned to specific areas, determined mostly by geographical location. For example, we all know that the area code 212 is assigned to New York City. The office codes are assigned to specific central offices in that area code. With the new FCC mandate, a user who moves from New York City to North Carolina can keep the 212 telephone number.

This is actually easier for wireless networks than it is for wireline network operators. The wireline network does not use databases to determine the location of a subscriber for incoming calls. With local number portability, database lookups will be required for all calls within an area code/office code (NPA/Nxx) if even one number is "ported."

Already, cellular networks search databases to learn the location of a cellular subscriber. Number portability will require an additional database search before a call can be connected, possibly extending the call setup time. The FCC already has mandates regarding call setup time for 800 numbers, and it is very likely that requirements will be set for all call setups as number portability starts to unravel.

Service portability is achievable through the use of smart cards, intelligent networking, or both. Network databases containing subscriber information can be made accessible to all cellular providers through SS7. These databases are in use today for roaming and location management.

Smart cards are used throughout Europe (as part of the GSM network). These cards, called Subscriber Identity Modules (SIMs), contain subscriber information and are inserted into GSM telephones through a slot on the bottom of the telephone. They are usually about the size of a credit card and can be used in any GSM phone. This is also being used in some PCS networks here in the United States.

The advantage to using this type of system is that users can buy or rent cellular telephones from anyplace, without activation. The phone becomes activated when the card is inserted into the phone, providing all of the information the network needs to perform authentication.

Before we look at the technologies used in the United States and abroad, it would help to understand how a cellular telephone works and how the network receives and originates calls. This next section will talk about the generic operations of a cellular network. Understand that different technologies may provide different procedures. I will point these differences out as we go along.

7.2.1. The U.S. Network

Within the United States, there is only one kind of analog network. This is referred to as the Advanced Mobile Telephone System (AMPS) and is presently deployed throughout North America. Analog systems are quickly being replaced with digital systems.

The analog system is limited as to how many subscribers it can accommodate. Each frequency is capable of handling one call. There is a limit to how many frequencies there can be in an area (or cell) at the same time, which places a limit on how many subscribers can be supported within a cell at one time. This is one of the reasons the analog cellular system is being replaced with digital one.

There are two technologies currently being investigated by cellular network providers looking to convert to digital. TDMA was introduced in 1992 and is deployed in several networks throughout the United States. It is an air interface technique also used in GSM networks in Europe.

CDMA is a more recent solution and is quickly gaining popularity for a number of reasons. CDMA is being touted as being capable of handling more transmissions per frequency, higher data rates, and more secure transmission than TDMA. CDMA also provides additional control of power levels when transmitting, offering extended battery life to cellular telephones and less interference between cells. There are several differences in the way TDMA and CDMA operate, discussed in Sec. 7.2.3.

GSM is a European technology, now being deployed in U.S. networks. GSM development started in 1982, but it was not until 1992 that GSM was first deployed in Germany. GSM is a standard defining the entire network, not just the air interface. It addresses signaling, control of different entities within the network, and network management. GSM is discussed in more detail in Sec. 7.2.6.

Cordless telephone technology works in similar fashion to cellular but on a much smaller scale. A base station must be used within close proximity (creating a "pico" cell) to the handset and provides an air interface. The air interface can be analog or digital, depending on the standard used.

In the rest of the world, there are several choices for cordless telephone technology, most based on Cordless Telephony 2, or CT2. This is a second-generation standard defining the air interface for wireless communications at pedestrian speeds. This qualifies the standard for use in PBX systems as well as cordless telephones.

CT2 originated in the United Kingdom in 1992 and has been adopted by the European Telecommunications Standards Institute (ETSI) as that nation's cordless telephone standard. There are several other standards derived from CT2 that are used in Europe and the rest of the world. These are discussed in the next section. CT2 is also used here in the United States.

There are several new U.S. standards under development that are derived from CT2 and other standards. These are listed below:

- Personal Communications Interface (PCI) (based on CT2)
- Wireless Customer Premises Equipment (WCPE) (based on the Digital European Cordless Telecommunications standard, or DECT)
- Personal Access Communications System (PACS) (based on the Public Handphone System, or PHS, and WACS)
- Wireless Access Communication Service (WACS) (developed by Bellcore)

Some of these cordless telephone standards may become applicable to PCS solutions. For example, one company on the East Coast currently offers a service using both cordless and cellular telephone technology. The user gets a cordless telephone (based on CT2) that works as a portable telephone around the home. However, as soon as the user roams outside of the base station's range, the cellular network picks up the signal.

The handset is a dual-mode handset capable of working in both CT2 and AMPS mode. The service has been fairly popular, although it is currently limited to specific service areas.

For signaling, U.S. networks use the Electronics Industry Association/Telecommunications Industry Association (EIA/TIA) standard IS-41. This uses the services of either X.25 or SS7. IS-41 provides the procedures and protocols to allow Mobile Switching Centers (MSCs) to communicate with the other network entities to provide subscriber information and access to databases. As mentioned earlier, GSM defines the signaling protocols as well, providing one standard for the entire network.

7.2.2. The International Network

International cellular networks are far from compatible with those in North America. The original intentions of the International Telecommunication Union (ITU) was to create a universal network, allowing subscribers to seamlessly roam from network to network, country to country, using their cellular telephones. Unfortunately, given the solutions under development today, and the choices of both U.S. and international cellular providers, it is unlikely we will ever see such a network.

Europeans do enjoy seamless roaming in European countries, but outside of Europe, countries have chosen their own standards. Japan for example uses a standard proprietary to their country. It is not used anywhere else. The Middle East uses a variety of different technologies as well, making seamless roaming from the Middle East to Europe impossible. Of course, the United States has followed its own standards, making U.S. cellular telephones useless overseas.

Much of this difference lies in the frequencies allocated to cellular use. The FCC has no jurisdiction in other countries and does not influence how frequencies are allocated outside of the United States. Different countries have different needs and must follow their own requirements for frequency allocation.

Europe's cellular industry is unregulated, with competition encouraged. There must be at least two cellular providers in each region, which is different than in the United States, where only two service providers are allowed per area. This, too, is under change after the passing of the Telecommunications Act of 1996. Now wireline companies are competing for wireless subscribers, and wireless companies are looking at the local telephone service market.

There are still some analog cellular systems used internationally, although most of the networks have been converted to digital. The international cellular industry has grown much more quickly than the U.S. market because many countries do not have an existing advanced wireline communications infrastructure, forcing them to develop either a wireline network (at considerable cost) or deploy a cellular one. In many regions, it is far more cost effective to provide cellular services than try to maintain wireline infrastructure. The two international analog systems are Nordic Mobile Telephone (NMT) and Total Access Communication Systems (TACS). They are used mostly outside of Europe.

The most widely used digital cellular standard in the global community is GSM. This is actually a European standard, but it has been adopted for use in many other regions of the world, including parts of the United States. GSM provides a fully digital cellular solution.

The air interface used for GSM is TDMA. There are some differences between the TDMA used in the United States and the version used in Europe. Most of these differences lie in the frequencies used and the number of transmissions that are multiplexed onto a signal channel.

GSM is capable of supporting eight transmissions per channel, while U.S. networks support three transmissions per channel. TDMA can support more than this, depending on a number of factors, such as channel spacing. The United States uses 30-kHz channel spacing while Europe uses 200-kHz channel spacing, affecting the amount of bandwidth available.

There are many cordless telephone standards used throughout the world. In Europe, the DECT standard is widely used. This was derived from CT2 and adopted by ETSI as a standard. Japan uses its own standard, PHS, which is not compatible with any other cordless telephone standard.

Before discussing the various cellular solutions, we need to understand how cellular systems work, and how the network elements perform. There are differences in the various technologies, but for the most part, there are more similarities. We will first discuss how cellular networks operate and then look at the differences between the various solutions.

7.2.3. Cellular Operations

To understand cellular technology, you must first understand how a cellular network operates and the entities used within the cellular network. Figure 7.2 is the basic model for a cellular network. Some technologies, such as GSM, add additional entities, providing different features to the network (explained later).

Cellular networks are deployed according to markets. The wireline industry deploys their services according to service areas, or Local Access Transport Areas (LATAs). These were defined by the courts according to

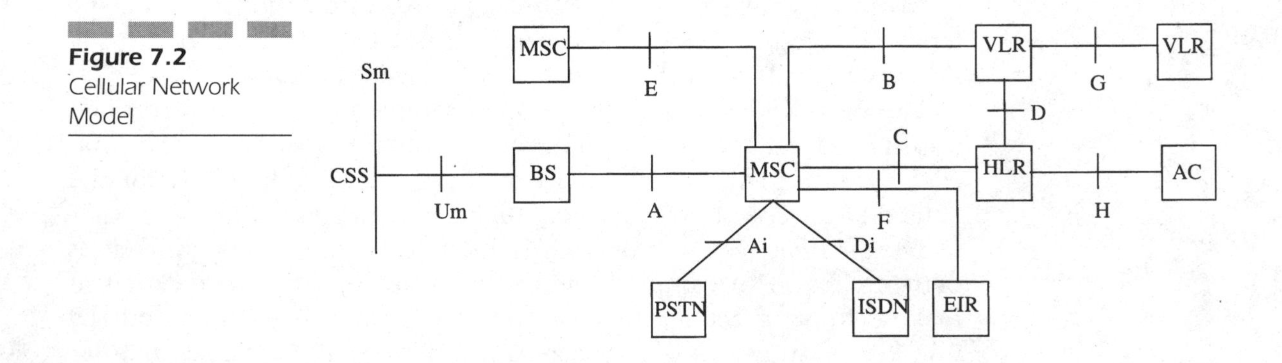

Figure 7.2 Cellular Network Model

demographic data. In the wireless industry, the areas are called Metropolitan Statistical Areas (MSAs) and Rural Statistical Areas (RSAs). There are some 305 MSAs and 482 RSAs as of this writing.

New PCS networks have been divided differently than cellular networks. Metropolitan Trading Areas (MTAs) and Rural Trading Areas (RTAs) are defined for PCS networks and do not cover the same territories as MSAs and RSAs. MTAs and RTAs are discussed in Sec. 7.3, but you should understand that they are two different definitions.

Two carriers are allowed to provide cellular services within each MSA or RSA. Frequencies are allocated to both of the carriers in blocks. The two blocks of frequencies within a market area is labeled as system A or system B. Cellular phones must be able to work in both systems, regardless of the technology used in that network. For example, if system A is an analog AMPS network, and system B is a TDMA IS-54 network, subscribers must be able to use their phones on both systems.

There are 21 setup, or control, channels within each frequency block. They operate in two directions, forward and reverse. The forward channel is used to page the mobile unit, while the reverse channel is used by the mobile unit to request a call setup.

When a cellular phone is powered on, it will search the 21 control channels for the strongest signal. Each control channel is associated with a cell site. When the cellular phone determines which signal is the strongest, it sends a call request to the cell site over the control channel. The cell site will then determine which voice channel is available and send a signal to the cellular telephone identifying the frequency. The cellular telephone can then begin the connection process on the voice frequency.

The heart of the cellular network is the MSC, which manages the routing of calls within the cellular network. It also controls handoffs between cell sites, access to system features, and access to network databases. There is typically an MSC per MSA/RSA. The MSC also coordinates handoffs between cell sites, manages mobile paging (the process used to alert a cellular phone), processes registrations from cellular phones when they are powered up, and provides and/or manages connections to the Public Switched Telephone Network (PSTN).

The MSC provides access to databases as well. All cellular networks have a Home Location Register (HLR). This database provides information about all subscribers in that home area. The HLR is owned by the cellular company providing service in that MSA/RSA. The subscriber's cellular phone is encoded with an equipment serial number, a Mobile

Identification Number (MIN), and a cellular telephone number. This information is stored in the HLR.

If you have owned a cellular telephone, you know you had to choose the area you wanted to be your home area. Any calls made outside of this area use roaming services, which may be provided by the same cellular company or by a competitor. The data regarding your telephone and the services you pay for are kept in the database located in that home area. When you roam, the cellular company servicing your call must access the HLR in your home area to determine who you are and how to handle service to your phone.

All MSA/RSAs have access to a Visitor Location Register (VLR) as well. This is also a location database, but it provides a little more detail than the HLR regarding your location. The HLR only knows the last MSC servicing your telephone. This information is maintained until updated by an VLR. The VLR knows the last cell site to provide you service. Here is how it works:

When a call is directed to your cellular phone, the telephone company handling the call routes the caller to your home MSC. The MSC will then query its HLR to see where you were last located (which MSC you last made a call from). If the HLR says you were in a different network, the MSC forwards the call to that MSC.

When the call arrives at the remote MSC, the MSC will query its HLR. When it does not find your MIN in the database, it looks in the VLR. The VLR will identify whether or not your phone is active in that network, and if so, which cell site is currently servicing your phone. This information is kept up to date; however, the VLR database will remove the record after a predetermined period if there has been no activity on your cellular phone.

While your cellular phone is idle, it periodically sends out a signal to the cell site to let it know the phone is still active. This signal will include the MIN assigned to the phone so that the information can be stored in the VLR for that area. If you move to another network, serviced by a different MSC, your home HLR is updated to reflect the new MSC. It is the responsibility of the VLR to provide this update information.

The VLR is a very dynamic database. Data is not stored forever since this would quickly fill the database beyond capacity. Instead, if a record ages beyond a specified time period without updates, the record is deleted.

The data in the HLR consists of the last VLR to provide registration information regarding the cellular phone. It provides the SS7 point code and subsystem number of the VLR (both addresses used in the SS7 protocol), as well as the subscriber data such as MIN, Equipment Serial Number (ESN), and mobile state (active/inactive).

The cell site consists of two parts, the Base Station Controller (BSC) and the Base Station Transceiver (BTS). The BTS is the radio transceiver and the antennas used at a cell site. The combination of antennas and transceivers is called the Base Station Subsystem (BSS). The antennas are connected to a system that allows the antennas to be switched to different transceivers. This is controlled by the BSC.

Since there are several components in the BSS, we will talk about them one at a time. First, we will talk about the various antenna systems used. Antennas can provide coverage in numerous ways. Omnidirectional antennas provide coverage in a uniform circular pattern.

Directional antennas provide coverage in one specific direction. These antennas are good in areas where signals need to be directed in a specific direction, in cities where buildings prevent a uniform circular pattern, or against hills and mountainsides where omnidirectional patterns would be significantly blocked on one side. Directional antennas can also allow transceivers to operate at lower power levels because the signal is more concentrated. This is only an advantage in systems such as CDMA where power levels can be fluctuated depending on signal strength.

Sectorized antennas are divided into three to six different sections. Each section emits a signal, in varying degrees of coverage. This provides wide coverage and allows signals to be concentrated in each sector rather than spread over an omnidirectional pattern.

Smart antennas use sectorization techniques but further divide the sector into single signal beams. The signal strength of cellular phones is monitored, and the beam receiving the strongest signal is used to service the cellular phone. These systems use switched beam technology, which allows the antenna controller to switch from beam to beam. This provides the function of a directional antenna with the coverage of an omnidirectional antenna.

Another advantage to this type of antenna technology is that it provides stronger signals with less possibility of interference. This is the same as directional antennas which are able to concentrate their signals in one direction, without interfering with surrounding signals.

For small coverage areas (microcells and smaller), a new technology called Remote Antenna Driver (RAD) and Remote Antenna Signal Processing (RASP) is being used. A RAD is placed in the network, on top of street lamps and telephone poles. This small antenna communicates with a RASP over cable television facilities. The RASP is placed in opera-

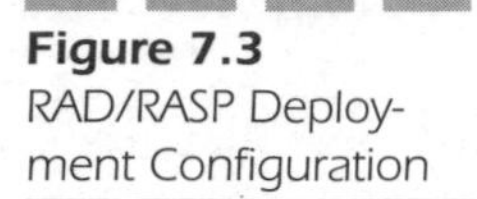

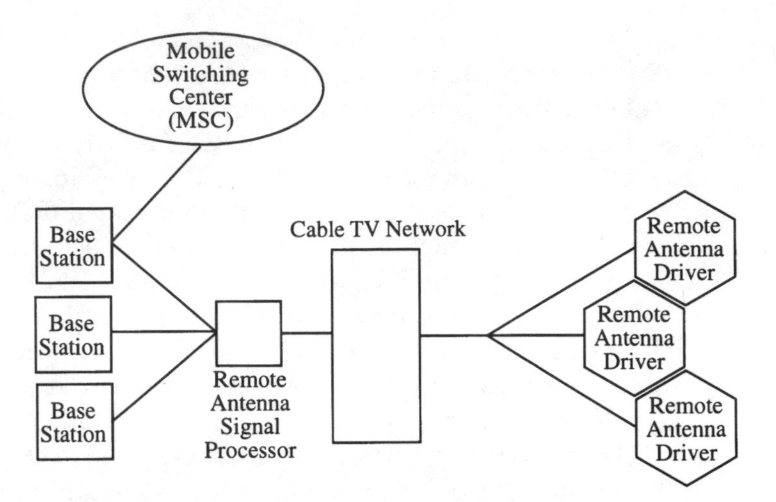

Figure 7.3 RAD/RASP Deployment Configuration

tions centers and translates signals from cellular phones for transmission over the wireless networks (see Fig. 7.3). This is an attractive solution for cable television operators who are looking to move into wireless communications using the existing infrastructure. It can also be used to deliver cable television signals to homes without cable using wireless receivers in the home.

Multisubscriber units can be installed in apartment buildings or campus environments. These units allow standard two- and four-wire equipment to be connected via the cellular networks. All users are then multiplexed over the single air interface (acting as a concentrator). Single subscriber concentrators can be used to attach multiple telephone extensions to one air interface. This is how telephone companies in Mexico are delivering telephone service to residents where no telephone infrastructure exists.

Using cellular networks to deliver residential telephone service has demonstrated savings of 36 percent over wireline services. PCS may use the same techniques to provide wireless services to residential areas.

The cell site antenna system is attached to the base station transceivers. These transceivers are used to transmit and receive calls over the air interface. Each transceiver can support one or more multiple transmissions per frequency depending on the technology used over the air interface. In the United States, the TDMA solution supports three transmissions per frequency, while the same air interface (TDMA) in Europe supports eight transmissions per frequency (GSM).

The BSC is responsible for call processing and handoff procedures under the direction of the MSC. The BSC also provides audio compression and decompression. The handoff procedure is a complex procedure used when a cellular phone moves from one cell to the next. In CDMA networks, a soft handoff is used, where both the old and the new cell site handle the call at the same time. This is explained in more detail in Sec. 7.2.7.

The BSS is connected to the MSC. One MSC controls many BSSs. Remember that there is usually one MSC per MSA/RSA (in most situations). Now that we understand all of the components of a cellular network, we need to understand how they all interconnect and communicate with one another. Figure 7.4 shows the relationships of all of these components and how they are interconnected.

We have already talked about the air interface. This is the interface from the BTS to the cellular phone (or handset or cellular telephone or whatever you want to call it). The air interface can be analog or digital and can use a variety of technologies to support it. We will be looking at the differences between these various solutions in a moment, but for now know that there are differences between them, and they are not

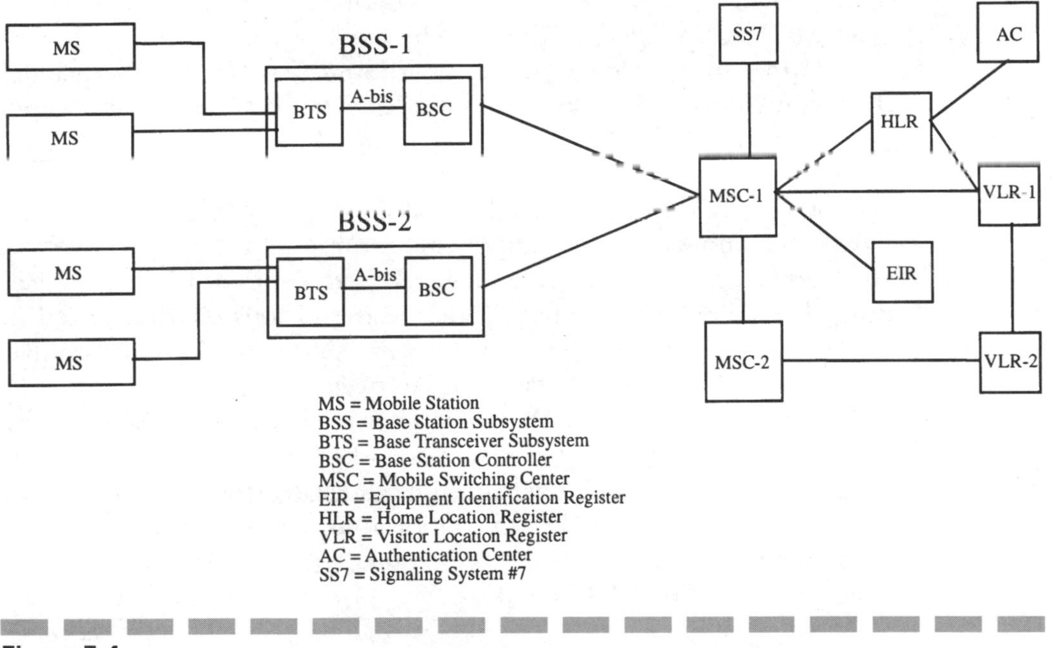

Figure 7.4
Cellular Network Model

always compatible. Cellular phones are usually manufactured as *dual mode*, meaning they can operate over an analog air interface or a digital one.

From the BTS, there is a communications link to the BCS, which is the controller. As we mentioned earlier, the BCS can be collocated with the BTS or it can stand alone in the network, controlling several BTSs. The interface between them will depend on the strategy deployed. In a GSM network, the SS7 network provides services for the application protocol Base Station Subsystem Mobile Application Part (BSSMAP). In U.S. networks, there is probably an X.25 link using SS7 TCAP and IS-41 signaling protocol.

From the BCS, a link to the MSC provides channels for both voice and signaling data. The signaling data is usually run over an X.25 or SS7 link, with IS-41 protocol supporting the cellular signaling application. The voice link may be an ISDN line, as is the case in many GSM networks.

In the United States, the signaling application used is IS-41. This protocol runs on SS7 Transaction Capabilities Application Part (TCAP) and defines the messages used to control handoffs and channel assignment. IS-41 does not define the air interface, and it works independently of the rest of the network. The air interface is defined by other standards such as TDMA or CDMA (depending on which solution is chosen).

In European networks, GSM defines the entire network. This includes the air interface (TDMA) as well as the links between all of the components. This is why many PCS vendors favor GSM over other U.S. solutions. GSM provides a standard for the entire network, not just a portion of it. We will discuss GSM more later on in this chapter.

Now we know how the components communicate with one another, and we have seen an overview (as confusing as it was) of the cellular network. The following sections will define the various standards used in the air interface and the signaling network. Let us first look at a cellular call and the procedures that take place to make that call.

The first and most important step to being able to make and receive cellular calls is registering with the network. When a cellular telephone is powered up, a signal is sent from the cellular telephone to the cellular network. This signal provides registration information, which is stored in the home HLR and VLR and, if the cellular telephone is in another network, in the visiting VLR as well.

The registration is sent to the MSC, which manages the registration of all cellular phones in its network. The MSC examines the MIN to determine whether or not the cellular phone should receive service. The MSC forwards the message to the VLR. The VLR updates an existing record if one exists. If there is no existing record for this MIN (remember the VLR is dynamic), a record is created. The VLR notifies the home HLR and requests a service profile to be used for the new record. The HLR sends the profile after authentication has been completed (a check to make sure there are no flags on record to deny service to the specified MIN).

There is always a chance that the cellular phone is registered in another VLR somewhere. The HLR knows the last serving MSC and can determine if a record exists in another VLR somewhere for the specific MIN. In this case, the home HLR must send a message to that VLR, instructing it to flush the record from its database.

When you dial a cellular telephone number, the office code of that number identifies the MSC which is registered as the "home" MSC for the subscriber. The subscriber may or may not be in that network, and the MSC must determine how to route the call. When the MSC receives the call, it examines the called number and queries its HLR. The HLR will identify the last MSC to serve the cellular phone. If the last MSC was the home MSC, the MSC can query the VLR to determine exactly which cell the cellular phone is now in. If it is registered in another MSC, the home MSC must transfer the call to the serving MSC.

Before the transfer can take place, the home VLR must determine how to route the call to the now serving MSC. The home VLR queries the serving MSC to determine how to connect the call and receives a temporary local directory number (TLDN) from the serving MSC. This TLDN is entered into the VLR, which will then update the HLR for future calls. The home HLR then sends the TLDN to the home MSC, which forwards the call to the TLDN.

The VLR in the visited area identifies which cell is serving the cellular phone and determines whether or not the cellular phone is active or inactive. If it is active, the MSC sends a signal out to the BSC requesting that the cellular phone be paged. The BSC will order the BTS to send a paging signal out to the cellular phone on the control channel for that cell. The paging signal will tell the cellular phone which frequency to use to receive the call on.

When the cellular phone receives the paging signal, it switches to the proper frequency and sends confirmation to the BTS, which in turn sends confirmation to the BSC. The call can now be routed from the MSC all the way through to the cellular phone. All of this does take time, longer than wireline services would. You will notice a delay in the call setup when you dial a cellular telephone number, especially if the person you are calling is out of his or her home area.

7.2.4. Time Division Multiple Access

TDMA first became available in 1992. TDMA is a digital air interface technology which allows cellular network operators to multiplex multiple transmissions over one radio frequency. This provides them with increased subscriber support using the available frequency spectrum. Estimates of deployment show some 2.5 million subscribers in 22 different countries currently use TDMA-compatible cellular phones.

In the United States, TDMA-compatible cellular phones must also be AMPS compatible so that the same cellular phone can be used in both AMPS and TDMA networks. This is known as dual mode (analog and digital). These phones are capable of detecting whether or not they are in an analog network, in which case they transmit analog signals, or a digital network, in which case they transmit in digital mode.

TDMA currently supports three digital transmissions over one frequency. This is a big boost to existing networks in dense subscriber areas as they struggle to support all of their subscribers over the already limited frequency spectrum. The TDMA standard defines support for up to 10 transmissions per frequency, but this has not yet been proven in any network.

In GSM networks, TDMA is divided into eight time slots rather than three. This is why GSM can support more subscribers per channel than U.S. networks. The difference lies in the channel spacing. In U.S. networks, AMPS networks use 30-kHz channel spacing, while in GSM networks, 200-kHz channel spacing is used. In the United States, it is important to follow the 30-kHz channel spacing for interworking between digital TDMA and analog AMPS networks.

When a cellular phone requests service from a cell site, the cell site will identify the frequency the phone should be transmitting on (remember our earlier discussion about cellular operations) as well as the time slot to be used. This time slot is then assigned to the cellular phone and is not used by any other cellular telephone in the cell. The phone does not transmit anything until its time slot becomes available (usually every 4.615 ms). The phone is idle the rest of the time. This is known as bursty transmission. There is a big difference in frequency efficiency between TDMA and CDMA, discussed later in this chapter.

The standard which defines TDMA was developed by the EIA/TIA. Known as IS-54, this standard defines the air interface procedures. This includes the method used for accessing a cell site transceiver and digitizing the voice transmission. A newer standard, IS-136, is compatible with IS-54 and provides a number of enhancements over IS-54 for PCS networks.

The new IS-136 standard supports macrocellular service for large geographical coverage with low subscriber density. It also provides procedures for in-building networks, including "fixed wireless" networks. Fixed wireless networks provide cellular network support using fixed telephones and data terminals instead of portable handsets. These are popular where cabling is difficult or impossible.

IS-136 also supports seamless interworking with existing AMPS networks. This is critical to the success of PCS networks since they operate at different frequencies than AMPS networks. Cellular phones used for PCS networks operate in the 2-Ghz range, while AMPS networks operate in the 900-MHz range. This means dual-mode phones must be able to operate in both frequency ranges to support roaming between AMPS and PCS networks.

Some new features supported in IS-136 not found in IS-54 include a message waiting indicator in the handset, calling number identification, alphanumeric paging capability (short messaging service), and sleep mode. The sleep mode automatically shuts down the phone when idle but periodically reactivates to check for messages.

It should also be mentioned that TDMA is used in GSM networks as well. If TDMA is so widely used, why do not all network providers select TDMA as their air interface? One of the fundamental problems in debate today is cellular telephone interference. Cellular telephones are notorious for interfering with aircraft navigational equipment and hearing aids, and while this is yet to be substantiated, there are reports of TDMA interfering with pacemakers.

The problem lies in the electronics required to support time division multiplexing. Transmissions must be multiplexed into time slots. This requires a timing system. Electronic clocks are run by crystals, which oscillate at specific frequencies. This oscillation emits radio frequency interference (RFI), which can interfere with the operation of other electronic equipment. To understand the effects of RFI, try placing your AM receiver antenna next to a fluorescent light bulb.

In CDMA phones, oscillators are not needed (unless they are working in dual mode, in which case an oscillator is needed). This means that CDMA cellular phones are less likely to interfere with other electronic equipment. There are a number of studies under way to determine exactly what effects there are from operating cellular telephones. None of the reports have been validated, and while there is a lot of concern over the impact on health and safety, there is no proof that cellular telephones present any health hazards, regardless of the technology used.

7.2.5. Coded Division Multiple Access

The EIA/TIA adopted the CDMA IS-95 standard from Qualcomm, a company which manufactures CDMA cellular equipment. There are many differences between CDMA and TDMA. The biggest one is in the use of the frequency spectrum.

In CDMA, several transmissions are sent over the same frequency without multiplexing. Instead, a unique digital code is added to the digitized speech for each transmission. Handsets receive all of the transmissions being sent over one channel and use microprocessors to decode each transmission and find the correct code. All other transmissions are then ignored.

While this function requires additional processing power in the cellular telephones, the cost of microprocessors has decreased, making CDMA affordable as a cellular solution. The technique is not actually new; it has been used for military radio transmission for years. The method of spreading transmissions over the entire frequency spectrum is known as spread spectrum technology.

There are a number of advantages to CDMA. The IS-95 standard defines procedures for a complex power control method, designed to save on battery life and help prevent cochannel interference. As a CDMA phone is transmitting, the receiving cell site is constantly measuring the signal strength of the transmission. When the signal weakens, the transceiver in the cell site can send a power control message to the phone, instructing it to increase its power. If the signal increases, the same signal instructs the phone to decrease its power.

Another advantage of CDMA is the absence of crystals to support time division multiplexing. As we mentioned earlier in our discussion about TDMA, crystals oscillate, creating potential RFI problems for other electronic equipment. Initial reports have shown little or no such interference from CDMA cellular telephones.

The handoff procedure used in CDMA is also different from other technologies. The TDMA air interface uses the standard hard handoff discussed in Sec. 7.2.3. CDMA uses the hard handoff, and it also was a soft one.

When a cellular phone crosses cell boundaries, rather than using the hard handoff, the original cell continues to provide service to the phone. The new cell is also activated, and the phone operates over both cell sites until reaching enough signal strength that the new cell can take over. This is a complex technique, requiring both cell sites to be controlled by the MSC throughout the handoff.

The MSC is responsible for processing the transmission received from both cell sites and determining which cell site is in control at any one time. While complex, the technique ensures reliable transfer of service during a call. There is no noticeable degradation in the transmission quality during the handoff (in TDMA and AMPS networks, static and volume drops are common during handoffs). This is especially critical for data transmission.

CDMA also provides support for data services. In today's CDMA networks, data transmission rates of 9.6 up to 14.4 kbps are supported. Future enhancements promise data rates of 64 and 500 kbps. Packet switching is also supported over the CDMA interface, without specialized packet-switching modems. CDPD integration is supported as well.

Future enhancements to CDMA include video and high-speed data. This will be important to PCS network providers that want to offer enhanced features and services such as wireless videoconferencing.

Despite all of its features, CDMA hasn't been without its share of problems. Early deployment of CDMA experienced many problems when operating in areas with AMPS networks. An AMPS handset operates at 1 watt (W) and a fixed mobile station (installed in vehicles) can operate at up to 3 W. A CDMA telephone operates at 10 milliwatts (mW). AMPS units would interfere with and overpower the CDMA phones. This problem has since been rectified but has marred the reputation of CDMA nonetheless.

CDMA continues to be a debatable solution. Many companies are investing heavily in CDMA equipment, but there are just as many investing in TDMA. The next few years will determine who the winner will be. If CDMA is able to deliver what is promised, CDMA will clearly be the solution of choice for many wireless providers. If CDMA cannot deliver what it promises, the result will be financial disaster for those who invested in it.

7.2.6. Global System for Mobile Communications

As mentioned earlier, GSM development began in 1982 by the Conference Europeenne des Pasteet Telecommunications (CEPT). It was not until 1992 that the first GSM network was put into place in Germany. There were many influences in the development of GSM networks: the European economy, political pressure to create an international network, and the absence of services to support Europe's changing workforce.

GSM did not start out as a digital solution. ISDN was already being deployed in Europe and had seen tremendous success. The Post, Telephone, and Telegraph (PTTs) administrations of Europe had realized the many advantages of using ISDN over analog facilities and were very interested in expanding ISDN into their wireless networks. This was the principal reason behind making GSM a digital network.

ISDN is used for interconnecting MSCs, for both voice and data. Signaling in GSM networks uses another European technology, SS7. SS7 development had begun in the mid 1960s, and it was first deployed in Europe. Since GSM was all digital and SS7 was already serving the wireline network, it was a natural choice to use SS7 in the GSM networks as well.

ISDN and SS7 were the only existing technologies used as part of the infrastructure. All of the cellular-specific applications were developed specifically for GSM, including the air interface. TDMA is used as the air interface in GSM networks, which was adopted by the EIA/TIA for use in North America as well.

GSM provides a number of features now found in North American networks as well. Some other features of GSM are just now beginning to find their way into U.S. networks. Caller ID, call waiting, call hold, conference calling, and data rates of 2400 and 9600 bytes per second (bps) are supported by GSM.

There are several types of cellular entities used in GSM that are somewhat different from those in the United States. Terminal equipment (cellular telephones and modems) can be either fixed, portable, or handheld.

A fixed mobile station is one which is permanently mounted into a vehicle. They can operate at up to 20 W, which is more powerful then their U.S. counterparts. The increased wattage provides more coverage but would introduce problems with interference if this were attempted in U.S. networks.

Portable stations are typically bag phones, which are larger than the units which fit in your pocket, yet small enough to still carry around. They are referred to as bag phones because they are installed in their own carrying satchels and operate at up to 8 W.

Handheld stations are much like the popular cellular phones used in the United States. They are very small and can be carried in a pocket. These stations operate at up to 2 W. While they do not provide the power of portables or fixed stations, they are popular because of their very small size.

Another unique feature of GSM phones is the SIM. This is a credit card-like device which plugs into the GSM telephone and provides subscriber information to the network. Rather than program this information directly into the telephone (as is done in the United States), subscriber information is programmed into the SIM card, which can be used in any GSM phone. This feature allows GSM subscribers to rent phones and immediately activate them by simply inserting their personal SIM cards.

New applications for these SIMs are now being explored. European GSM providers are expected to deploy many new services, including services where credit is programmed into SIMs for airtime. When the airtime is used up, the card is no longer valid.

The components in the GSM network are similar to those found in U.S. networks. Security in GSM networks is a bit better than in U.S. cellular networks because of several added functions such as an authentication center, which is a database which checks the transmission for an encoded value to determine if the phone is valid or not. If the phone has been reported stolen, or if there are a number of suspicious transmissions (such as those covering a large geographical area), the transmission is either denied or flagged as suspicious for further monitoring.

The VLR also assigns a temporary phone number for the cellular phone when the subscriber changes networks (by roaming into a new area served by a different MSC) as a security feature.

GSM phones can also scramble transmissions, making it difficult to monitor GSM calls. This is an inherent security feature of GSM which many U.S. cellular providers favor. However, because GSM uses TDMA at the air interface, there are many limitations as well. It is yet unclear whether or not GSM will enjoy widespread deployment in the United States in new PCS networks.

7.2.7. Packet Switching over Cellular

Data transmission over cellular networks is a lot like using a modem on a wireline service. Cellular modems convert the data stream into a series of audible tones and transmit the tones over the cellular air interface. The air interface treats the transmission the same as voice. Naturally, there are many limitations to this approach, including speed and cost.

Cellular Digital Packet Data (CDPD) provides a cost-effective solution for data transmission over cellular networks. Using the Internet Protocol (IP), CDPD provides a very cost-effective way to transmit small amounts of bursty-type data through the cellular network. The cellular network does not treat CDPD traffic as voice and provides special handling for the data.

The most distinct difference seen in CDPD is that the data traffic is sent over idle voice channels. When a call setup is requested from a CDPD modem to a cell site, it is immediately identified as a CDPD-type call. This means the transceiver must have CDPD software to know how to handle the call.

The transceiver then finds an idle voice channel for the call and sends instructions back to the CDPD modem as to which frequency to use. The modem can then begin transmitting. This is just a quick overview of CDPD. Its deployment has been somewhat slow, but many networks have begun offering CDPD services.

Billing for CDPD is different from the model used for voice over cellular networks. Instead of billing by the minute (which becomes quite expensive in the case of data), CDPD providers bill by the amount of data sent. The billing model is cost effective for bursty-type data but is still cost prohibitive for large data transfers.

Typical applications for CDPD include telemetry, e-mail access, point of sale (cash registers in remote areas and credit card transactions), and dispatch operations. Major airlines are using CDPD for ticket counters in remote locations, and police departments have installed CDPD modems in patrol cars for accessing criminal records.

7.3. Personal Communications Services

PCS is really not a new technology but a new feature set for cellular services. There are many dramatic changes introduced into the network to support PCS services. First of all, the frequency range is higher than what is presently allocated for cellular in the United States. PCS networks operate in the microwave band, which is 2 gigahertz (GHz).

Many new service providers have jumped into the pool hoping to reap the profits that cellular operators have enjoyed over the last decade. These new players face many obstacles before they can begin to realize these profits.

The FCC held several auctions in 1995 and 1996, allowing new service providers to bid for market areas, instead of simply allocating the frequencies for cellular usage. Many of these service providers have paid millions of dollars for their market areas, without even having a network in place. They now face millions of dollars worth of investments to build new digital networks to support the PCS services they hope to provide.

Making things even more precarious for PCS operators is the fact that cellular providers have already saturated these same market areas, offering free calls on weekends and free cellular phones (with extended contracts). To compete against the cellular network providers, PCS providers will need to first build their networks and then be able to offer the same coverage with extended features.

Cellular operators have been able to build their networks slowly since there was no other service like theirs. They spent 10 years building their coverage and negotiating access agreements with other cellular providers so that their customers could use their phones anywhere. PCS operators will not have the opportunity to spend this much time building their networks. PCS networks must be able to compete with cellular on the first day the network is turned up.

7.3.1. New Network and New Services

The technologies for PCS are all digital. TDMA, CDMA, and GSM are all likely candidates for PCS networks, as are IS-54, IS-136, and the other standards already used in cellular networks. The difference between PCS and cellular networks will be in the services offered.

For example, paging services will allow you to use your PCS phone as a pager, with one unique difference. When you receive a page, you can either receive an alphanumeric message, or you can elect to connect to the party paging you. The caller will be placed on hold, while you decide whether to answer the page by connecting or to ignore the caller and receive the alphanumeric page instead.

For the sake of brevity, we will not go into the many services being promised by PCS providers. You can see these offerings today in the many advertisements already being printed. The objective of PCS is to fulfill the requirement for personal mobility.

Some PCS providers are looking to use SIMs, like those used in GSM, to deliver many of their services. Others will rely solely on databases used in Intelligent Networking (IN) to store subscriber data and activate features. Whatever technologies are used in PCS networks, there will be many changes taking place in a very short time frame.

7.3.2. GSM—To Be or Not to Be

There is much debate about the success of GSM in U.S. networks. Several PCS providers are already vowing to use GSM, yet these same providers are casting a wary eye towards CDMA.

CDMA is not yet a proven technology. To date, CDMA networks are still in trial and have not delivered the capacity promised by its developers. Many are skeptical about the ability of CDMA to deliver what has been promised. Qualcomm is the principal developer of this new air interface, and many analysts question whether or not PCS providers should put all their investment into CDMA based on Qualcomm promises.

If Qualcomm is able to prove CDMA in live networks, their success will be rewarding for all who invested heavily in CDMA equipment. If Qualcomm fails, these PCS providers will be in big trouble. Obviously, CDMA equipment cannot be used with TDMA equipment, which means if CDMA fails, the networks relying on CDMA equipment will have to be rebuilt.

7.4. Specialized Wireless Solutions

There are many applications for wireless networks. Cellular service is a complex subject, addressed in a number of books. I have left out a lot of details about cellular networks for the sake of brevity, but I highly recommend continued studies in this area. Wireless is the fastest-growing industry in telecommunications and definitely the future choice for many applications. Below is a summary of many of the applications for which wireless is a good solution.

7.4.1. One Number Service

Imagine having only one telephone number. People wanting to call you would not have to know your home number, work number, pager number, fax number, e-mail address, 800 number, voice mail number, and whatever other number you may have to reach out and touch someone.

One number service means that you have one telephone number, and no matter where you are, you can always be reached through this one number. The same number would provide you access to facsimile, e-mail, voice mail, and virtually everything else you use to communicate with. Databases would keep track of your whereabouts and automatically forward calls to the appropriate location, which could be your cell phone, office phone, home phone, or voice mail.

7.4.2. Data Access

Many already enjoy accessing databases and e-mail through their laptop computers from remote locations. Cellular modems are common throughout the United States but still operate through the switch network. Circuit-switched networks can only support data rates of 9600 bps.

New packet-switched services, such as CDPD, are appearing in many networks. This allows cellular subscribers to access data at higher data rates than conventional cellular modems. The data rates are increasing as newer digital technologies are deployed over the air interface (such as CDMA).

7.4.3. Alarm Services

Alarm companies find wireless an excellent tool for monitoring locations. The problem with security systems attached to phone lines is that the phone line can be cut, disconnecting security companies from their equipment. The telephone lines are often configured in such a way that the security company has no indication that the line has been cut.

With wireless systems, the line cannot be cut. The connection is a cellular transmitter located in a secure location. Wireless alarm systems are already common in many large installations, allowing alarm monitoring companies to actually connect an audio channel when an alarm has been tripped, listening in on the burglars who usually have no indication the alarm has been activated.

7.4.4. Telemetering

Utility companies spend millions each year in salaries for personnel who must read meters at residences and commercial buildings. Their meters can be attached to cellular units and polled at any time. Utilities would no longer need to dispatch meter readers to collect billing information. They could poll meters and receive usage data in real time.

7.5. Chapter Test

1. How many transmissions are supported on one channel in U.S. TDMA (IS-54)?
2. Cellular networks in the United States use a seven-cell pattern. Are frequencies duplicated within the seven-cell pattern or are they different in each of the seven cells? (Hint: frequency reuse.)
3. Name one security feature used in GSM to prevent cloning of telephones.
4. Which database in the cellular network identifies the cell serving a cellular number?
5. When is the HLR updated with location information?
6. What two functions are found in the Base Station Subsystem (BSS)?
7. How many setup channels are there in U.S. networks (within one cell)?
8. PCS networks will use new technologies never before used in cellular networks?
   a. True
   b. False
9. In GSM networks, the subscriber places a credit card-like device into the telephone to provide information such as their subscriber identity to the network. What is the name of this card?
10. In CDMA networks, the handoff procedure is different. What is the handoff procedure in CDMA networks called?

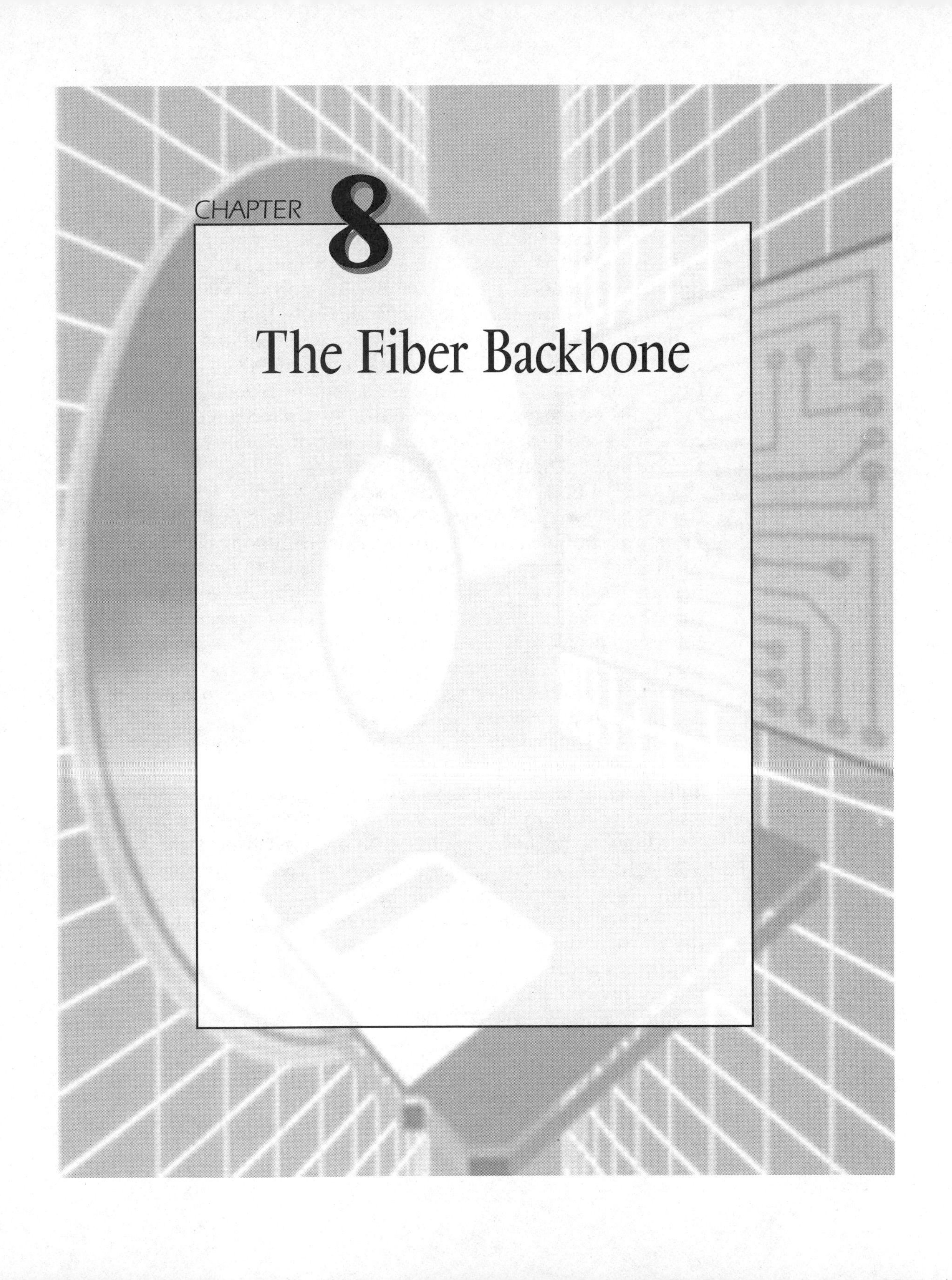

CHAPTER 8

The Fiber Backbone

8.1. From Copper to Fiber

Synchronous Optical Network (SONET) provides bandwidth in the gigabits, far beyond the capacity of the digital hierarchy. In fact, a DS1's capacity is 1.544 Mbps, based on 64-kbps channels. An OC1 is the lowest signal in the optical hierarchy, and it supports 51.840 Mbps of bandwidth. An OC48 supports 2.488 Gbps, the equivalent of 1344 DS1s.

Before SONET was developed, fiber equipment and their interfaces were highly propriety. This prevented companies from connecting their fiber equipment to other vendor's equipment. SONET changes this by providing a standardized interface that all vendors adhere to. Telephone companies, cable television operators, and power utility companies have already invested heavily in SONET networks.

SONET was developed by the Exchange Carriers Standards Association (ECSA) for the American National Standards Institute (ANSI). The European equivalent to SONET is the Synchronous Digital Hierarchy (SDH). There are subtle differences between the two. SONET got its beginnings around 1984, when the ECSA was asked by the Interexchange Carriers Compatibility Forum (ICCF) to create a standard for connecting fiber optic transmission equipment from different vendors together. This was important because many new carriers were trying to connect to existing fiber networks but were unable to interwork their equipment because of the lack of standards.

Bellcore extended the concept into standardization of all fiber optic network elements, including multiplexers and Digital Cross Connects. Today's standards cover all aspects of fiber optics in both public and private networks. Many universities are using SONET as their campus backbone (such as the University of North Carolina, Chapel Hill).

In Chap. 2, we discussed the business reasons for migrating away from copper to fiber optic facilities. In large part it is because of economics that in the United States in most cases only the backbone, not the entire network, has been converted. However, as time evolves, and facilities reach their maximum capacity, they too will be replaced by fiber optics.

The European countries have an advantage over their U.S. counterparts. In many parts of Europe the infrastructure is sparse. There is not as much investment at risk, allowing the European Postal, Telephone, and Telegraph (PTT) operators to build an entire fiber optics network at low risk. This is one of the reasons the Europeans seem to have a more advanced telephone network.

If fiber optics is so wonderful, why do telephone companies not just make the investment, taking advantage of all of the features fiber optics would provide? Decisions like this one are not made quite so easily. Investments in the public network are based on the cost per subscriber and the length of time it will take to recoup that cost balanced against the cost of the existing infrastructure. Telephone companies must scrutinize costs in their network closely, since margins fall with stiff competition. This is especially true in local loop service.

With the new telecommunication laws in effect, telephone companies will need larger margins to survive. This will be difficult with fierce competition. Making it even more difficult will be companies competing for the local loop who already have their own fiber optics network in place. This is the case with many cable television operators and electric utility companies who are looking to provide local telephone service.

These companies do not face reinvestment in their networks for many years because they are already using state-of-the-art equipment and fiber optics throughout their network (not just the backbone). This does not mean the local telephone companies will be put out of business by these new service providers. While the electric and cable television companies have the fiber optics backbone, they do not have the customer service departments, billing systems, switching architecture, and alliances that the telephone companies established many years ago. This alone will keep the local telephone company in business for decades.

The move to fiber optics is still important, and many companies have already begun the migration.

8.1.1. Existing Digital Transmission Overview

Looking back at the existing digital hierarchy, you can see some distinct disadvantages. One of the major disadvantages is the arrangement of the transmission frame. The digital hierarchy in North America uses a basic building block of 64 kbps. This is referred to as a digital signal 0 (DS0).

The next level from the DS0 is the DS1. The DS1 comprises 24 DS0s. These are arranged into a bit stream of 24 contiguous channels, followed by a framing bit (refer to Chap. 2 and our discussion on digital transmission). The DS1 level is typical between subscribers (large corporate subscribers usually) and the local exchange.

Between the local exchange and say, a tandem office, there is most likely a DS3. The DS3 is equivalent to 28 DS1s. Remember that our building block to the DS1 was the DS0. The DS1 was formed by 24 contiguous DS0 channels, followed by a framing bit. Now, the DS1 is placed into a part of a DS3 bit stream. However, there must be framing bits within the DS3 as well so that the individual DS1s can be extracted properly.

The transmission equipment used in the digital network work on these framing bits to determine where the frames begin and end and on framing bits within the DS1s to determine where the payload is located (individual DS0s). This means that if you were to connect a protocol analyzer to the middle of a link for a DS3, your protocol analyzer would have to first demultiplex the DS3 down to individual DS1s and extract the framing bits to reach the individual DS0s within the frames.

SONET uses a different approach. Rather than use a framing bit that is position sensitive (it must always be located in the same position of the bit stream), SONET uses a pointer. The pointer identifies where the payload is located within the SONET frame. Another unique feature of SONET is its ability to deal with jitter and wander. These are two characteristics of any digital communications.

In digital transmission, whenever timing slips, the bit stream becomes out of synch with the rest of the network. This means that a receiving node can no longer decipher the bits because they are being received during the wrong time intervals. The framing bits can no longer be located properly. Think of a poor video, where you see someone's lips moving, but the voice is delayed just enough to be annoying. The sound is out of synch with the picture.

In SONET, the pointer is changed when the payload shifts location within the SONET frame. If the timing gets out of synch, and the bits begin to wander or jitter, the payload may actually shift position within the frame. The pointer takes this into account and can actually change (on a node by node basis) to reflect the new position of the payload. This makes SONET more tolerable of these types of digital phenomena.

Another factor in digital transmission is the equipment required to support the digital signal. Copper wire is capable of maintaining a digital signal for a limited distance. Keep in mind that digital transmission over copper is actually an electrical signal. The thickness of the wire determines the resistance, as does the distance of the wire. The object is to provide as little power as possible (in the form of amperage) and space repeater equipment as far apart as possible.

As luck would have it, underground cables are spliced in access holes about every mile. If you look around any major metropolitan city, you will find these holes everywhere. They are actually cable vaults, used to house repeater equipment, splices to other cables, and any other transmission equipment. The repeater is actually a regenerator (*repeater* is a term used in the analog world, where an analog signal is actually repeated as received). A regenerator receives a digital signal, cleans it up, and regenerates the original wave form.

Fiber optics can transmit much further than copper, without regenerators every mile. Multiplexers and other network transmission equipment are not necessary with fiber optics, making it a more economical choice. However, remember the investment already in place, and keep in mind that accountants do not like replacing new equipment that has not been fully depreciated. When the time comes, fiber optics can be placed in very small conduits (much smaller than existing copper), can be bent much more easily than copper cables, and is not susceptible to noise and external interference.

Another difficult problem with digital transmission is clocking. Clock signals are vital to successful digital transmission. In asynchronous networks, the clock signal is inserted into the bit stream and used by the distant end to synchronize its clock. In other words, when a DS1 is sent to a distant switch, some of the bits inserted are used for clock synchronization. When the distant node receives the DS1, it looks for the clock bits and uses them to make sure its clock is ticking in unison.

This is an oversimplified description of clock synchronization, but it is important to understand that a DS0 signal is no good without a synchronized clock source. Both the source and the origination must be in synch, or the transmission will fail. The telephone companies have deployed clock networks that actually use the Global Positioning Satellite (GPS) or LORAN-C systems for synchronization.

8.1.2. SONET—The Solution

SONET is the first international transmission standard. Unlike the digital signaling hierarchy used today, SONET multiplexing is the same worldwide. Equipment vendors are no longer faced with various "country flavors" when it comes to transmission facility interfaces.

This is a huge relief to service providers that must interconnect to international networks. Today, equipment vendors must provide equipment that is capable of mapping DS1 traffic to E1 interfaces. This presents a challenge because there is no one-for-one correlation between the two. E1 supports 32 channels, while DS1 only supports 24. There are many other issues related to these types of facilities.

Defining an international standard for fiber optics has led to the quick deployment of SONET/SDH. All companies can interconnect their networks without looking to vendors to provide "special" interworking solutions. Vendors can concentrate on developing a good SONET/SDH interface without worry about investing research and development money for small countries.

SONET is already appearing in networks everywhere. Cable television companies began using SONET for their transmission backbones, feeding video signals through fiber optics to hubs, which then fed the video signals to hundreds of homes via coaxial cable. Electrical utility companies have been using fiber optics to send data from one switching center to the next. They started leasing fiber out to other companies to use for their data networks and will soon be sending voice over the fiber.

Fiber was an important choice for power utilities. Copper works like an antenna, and if placed in areas such as high-tension power lines, would be unusable for data transmission. Fiber can be placed right next to high-power transmission lines, without any interference. There is no electrical signal in the fiber, only light.

Another advantage of SONET is the fact that it is synchronous. This means that it requires an external clock source rather than an internal clock bit in the actual transmission stream. A clock synchronization network is mandatory for any SONET facilities because both the originator and the destination must be in synch with one another.

This is accomplished by every telephone office building an internal clock network. The clock generators are highly accurate clock cards (usually rubidium oscillators) capable of maintaining highly accurate clock signal (not your typical Timex). These clocks are then connected to a master clock signal. The master clock signal for all of the telephone offices must be the same. This is achieved by receiving a clock signal from GPS or LORAN systems.

The clock systems used in these networks are capable of receiving multiple clock signals from various satellites, determining which clock signal is the most accurate, and maintaining clock synchronization should the main source be lost. Clock systems are discussed in Chap. 2 as well.

One of the biggest advantages to SONET is the availability to get to individual bit streams anywhere along the transmission path. This is not true of asynchronous digital transmission, where framing bits are embedded at various multiplexing levels, requiring the entire bit stream to be demultiplexed down to the lowest level (DS0) before the individual signals can be accessed. In SONET, a pointer can be used to identify where in the frame the payload is located.

One option for companies who want to maintain their own private network is to lease "dark fiber" from the telephone companies. This is analogous with leasing "dry" copper pairs to interconnect offices. Dark fiber uses customer-provided equipment at both ends. The telephone company provides only the fiber optics cable. Subscribers can then provide whatever equipment is needed and maintain the fiber in their own private network. Many of the power utilities lease dark fiber to major corporations and even to other telephone companies.

8.2. SONET Overview

SONET can be deployed as either a bus topology (point-to-point) or in a ring configuration. While there are many advantages to the ring topology, it is more difficult to manage from the protocol's perspective. This part of the standards is still improving as more and more companies deploy self-healing rings.

The ring provides an additional level of reliability to any network, providing alternate routing in the event of failure. If a failure occurs, SONET nodes reverse the direction of traffic using the other ring. Figure 8.1 illustrates how these loops can occur within any node, reversing traffic over opposite rings.

8.2.1. SONET Network Nodes

There are just a few devices used in the SONET network compared to many different devices needed to support the digital signal hierarchy. In conventional digital transmission facilities, there are cross connects, channel banks, multiplexers, repeaters, and many other devices used in the outside plant. With SONET, the equipment list is reduced, representing reduced operating cost to network operators.

The Add-Drop Multiplexer (ADM) can sit anywhere in the SONET network. It has the ability to drop any Synchronous Transport Signal (STS) frame from within an optical carrier and route the STS frame onto another optical carrier or out to a non-SONET device. The multiplexer may sit in a central office where circuits need to be "plugged into" a SONET facility (which is where the "add" function fits). Figure 8.2 shows the basic function of an ADM.

A SONET terminal can receive signals from a variety of digital facilities

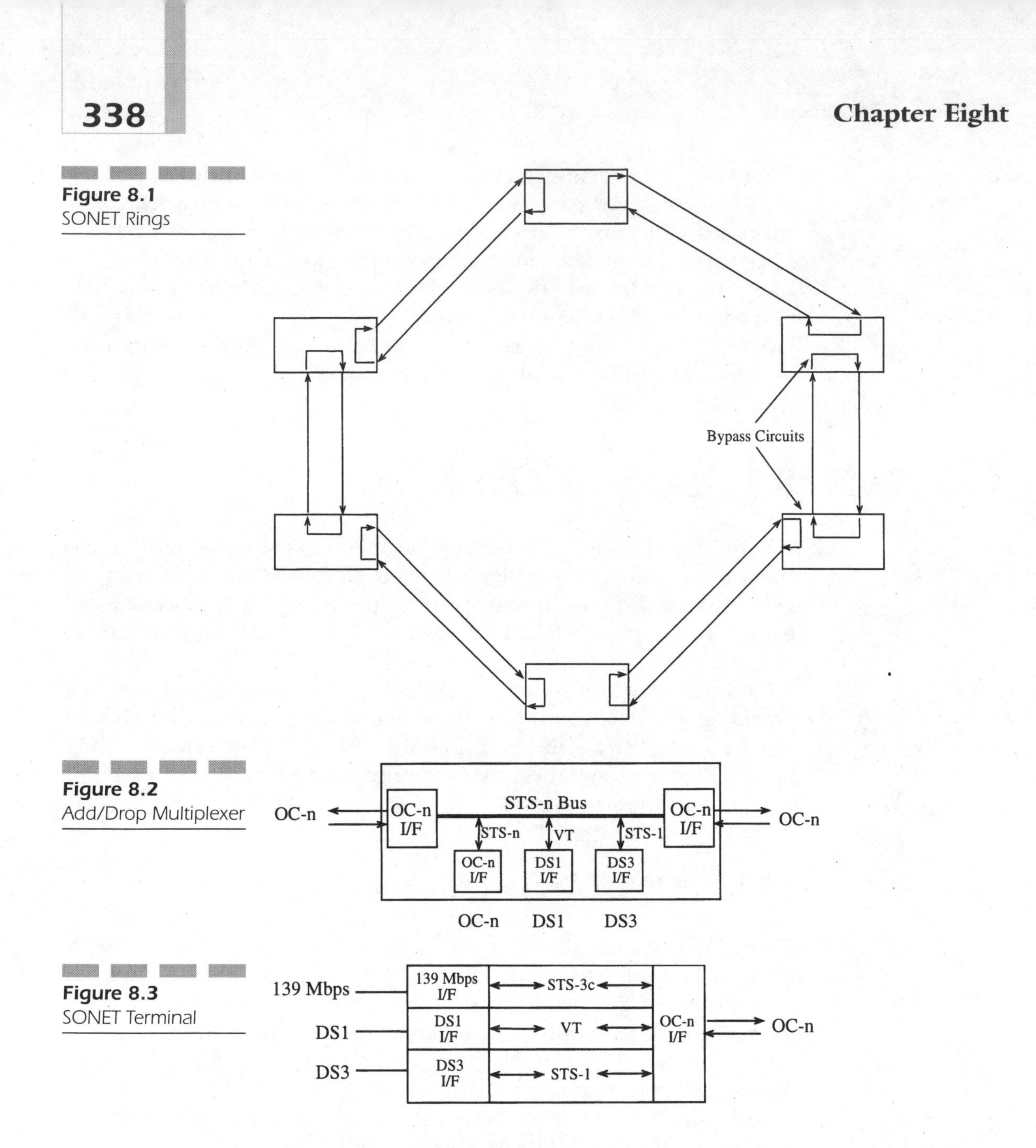

Figure 8.1
SONET Rings

Figure 8.2
Add/Drop Multiplexer

Figure 8.3
SONET Terminal

and output a SONET optical carrier (see Fig. 8.3). It cannot pass traffic straight through or add or drop traffic from the optical carrier. The terminal is usually at either end of a SONET point-to-point configuration, while the ADM is typically placed as an intermediate device (between terminals).

Digital cross connects are replacements to the old patch panels, where technicians used patch cords to manually connect one facility to another. The digital cross connect provides the same overall function, but there are no patch cords. A terminal is used to access software within the digital cross connect and configure which facilities are to be patched to which facilities.

The digital cross connect is a real cost saver for all network operators because it simplifies and speeds the task of placing facilities into service and mapping them over the existing cable plant.

Synchronization in SONET is achieved through a hierarchical clock distribution network. The various levels in the clock distribution network are referred to as stratums, with Stratum 1 being the most accurate signal available. There are five stratum levels.

As signals are passed from one node to another within the exchange, the signal starts to degrade. As the signal becomes less accurate, it drops to a new stratum level. You could not use a lower stratum signal to feed a device requiring a higer stratum signal (for example, you could not use Stratum 3 to drive a device requiring Stratum 2).

Stratum 1 timing signals are usually generated by cesium clocks, which are traceable to Universal Coordinated Time (UTC). The GPS or LORAN-C is used to receive these reference signals, which are then distributed to clock cards for reference. The reference provides highly accurate timing phases (like the swing of a pendulum), to be used by the Stratum 1 signal being generated. Figure 8.4 shows the clock hierarchy in relation to SONET nodes.

Figure 8.4
SONET Clock Synchronization Distribution

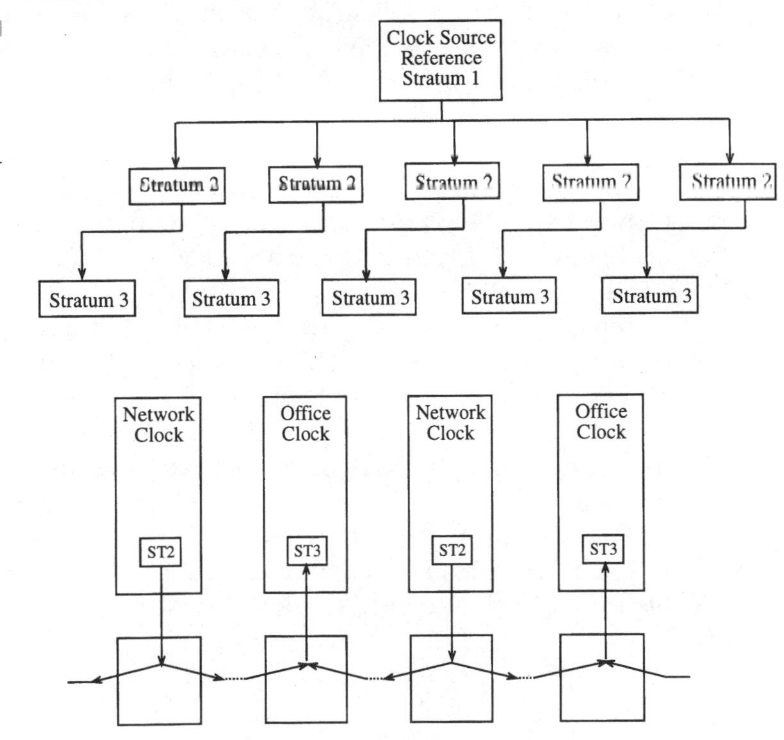

TABLE 8.1

SONET/SDH Hierarchy

North American Designation			
Electrical Signal	**Optical Signal**	**Data Rate (Mbps)**	**CCITT Designation**
STS-1	OC-1	51.84	
STS-3	OC-3	155.52	STM-1
STS-9	OC-9	466.56	STM-3
STS-12	OC-12	622.08	STM-4
STS-18	OC-18	933.12	STM-6
STS-24	OC-24	1244.16	STM-8
STS-36	OC-36	1866.24	STM-12
STS-48	OC-48	2488.32	STM-16

8.2.2. The SONET Protocol

The basic building block for a SONET frame is the 51.84-Mbps frame, as shown in Table 8.1. All of the rest of the SONET signal levels are built in multiples of this STS-1 level. In the table, you will notice two identifiers for each level. There is the synchronous transport (STS) side, which is the electrical input into the SONET multiplexer, and the optical carrier (OC), which is the optical output from the multiplexer.

Another unique feature of SONET is its ability to accept input from a variety of different types of sources and multiplex them into SONET frames. Figure 8.5 shows various types of input, including varying levels of digital signals, all multiplexed into SONET frames at whatever bandwidth is available.

Yet another unique feature of SONET is the ability to extract a single transmission from the SONET facility anywhere in the network. In other words, if you are troubleshooting a SONET facility and suspect a particular node may be causing the problem, you can attach a monitor to the facility and read the transmission from just that node. This would be impossible to do in, say, a T-3 because the entire facility would have to be demultiplexed down to a T-1.

The SONET system is divided into four layers. Each layer is responsible for specific functions. The four layers are:

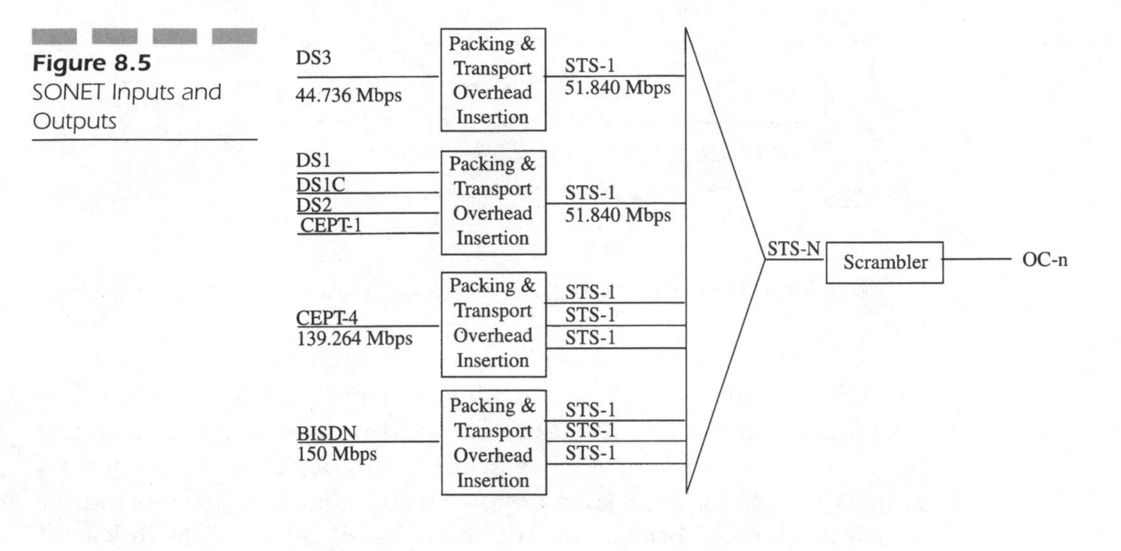

Figure 8.5
SONET Inputs and Outputs

- Photonic
- Section
- Line
- Path

The photonic layer converts electrical signals to optical signals for output as an OC. This layer also performs the reverse, converting optical signals to electrical signals. The photonic layer is also responsible for managing the optical pulse shape, wavelength of the light, and power levels. SONET uses single-mode fiber but may vary in light strength and type depending on the distance to be covered.

The section layer defines the basic SONET framing to be used. It is responsible for communicating between adjacent signal regenerators as well as between regenerators and SONET terminals. It manages the transmission of multiplexed frames across the fiber, providing framing, scrambling, and error monitoring (of the section layer only). Section overhead is found in the first three rows of columns 1, 2, and 3.

The line layer provides synchronization and multiplexing of STS-1s between adjacent network nodes. Each node "terminates" the line layer, which is why they are referred to as line-terminating equipment. Maintenance at the line layer and protective switching is also managed here.

The path layer provides end-to-end transport of SONET frames. A logical connection is established between source and destination path-

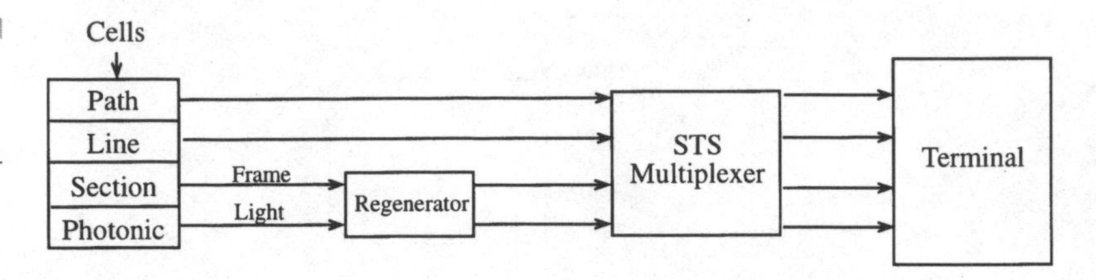

Figure 8.6
Logical Hierarchy of SONET

terminating equipment (a path is a logical connection between these two points).

The layers are somewhat similar to the various functions in a Local Area Network (LAN). If you remember our discussions of LAN protocols, you will remember that there is the MAC layer, the LLC layer, and then the network layer (which is usually another protocol altogether). Routers use the network layer protocol information, while ignoring the data link protocol. Bridges, on the other hand, use the data link layer (MAC and LLC), while ignoring the network layer.

Figure 8.6 shows how the various layers in the SONET hierarchy communicate with various components in the network. The header for the path layer is read only by SONET terminals and is treated as part of the user data by regenerators and multiplexers. Likewise, the line layer is only used by multiplexers and SONET terminals, and so on.

The various layers also have associated overhead. This overhead is appended to the user data and used by the various devices in the network, depending on the functional layer they represent. There is no overhead associated with the photonic layer. This is where the signal is output as light and is analogous to the physical layer of any other protocol. Whatever is passed through all of the layers above eventually makes its way to the lowest layer, whose job it is to transmit what it receives. This is the photonic layer.

The section layer does come with overhead. The section overhead defines the SONET frame. The following parameters are part of the section overhead:

- *A1 and A2.* Framing bytes used to synchronize the beginning of a frame.
- *C1.* STS-1 identifier used to identify individual STS-1s within an STS-*n* frame. This is used when multiple STS-1s are interleaved into a larger STS-*n* frame (such as an STS-3). It is a binary number, assigned during multiplexing. The first STS-1 is assigned the value of 1 (0000 0001), while the next STS-1 is assigned the value of 2 (0000 0010). Interleaving is explained a little later.

- *B1.* Parity byte used to provide even parity. It is used to check for transmission errors in a section. The value is calculated from all of the bits in the previous STS-*n* frame, after the previous frame was scrambled. The value is then placed in the first STS-1 of an STS-*n* frame before it is scrambled. This parameter is also referred to as the bit interleaved parity code (BIP-8) byte. Each parity byte is for error checking previous STS-*ns*.
- *E1.* An optional 65-kbps voice channel used between "section terminating" equipment.
- *F1.* 64-kbps channel reserved for user purposes.
- *D1—D3.* These bytes are used as the section data communications channel (DCC). The 3 bytes provide a 192-kbps channel, which is used for alarm monitoring, control, and administration. There are a number of protocols used to communicate over the DCC. The DCC protocol stack aligns with the ISO OSI Model, as shown in Fig. 8.7. The next layer of overhead is the line overhead. The line layer points to the path overhead position in the frame. Pointers are used, rather than assigning fixed positions in the SONET frame, to allow for timing deviations which may shift the bit positions within the frame.

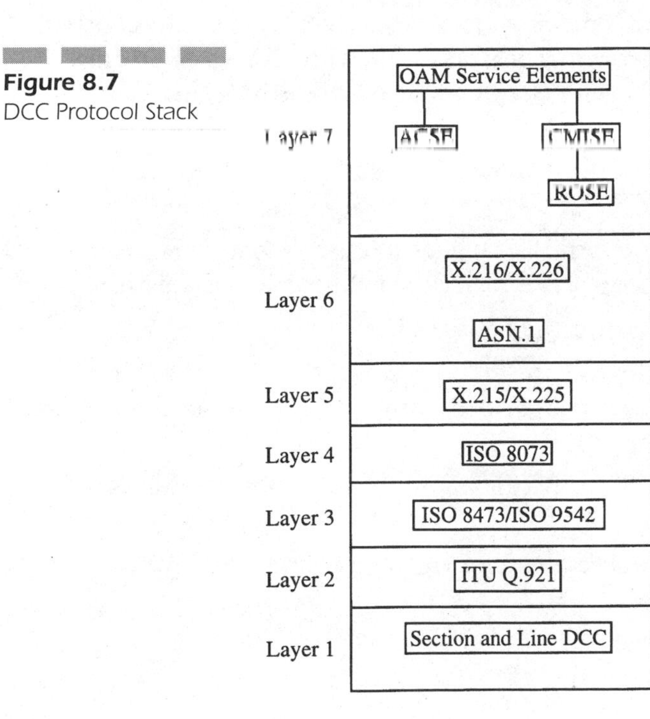

Figure 8.7
DCC Protocol Stack

- *H1—H3.* Pointer bytes that are used in frame alignment and frequency adjustment of the payload.
- *B2.* Parity byte used for error monitoring at the line level. This is the same as the B1 parameter except that this is for checking transmission errors at the line level. The B2 parameter (and the BIP-8 code) is provided in all STS-1 frames within an STS-*n* (which is also different from the B1 parameter, which is only placed in the first STS-1 of an STS-*n*).
- *K1, K2.* Signaling bytes used between line-level switching equipment. A bit-oriented protocol is used on this channel-to-control Automatic Protection Switching (APS). APS provides for an alternate path when there is a failure. The process is explained in more detail later in the chapter.
- *D4—D12.* 576-kbps data channel for alarms, maintenance, control, monitoring, and administration at the line level.
- *Z1, Z2.* These bytes are used with BISDN applications at the User-Network Interface (UNI). They provide information on errored blocks of data (detected by BIP-8). Only Z2 of the third STS-1 within an STS-*n* is used.
- *E2.* 64-kbps voice channel (Pulse Code Modulation, or PCM) for line-level orderwire. This provides a voice channel that can be used by maintenance personnel. It works somewhat like an intercom, providing a 64-kbps voice channel over which technicians can communicate. The path overhead appears ahead of the data and indicates the demultiplexing format through the C2 byte. The path overhead parameters are:

  J1. 64-kbps channel; sends a 64-octet string at repetitive intervals. Receiving terminals use it to verify the integrity of a path.

  B3. Parity byte used at the path level.

  C2. Signal label that is used to indicate whether or not a line connection is complete with a payload (equipped) or complete without a payload (unequipped).

  G1. Used by path terminating equipment to exchange status information.

  F2. 64-kbps channel reserved for the user of the path level.

  H4. Indicates multiple frames were needed for the payload (segmentation indicator).

Z3–Z5. Reserved for future use.

8.2.3. SONET Framing

Now that we have covered all of the overhead presented in a SONET frame, look at Fig. 8.8 and see if you can identify all the various layers of information. To present all of this information in one straight contiguous line would be rather difficult, so SONET frames are depicted in rows and columns. The actual transmission is a rather fast string of serial data.

The STS frame consists of 9 rows of 90 columns (each being 8 bits) for a total of 810 bytes. Byte transmission order is row by row, starting with the top row and moving from left to right. These frames are transmitted at the rate of 8000 frames per second, for a transmission rate of 51.840 Mbps.

The first three columns of each frame are overhead, while the remaining 87 columns are the Synchronous Payload Envelope (SPE). The SPE has its own format. There are 9 rows and 87 columns (transmitted row by row, left to right). The first column contains the path overhead, while the remaining 86 columns are for the payload.

A unique feature of SONET is its ability to allow the payload to

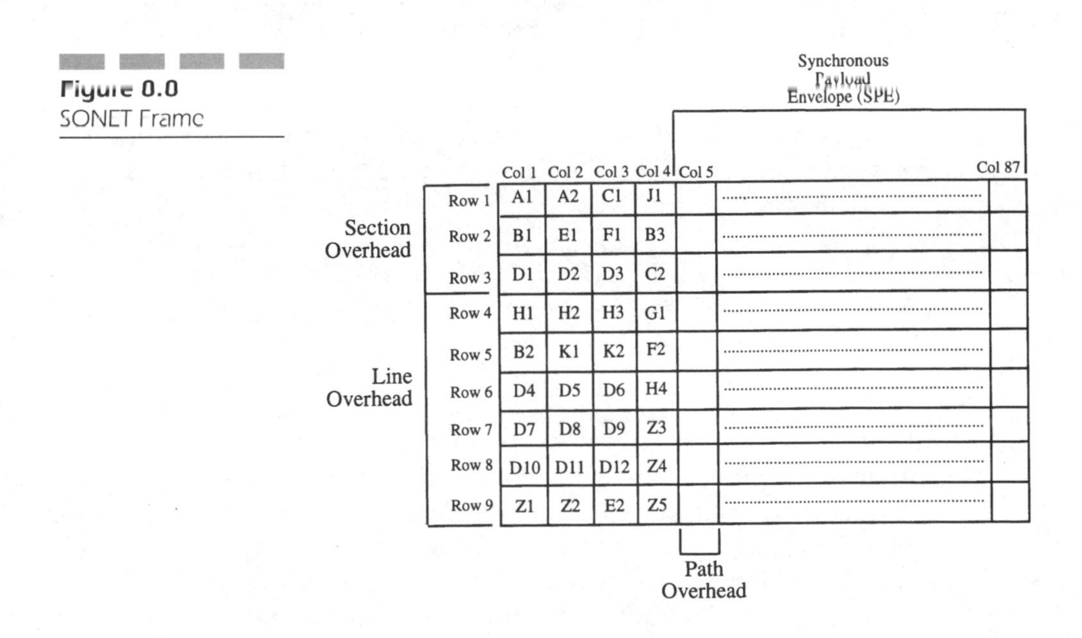

Figure 8.8
SONET Frame

"float" within an STS-1 frame. For example, if the payload is traveling a bit slower than the STS-1 itself, SONET allows for the slippage. This is accomplished by adding bits into the SPE portion of the STS-1 to compensate for the slippage. For example, if the payload slips in time, SONET will insert a byte at the beginning of the SPE. A pointer is then used to identify where the data within the SPE begins. This is adjusted at each node as the payload (user data) shifts within the SPE.

Both positive and negative bit stuffing are used to compensate for payload shifting. The SPE portion of the STS-1 is decoupled from the rest of the frame. This also allows the payload from any STS-1 to be dropped from any STS-*n* (or examined) for maintenance purposes.

In negative bit stuffing, the payload is traveling faster than the STS-1. A byte of the actual payload data is inserted into the H3 field of the transport overhead, giving it a 1-byte advance.

8.2.3.1. Virtual Tributaries SONET was designed to carry 50 Mbps of data, but it was also designed to support existing digital facilities. A digital signal of any level could be mapped into a single STS-1, but this would be a waste of bandwidth.

Virtual tributaries allow multiple digital signals to be carried within one STS-1. There are different sizes of VTs (or types), each type supporting different digital signals (such as DS1, DS3, etc.).

The smallest VT is called VT1.5 and provides a data rate of 1.728 Mbps. This is used to support DS1s. A DS1 can be supported using a 27-byte structure. If you multiply the number of bytes by 8000 frames per second, multiplied by 8 bits per byte, you get 1.728 Mbps data rate ($27 \times 8000 \times 8 = 1.728$). I mention this formula so you can see that deriving these bit rates is not at all mysterious. This is the only math used in this book.

There are four types of VTs:

- VT 1.5
- VT 2
- VT 3
- VT 6

Table 8.2 shows the capacity for each virtual tributary. They are often combined into groups to meet traffic demands.

Virtual tributaries are placed into groups to allow mixes of VTs within one STS-1. There can be seven different VT groups within an STS-1, each group using 12 columns of the SPE. A group cannot consist of mixed VT types. All of the VTs within a group must be of the same

TABLE 8.2
VT Capacities

VT Type	VT Rate (Mbps)	# of Columns	Possible Payloads
VT 1.5	1.728	3	DS1
VT 2	2.304	4	CEPT-1
VT 3	3.456	6	DS1c
VT 6	6.912	12	DS2

type. The capacity of a group depends on the type of VTs being supported. The following are the VT group types and their capacities:

- Four VT1.5s
- Three VT2s
- Two VT3s
- One VT6

Figure 8.9 shows how all of these VTs are sent into a multiplexer and then sent out as one STS-1.

Each of the inputs to the SONET unit is placed into a virtual tributary, which is then combined with other VTs and sent over the electrical interface as an STS signal. The multiplexer then combines all of the various STS signals and sends them to a scrambler, which converts the electrical signals into an optical signal.

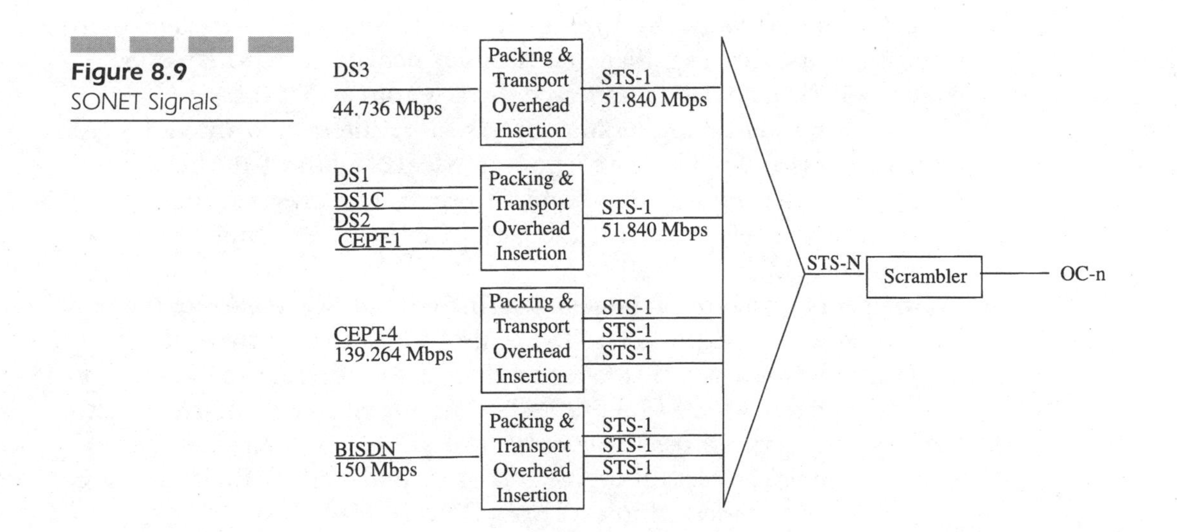

Figure 8.9
SONET Signals

There are two modes that can be used for VTs, locked and floating. If maximum efficiency of the 64-kbps DS0 structure must be maintained, locked mode is used. There is a one-for-one mapping, and the DS1s are not decoupled from the STS-1. This means that payload pointers are not needed and are available for payload instead. Locked mode of operation does not allow switching of VTs.

VT switching requires add/drop capability. A node is able to add or drop any VT within an SPE, switching it from STS-1 to STS-1. This function requires the VT to be in floating mode so that individual VTs can be accessed. This also requires the use of payload pointers so that the node can determine where the payload begins.

8.2.3.2. Byte Interleaving Byte interleaving is the process used when sending multiple STS frames in an STS-*n* frame (for example, three STS-1s in one STS-3 frame). The bytes are interleaved into the STS-3 so that they can be removed from the STS-*n* individually. For example, in the case of an STS-3, the first byte of the first STS-1 is placed in the SPE of the STS-3. This is then followed by the first byte of the second STS-1, which is then followed by the first byte on the third STS-1. Three STS-1s can be interleaved into an STS-3.

The process is then repeated with the second byte of the first STS-1, and so on. The STS-3 can then be byte interleaved into an STS-12. Multiple STS-3s would be interleaved in the same fashion. The first byte of the first STS-3 is placed in the first byte of the STS-12, and so forth. It is byte interleaving that allows access to any portion of an STS-1 without demultiplexing the entire STS-*n*.

There are two modes of byte interleaving, single and two stage. With single-stage interleaving, there are no intermediate steps to larger STS-*n* frames. All STS-1s can be interleaved directly into an STS-12.

Two-stage interleaving requires STS-1s to be interleaved into an STS-3 first, which is then interleaved into an STS-12. Figure 8.10 shows both methods of interleaving, single and two stage. The only requirement of single-stage interleaving is that the pattern must be the same as for two-stage interleaving.

In Fig. 8.10, the boxes represent the first byte of STS-1s. Notice the pattern when interleaved into an STS-12. Also notice that in two-stage interleaving, the same pattern is followed. This is so individual STS-1s can be found no matter what STS-*n* frame they are interleaved to. There are also bytes in the overhead used to correlate the STS-1s when they are received. The C1 byte of the section overhead is used to identify individual STS-1s when they are interleaved into a larger STS-*n* frame.

Figure 8.10 Single-Stage and Two-Stage Interleaving

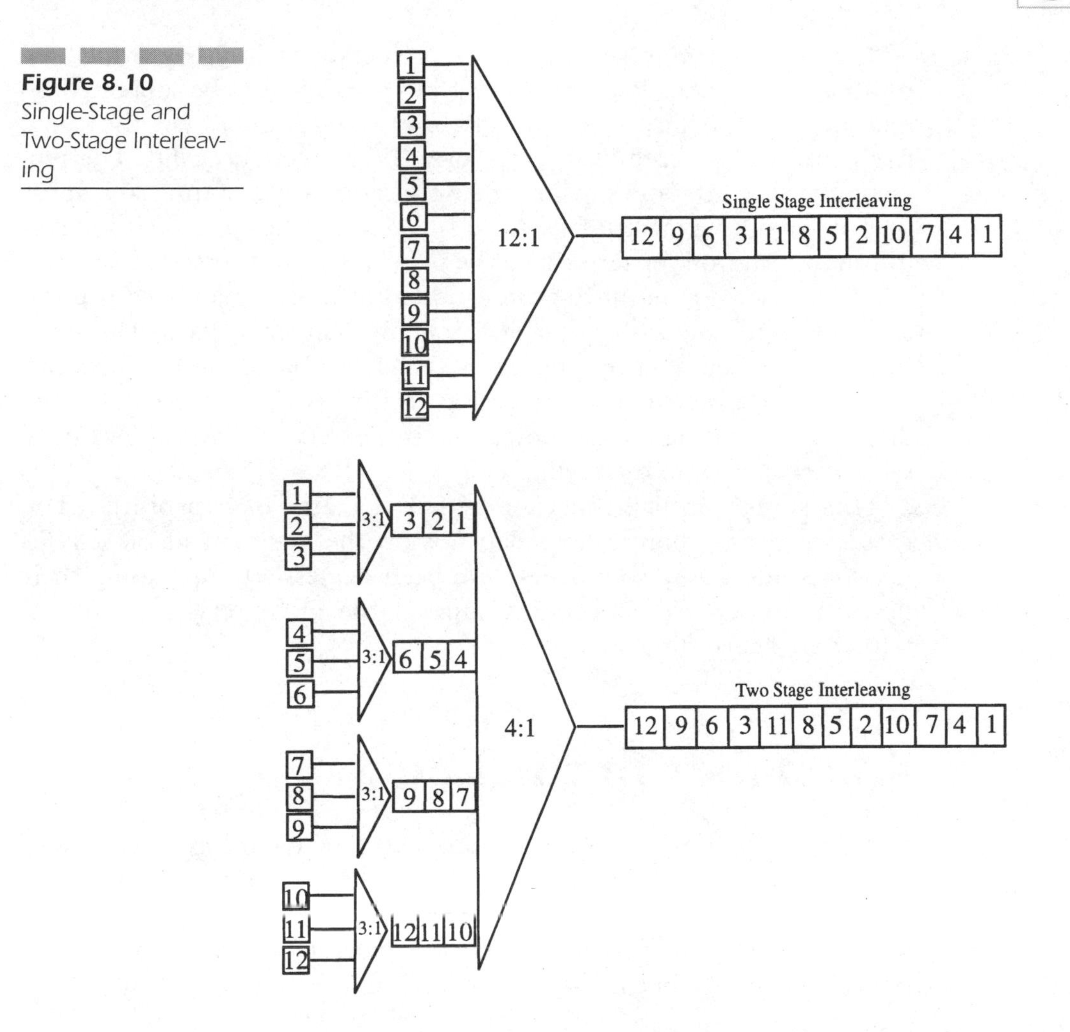

8.2.3.3. Automatic Protection Switching An APS is part of the self-healing aspect of SONET. There are two types of APS, 1 + 1 and 1:*n* protection switching. With 1 + 1 protection, there are two paths for optical signals. The working path has a backup path, which is referred to as the protection facility.

The working facility is mirrored over the protection facility as if there were two different cables to the same termination point. Everything that is transmitted over the working facility is duplicated over the protection facility. If a failure is detected at either end of the two facilities, the protection facility is immediately used. Since the same signal is already going over the protection facility, there is no "switchover" requirement.

With 1:n protection switching, one protection facility is provided for multiple working facilities. Up to 14 working facilities can be supported by one protection facility. If any of the 14 working facilities fail, the signal from that working facility is placed over the protection facility. The protection facility can only support one working facility at a time, so if more than one working facility fails, there is a prioritization scheme for determining which working facility will be switched to the protection facility.

APS is used for point-to-point and point-to-multipoint configurations. With a ring topology, a different mechanism is used. The same bytes are utilized for ring protection switching, but the procedures carried out are different. The protection facility is the other ring in the dual ring configuration, and any node in the network can automatically switch traffic over to the backup ring.

This is just a quick overview of SONET and its capabilities. The future of telecommunications depends on the deployment of SONET networks, and many companies have been aggressively upgrading their networks to SONET networks to support the many services their customers are demanding.

8.3. Fiber in the Loop

In Chap. 2, we mentioned the various solutions for bringing fiber into the local loop. In this section, we will look at the various options available and the differences between them. All of these solutions are currently being deployed in various telephone company networks. The differences lie mostly in cost and bandwidth.

8.3.1. Current Implementation Plans

While new technologies are being developed to increase the capacity and bandwidth of our telephone network, work is also under way to find a replacement for the subscriber loop. This is the point between the central office and its customers. The objective is to find a technology which can continue to support existing analog services while also having the capability to expand and support newer broadband services as they are deployed.

The primary targets for these technologies are residences and small businesses. Large businesses are more likely to use special dedicated circuits such as BISDN or Frame Relay for their connections.

There are a number of Digital Subscriber Loop (DSL) technologies to fill this requirement. Telephone companies are busy testing these now to determine which is the best solution for them. The biggest problem is finding a technology that will support video, data, and voice over the existing wiring. Twisted-pair cable used in residential service does not support high-speed data or video services, so this has become a major issue. On the other hand, fiber optics is expensive to deploy, making the choices difficult.

Pacific Bell embarked on an aggressive project in 1995, changing the twisted-pair wiring to coaxial cable in all of their residential areas. This project also includes replacing the cable from the central office to the subscriber areas with fiber optics. This is known as Fiber to the Curb (FTTC). An Optical Network Unit (ONU) serves as the interface between the central office and the subscriber. The ONU provides optical-to-electrical conversion, voice conversion to PCM, and multiplexing functions.

The ONU serves up to 24 customers. The intent is to serve as few customers as possible, providing more bandwidth on the upstream channel for videoconferencing and other interactive services. ONUs connect to a Host Digital Terminal (HDT), located in the switching equipment in the central office. The HDT can also be deployed at remote sites. The function of the HDT is to control multiple ONUs and act as an interface to the rest of the Public Switched Network (PSTN).

By using FTTC and coaxial cable to the residence, telephone companies have positioned themselves to provide almost any service subscribers need. Coaxial cable is capable of supporting high-speed data as well as video and voice, and with technologies such as ATM in the network, the telephone and the cable companies can support them all in one network.

There is also a cost benefit. Once deployed, there are no repeaters, network interface modules, or amplifiers to service. However, this also represents additional costs to telephone companies, which already have these elements in their outside plant. All of the outside plant must be changed to support FTTC, which raises the deployment cost significantly.

Another alternative to FTTC is Hybrid Fiber Coax (HFC). This is a favorite among cable television operators because they already have a fiber optics backbone in place. However, HFC uses a bus topology, with drops to coaxial facilities serving many users at once. The voice has to be multiplexed using frequency multiplexing because many conversations share the same bus. This represents a cost to the cable companies that want to use their existing networks for voice and data.

HFC is costly to deploy in the network because of the amount of outside plant that must be replaced. This is one reason why the telephone companies have been slow to support HFC. They already have a substantial infrastructure to deal with, and it is too costly to try to convert it all. In contrast, cable companies are building their voice and data infrastructure from scratch, making it easier for them to invest heavily in fiber networks and coaxial drops to their subscribers.

The other advantage of HFC is that it supports both analog and digital signals. Cable companies can migrate to digital technologies much more slowly, while still providing the services they want to their subscribers. The next generation of HFC will support higher bandwidths but will require ATM switches as well as special digital set-top boxes. The newer version of HFC will support hundreds of video channels as well as voice.

With HFC, the fiber does not penetrate as deeply into the network as is the case with FTTC. The fiber is run to distribution points, which may serve hundreds of customers. From the distribution point, coaxial drops are used to the subscriber. For this reason, HFC is sometimes referred to as Fiber to the Loop (FTTL).

Work is under way to bring fiber closer to the home using HFC technologies. The problem with the current implementation is that distribution points must support hundreds of subscribers. These subscribers must share the same upstream channel, which means videoconferencing and other similar applications cannot be supported. If the fiber is brought closer to the subscriber, fewer subscribers will be served by a distribution point, providing more aggregate bandwidth per subscriber on the upstream channel.

Asymmetric Digital Subscriber Loop (ADSL) uses the existing twisted pair at a residence. It does require changes in the outside plant. A new line card supporting ADSL must be installed in the central office switching equipment, and all repeaters must be changed. Special set-top converters are required to convert video signals from digital back to analog for viewing.

ADSL splits the bandwidth of a copper loop into three channels:

- 1 downstream (to subscriber) 1.5- to 6.1-Mbps channel
- 1 upstream (to the central office) 16- to 640-kbps channel
- 1 voice channel

The downstream channel is capable of supporting near-VHS-quality video, using Motion Picture Experts Group (MPEG) compression. However, it is unlikely that broadcast-quality video will be supported for

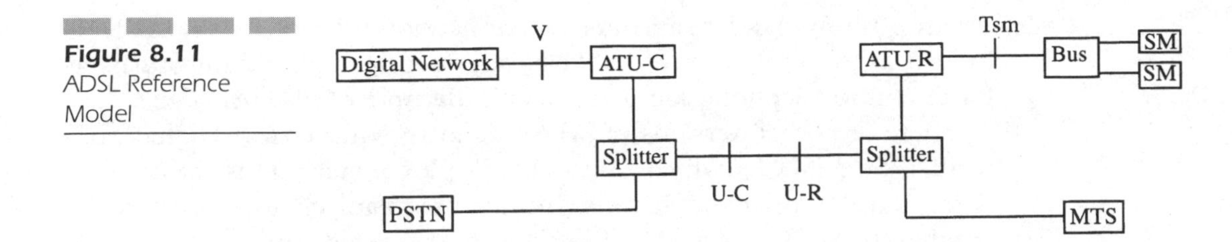

Figure 8.11 ADSL Reference Model

some time to come. If broadcast video is to be supported, the outside plant may prove inadequate due to its high noise potential.

A modem at the customer premise is used to multiplex these channels over the twisted pair. The modem is the line termination and is placed outside the home where the outside wire meets the house. If the service is more than 18,000 feet from the central office, a repeater is needed to support the signal. This adds additional cost to the service but is still cheaper than FTTC.

The modem uses Discrete Multitone (DMT) for modulation, a technique developed specifically for ADSL. This is only necessary if fiber is not used. The in-house wiring is sufficient if DMT is used, even allowing for a private Local Area Network (LAN) within each residence. In the event there is a power outage, Plain Old Telephone Service (POTS) service is guaranteed by the ADSL modem. Figure 8.11 shows the reference model for ADSL, depicting the various functions required and the interfaces to those functions.

The ADSL Transceiver Unit/Central Office (ATU-C) is the line card used in the central office to provide the ADSL service to subscribers. This is used in the central office switch and does not imply a new piece of equipment to the telephone company (other than a new type of line card). The opposite end of this circuit is the ADSL Transceiver Unit/Remote (ATU-R), which is the subscriber side of the line. The ATU-C converts analog signals received from within the switching network to an ADSL signal before transmitting to the remote end. The ATU-R provides the same functionality at the other end of the circuit.

The splitter can either combine or separate signals, depending on the direction of transmission. The digital network can be any kind of service. For example, connections to a video server would be through the V reference point to the digital network.

ADSL was first looked at for video-on-demand services. When the telephone companies discovered that there was no great demand for 200 channels, or video-on-demand, they quickly dropped the trials and began pushing ADSL as an Internet connection solution. Today, ADSL is looked at as an alternative to ISDN, supporting higher bandwidths at lower cost to telephone companies (typically under $1000 per home).

There are other versions of ADSL. One is Symmetrical Digital Subscriber Line (SDSL), which offers full duplex service. This means that there is a downstream channel and an upstream channel with equal bandwidth. SDSL provides T1/E1 speeds over twisted pair.

High-bit-rate Digital Subscriber Line (HDSL) provides higher bandwidths than ADSL but at a higher cost. Introduced in 1992, HDSL was used to deliver T-1 leased line services to businesses. The intent of the telephone companies was to use HDSL as a replacement for all of its T-span carrier equipment in the outside plant. This would lower their operating costs and provide more reliable service.

HDSL modems are cheaper to build than ADSL modems because they are not as complex. However, the overall deployment cost of HDSL is higher than ADSL. While the bandwidth of HDSL is higher (6 Mbps downstream), some are still skeptical that HDSL will win over ADSL. There still are no requirements for that much bandwidth in the home.

The jury is still out on which of the fiber-in-the-loop solution will be the winner. There is still time to weed out the weak entries and name the winner. But applications are changing, and as applications change, the ideal solution changes as well. It is hoped that the marketplace will find its calling, and the solution will be there.

8.4. Chapter Test

1. What is the photonic layer responsible for?
2. The SONET building block is based on ________ STS frames.
3. SONET uses what type of fiber?
   a. Multimode fiber
   b. Single-mode fiber
   c. Both multimode and single-mode fiber
   d. None of the above
4. Automatic Protection Switching (APS) uses two modes; what are they?
   a. $1 + 1$ and $1 - 1$
   b. $1{:}n$ and $1 - 1$
   c. $1 + 1$ and $1{:}n$
   d. $1 + 1$ and $1 = 1$

5. There are two types of byte interleaving, single stage and ______.
6. The smallest virtual tributary is:
   a. VT6
   b. VT2
   c. VT3
   d. VT1.5
7. Virtual tributaries are placed into groups. How many VT1.5s can be placed into one group?
8. How many VT1.5s can an STS-1 carry?
9. Positive and negative _______ are used to compensate for payload drift (floating) within a Synchronous Payload Envelope (SPE).
10. How many columns of payload are left in the SPE after overhead?

CHAPTER 9

ATM—Key to the Future

9.1. Integrating the Public Switched Telephone Network

Asynchronous Transfer Mode (ATM) standards are referred to as Broadband ISDN (BISDN) in the International Telecommunication Union (ITU) standards publications. BISDN is the next evolution of ISDN, providing higher bandwidth and faster speeds than ISDN. The ATM Forum has gone one step further and defined services for non-BISDN applications (data only). These applications are data-specific, and include applications such as bridging Local Area Networks (LANs) through ATM connections. BISDN standards define integration of data, video, and voice networks over one network.

Integration is certainly not a new concept. The whole basis of ISDN was to create one common interface to support all forms of traffic, including data and video. ISDN fell short of delivering for a variety of reasons, and its acceptance has been slow because of a lack of killer applications.

ATM has already suffered some of the same maladies as ISDN. Slick marketing has touted ATM as the network solution, providing more bandwidth to the desktop than anyone could possibly need. Which was exactly the problem. Very few companies have expressed a need, let alone an interest, in having more than 100-Mbps bandwidth delivered to their desktops.

So why was ATM developed? The original idea was fostered by the Bell Operating Companies (BOCs) long before divestiture. The idea of integrating their telephone networks generated the idea of ATM to replace the existing circuit-switched network. ISDN was the first step in integration, but as time evolved, the telephone companies recognized the need to increase the capacity of their backbone networks (as well as the local loop).

ATM was seen as the answer for backbone applications. By replacing the switched facilities with ones that made more efficient use of the bandwidth and supported all types of digital traffic, the telephone companies could consolidate their intraoffice facilities.

ATM standards began with the ITU-T, in conjunction with the American National Standards Institute (ANSI) and now Bellcore. In 1991, four companies joined together to expedite the standards development process. Adaptive, Cisco, NTI, and Sprint formed the ATM Forum, with the intent of creating implementation agreements to be used as standards until the ITUs work was complete.

An implementation agreement is an agreement among all member companies to implement a technology in a specific way. It is much like a standard but is not officially drafted as one. The ATM Forum quickly grew to over 500 members.

The ATM Forum also streamlined the voting process in hopes of getting consensus on various contributions more quickly than the unanimous rule used by the ITU. The ATM Forum uses a majority rule, using a simple show of hands rather than formal balloting. What the ATM Forum did not count on were the problems in getting consensus among 500 different companies. Some companies have joined together and implemented their own solutions because they did not like the ATM Forum decision. The ATM Forum is explained in greater detail in Chap. 1.

As I will explain a little later, ATM is part of BISDN. In fact, most books on BISDN mostly explain ATM. BISDN is a combination of ATM, the ATM Adaptation Layer (AAL), and various protocols such as Q2931 (signaling protocol derived from ISDNs Q931). I will continue discussing BISDN in this chapter, even though it is also discussed in Chap. 6, but want to make sure you understand that ATM can be offered without BISDN procedures (to support data transmission in a point-to-point network, for example).

9.1.1. The Reason for ATM

There are many issues with the current methods of switching. Circuit switching uses a fixed bandwidth, usually based on 64-kbps channels. It also uses fixed multiplexing, with each "conversation" dedicated to a specific channel. This means that when there is nothing to send, the channel is empty and wasted.

Each connection requires connection-oriented procedures, requiring extra processing at end and network nodes. Once a connection is established, it must remain "nailed up" until transmission is complete. This is suitable for voice and continuous data transmission but is a waste of bandwidth for bursty-type data applications.

Packet switching is more versatile than circuit switching, providing better use of the network. Packet switching also makes better use of the available bandwidth and uses simpler multiplexing methods. However, most packet-switching technologies require more processing at network nodes for routing and error detection/correction, introducing higher delays. This makes packet switching unsuitable for voice and other delay-sensitive traffic (including continuous data transmission and video applications).

ATM was found as the best of both the circuit- and packet-switched methods of transmission. It supports connection-oriented as well as connectionless applications, while meeting the requirements of delay-intolerant applications (voice and video). However, ATM is not a single solution; it is part of a total solution.

Within the public network, ATM is seen as a backbone technology, used to send traffic between central offices and tandem offices. This is important to understand because it has been a misconceived notion that ATM was intended for the desktop. This was never true, but many data communications companies hoped to sell product to provide that capability to meet the demands of corporate networks.

Indeed several companies have been successful selling ATM products for private networks. However, many have not, and it has become even clearer that while ATM may be delivered to the desktop, it will not be at the bandwidth seen in the public network for a variety of reasons.

First, where is the killer application? Most corporate network users do not need that much bandwidth. Second, the cost of replacing network components is high, especially when compared to options such as Fast Ethernet, delivering 100 Mbps to the desktop at minimal cost.

To the user, ATM alone is not enough. In order to send voice to the public network, there is a need for signaling. The signaling is used by the telephone network to set up a virtual connection end to end and to manage that connection. The signaling from the user interface is different from that used in the public network. By now, you should have become familiar with the terms *User-to-Network* Interface (UNI) and *Network-to-Network* Interface (NNI).

The signaling within the public telephone network is based on Signaling System #7 (SS7), while the signaling from the subscriber to the network is based on BISDN. Yes, BISDN is ATM, but it is also a lot more. The BISDN standards are a lot like NISDN, but they are based on packet switching (or cell relay if you like). ATM only describes one portion of BISDN, which is why I have isolated it by itself, instead of combining the two topics.

ATM can be provided by itself, for data applications or video applications, but without signaling, ATM to the subscriber will not be able to provide true integration, sending a mixture of traffic on demand to any point in the network. Even today, the ATM switches being deployed are being configured for point-to-point communications until the work on standards can be completed.

To understand the differences between cell relay, packet switching, and circuit switching, let us look at the basic concepts of cell relay (and ATM). In cell relay, there are no dedicated time slots. This is because it does not rely on channelized facilities (such as T-1). ATM does not use a dedicated route to a destination, which makes it look more like packet switching.

The addressing in ATM identifies a logical path, not a node address. This makes addressing much simpler and reduces the amount of overhead required. If every node in the network required a unique address, the header would require enough space to introduce millions of possible addresses. Look at the problems with Internet Protocol (IP) addressing now that the Internet has grown in popularity. IP has not run out of addresses quite yet and will not require as many addresses as ATM, but then the Internet does not have near as many nodes as the public telephone network, cellular network, and various public and private data networks.

9.1.2. From the Network to the Desktop

There are many who believed that ATM could be delivered right up to the desktop, providing 155 Mbps of bandwidth to every computer. This certainly is true, but the problem lies in finding an interest in this much bandwidth at the desktop. Most applications do not require this much bandwidth, and as a result there has been a lack of interest in this approach.

The ATM Forum has promoted ATM in the local network for many years and has been instrumental in developing implementation agreements that vendors could develop from. Many vendors are already providing equipment to support ATM in the local network, but the interest is still missing, and ATM to the desktop has never caught on.

Another factor has been the availability of 100-Mbps Ethernet. This is a more economical approach, requiring little more than software upgrades to existing equipment. The existing LAN cabling can support 100-Mbps (category 5 cable), and many routers can be upgraded inexpensively. Providing ATM in the local network would require changing all of the equipment in the network and possibly some cabling as well (if it is not category 5).

Many companies are looking at deploying 100-Mbps Ethernet in the local network and using ATM as the corporate backbone to the public network. This certainly makes sense and is certainly more economical than changing out the whole local network to support ATM.

9.1.3. From LAN to LAN

LAN Emulation Services (LES) allows two separate LANs to be connected transparently by an ATM network. LES is transparent to the applications as well as the LAN itself because it emulates the Media Access Control (MAC) layer of the LAN. This means that two Ethernet networks could be linked together and look as though they were one large LAN.

Companies with several networks would benefit from this feature, especially if they had geographically divided networks. Remote offices could be interconnected with the corporate office LAN using ATM LES. The applications and network software would not have to be changed to support ATM LES because of its transparency.

LES presently supports IEEE 802.3 Ethernet frames, IEEE 802.5 Token Ring frames, and a control frame used for address resolution, the registration of LAN emulation clients, and other network functions not related to the actual transfer of data.

For companies using channelized services to interconnect their facilities, ATM also supports Circuit Emulations Services (CES). This service allows ATM to interwork with DS0, DS1, and E1 leased lines. These are channelized facilities, typically found between the local telephone exchange and customer switching equipment.

Since ATM is emulating these facilities, idle channels are sent as cells. This does not seem efficient, but channelized facilities need to send these idle channels to maintain clock synchronization. ATM CES will have to provide clock recovery at the remote end. ATM does not presently support the requirement of alarms in channelized services. Alarms and measurements such as errored seconds are still undefined.

By providing both LES and CES, a company can merge their data network with their voice network, sending both voice and data through their own private ATM network. This is the goal of ATM services, providing integration of networks.

9.1.4. ATM Services and Applications

Now let us look at various ATM services and applications. There are four basic applications supported by our public telephone network (as well as private networks). They are voice (telephone service), data, cable television, and videoconferencing. Each of these applications requires its own set of unique requirements from the network.

For telephone services, all connections must be "nailed up" before transmission can begin. This means the network must provide connection-oriented services. Voice does not require a lot of bandwidth, so low bandwidth must be provided. Voice transmission is not tolerant of network delays, so the network must ensure no delays in transmission. Last but not least, the network must support a continuous bit rate.

Data is much more flexible. It can be sent in either connectionless or connection-oriented modes. It is not sensitive to variations in delay, or at least it is a lot more forgiving when delay occurs. Data can use both high and low bandwidth and is bursty in nature (rather than continuous). The bit rate is variable rather than constant.

Cable television relies on connectionless services. Cable television is unique because the transmission is mostly in one direction, from the service provider out to all of the subscribers. Digitized video requires high bandwidth and, depending on the compression method used, can transmit at a variable or constant bit rate. If some of the new compression methods are used such as (MPEG-2), digital video is sent at a variable rather than constant bit rate.

Videoconferencing requires a connection-oriented service. It is sensitive to any delay variation and transmits at both low and high bandwidth. Again, depending on the compression method used for the video portion, this application may also transmit at a variable bit rate.

Now let us look at some specific applications and see where they fit in the above descriptions. These applications are not absolute by any means, but they best classify the applications suited for ATM networks.

9.1.4.1. Voice Networks Voice networks are both private and public (the telephone company networks). This is where BISDN fits in. Many of the services in BISDN focus on voice applications and data integration. The signaling provided by BISDN (based on Q931 ISDN signaling) is designed for voice and data transmission over public networks.

Voice is sensitive to transmission delays for a number of reasons. If you have ever called someone overseas, you have probably experienced what is known as propagation delay found in satellite transmission. You hear a short period of silence, and just as you begin to speak, you hear the voice from the other end. Of course, they cannot hear you yet, because your voice has not reached them.

The result is an awkward conversation until you learn to pause before you speak to ensure you are not talking over the other party. This is only an inconvenience, but there are other more technical reasons why voice is delay-sensitive.

Voice-switching equipment relies on synchronized clock signals, which are derived from the constant bit stream found in channelized facilities. Even when there is nothing being transmitted, idle channels are filled with "flags" to maintain synchronization between nodes. In the case of ATM, these idle channels are emulated to maintain clock synchronization.

ATM has been modified to interwork with non-ATM voice networks, and it provides clock restoration at the destination (or remote end). While this may seem a waste of ATM bandwidth, it is a necessary function when dealing with voice switches. There are other devices used in the voice network which rely on these clock signals as well.

In addition to clock synchronization issues, voice requires a connection-oriented service, establishing a virtual connection between both parties. Conference calling introduces a different challenge, requiring transmission from one point to multiple points. This part of the standards is still evolving.

9.1.4.2. High-Speed Data ATM standards work (at least in the ATM Forum) has focused on high-speed data in hopes of deploying ATM in private data networks quickly. Most private networks will probably deploy 155-Mbps ATM facilities, which should be more than enough for the average corporate network. How fast is 155 Mbps? A 128-megabyte image would take about 1 second to transfer over an ATM 155-Mbps link. A 2-gigabyte drive would take around 100 seconds to download its entire contents to another drive.

The idea in most data networks is that not everyone needs to send data at the same time, in large amounts. For the most part, data networks must be able to deal with bursty traffic, where a user may be downloading a large file, but once it is downloaded, the network is dealing with smaller-sized files. The backbone of the network is hardly taxed with this type of traffic mix; it is different when several users try downloading large files at the same time.

Remote access to large databases and central mainframes could also require high-speed data facilities. In the retail industry, many department stores use an automated process where all of the cash registers report sales to a central computer in the store. Every night, the store's computer downloads all of the day's transactions to a large corporate server, which then compiles the information for corporate sales reports.

This type of application would certainly benefit from a high-speed solution such as ATM since many stores are downloading a lot of data at the same time. Each store may not need a large link (again, 155 Mbps is sufficient), but the corporate server would certainly benefit from a bigger pipe to retrieve all of the information from multiple computers.

9.1.4.3. High-Resolution Graphics Depending on the type of applications being run over the data network, 155 Mbps will be enough for any company. However, there are specialized applications that will require more bandwidth. Image files require a lot of bandwidth, with images reaching 128 megabytes and more in size. These types of files can overload a data network at 100 Mbps, especially when there are many of them to be moved around the network.

Architects, mechanical engineering groups, and medical imaging applications are a few examples of where 600 Mbps is more likely in the private network. These applications do not represent the majority of data network users, which means 600-Mbps ATM networks in the private sector will not be as common as many originally thought.

Marketing and print companies deal with large image files as well. Clients with ATM access could download large image files of products and marketing material to their service bureaus for printing. This would eliminate the need to transport these large files via hard drives and magnetic tape. It also gives the service bureau the added advantage of being able to send proofs over the network rather than printing them and sending them via a courier service to their clients.

9.1.4.4. Video and Audio There are many applications incorporating video and audio. ATM is a good fit for these applications. The 53 byte cell size was a compromise between the data users and the voice/video users, allowing real-time applications such as video and voice to be supported efficiently. ISDN attempted to fill the needs of many corporate users who want to deploy videoconferencing networks, but ISDN does not provide enough bandwidth to support broadcast-quality video.

The recording industry is already experimenting with the idea of receiving audio from remote studios and mixing the final cut of an album in one central studio. This has already been accomplished by at least one recording artist. Frank Sinatra recorded an album without ever performing with the orchestra. All of the orchestra cuts were sent over ISDN links to a main U.S. studio.

Some of the artists were from countries such as Italy and France, performing in studios close to home. This saved thousands of dollars in travel expenses and allowed the recording company to get some of the best orchestra players in the world to record for them based on their personal schedules.

There are many more applications such as this one combining video and voice. I mentioned this one because it was a creative and unusual use for digital networking. The industry is actively defining standards that will allow recording equipment to be connected to data networks (recording consoles, equalizers, amplifiers, and even effects equipment). This means that if a studio does not have a particular piece of equipment, it could use equipment from another studio by connecting to it over a network.

Another video application is the distribution of television over ATM networks. Development has already begun using ATM as the backbone network for cable television. Fiber is used to feed the signals to the curb (or in some cases to a neighborhood), and specialized set-top boxes are used to convert the video signals from digital back into analog.

Telephone companies have a special interest in the distribution of television through the telephone network. As they search for new services to offer their customers, video is a good addition. However, existing telephone switches and infrastructure will not support broadcast-quality video, which is where ATM fits in. With SONET and ATM, telephone companies can provide telephone services, cable television, and data services over the same facilities.

Telephone companies are not the only ones interested in this approach. Cable television companies and electric utilities are also interested in broadening the services they provide to their customers. By using their fiber optic networks and ATM, they can provide the same services, broadening the scope of their offerings.

The biggest challenge in video is not supporting constant bit rate transfers but supporting compressed video in a variety of formats. Compressed video no longer transmits continuous video frames. Instead, it sends video frames only when there is a change in the frame. For example, if the frame is nothing more than a blue background, and nothing changes for several seconds, compression waits until there is a change in the scene (such as movement or a new color introduced). Many unchanging video frames could exist, and if compressed, they would never be sent over the network. This causes video to be more of a variable bit rate, which can be supported by ATM Adaptation layer 2 (AAL-2).

Audio is sent separately from the video signal, which adds to the complexity of the protocol. The audio must be synchronized to the video frames they belong to, which means the protocol must be able to correlate audio cells with specific video cells.

The problem in distribution of video today is supporting a point-to-multipoint broadcast configuration, where there is one source transmitting to thousands of receivers. This will require multicast ATM switches, which are just now entering the market.

9.1.4.5. Interactive Multimedia Multimedia is not well defined. The purist definition is any application where video, audio, and graphics are combined. Interactive implies that the output of the application changes according to the user input. We have all seen interactive video games where the user fights the ugly step sisters and depending on the type of weapon chosen, can kill them in any number of ways (nice thought isn't it?).

There are a lot of different applications that could benefit from interactive multimedia. Simulators incorporating high-resolution images could be used to train physicians in certain types of surgeries, without ever requiring a body. By using virtual reality equipment, they could even use real medical tools to simulate working on a live person, viewing their "work" through specialized monitors.

Flight simulators are already using interactive multimedia to deliver sight and sound based on the movements and actions of the pilots. Driving simulators provide the same type of effect for those who want to learn to drive without ever hitting the road. With faster access to the Internet, multimedia games could be delivered right into our homes and played through specialized television sets (already being developed).

9.1.4.6. ATM Services There are several ATM services which support many of the applications we have already discussed. The services are:

- Cell relay services
- Frame Relay bearer services
- Connectionless services
- LES
- ATM video and audio services
- ATM CES

We have talked about some of these services already, but for clarity I will cover each one again. Cell relay services provide point-to-point data transfer, without the services of the AAL. Cells are generated at the source, which means the source must be ATM equipment. It is believed that this will be the first usage for ATM networks, primarily supporting data transfers. However, it has become obvious that this may not be the case since many companies choose alternatives such as Fast Ethernet and Fiber Distributed Data Interface (FDDI). The ATM layer sends these cells over the network exactly as they were received and when they were received, making this the true ATM application.

Frame Relay bearer services requires the functions of AAL-5. This is an interworking solution for Frame Relay connections to the ATM network. The class C functions of AAL-5 (and the various sublayers of the AAL) will perform checks on the Frame Relay data to ensure that the length of the frame does not exceed what was negotiated at setup time and that the Data Link Connection Identifier (DLCI) is active and assigned.

This service may be used to interconnect two geographically separated Frame Relay networks. It may also be used to connect Frame Relay devices to ATM devices.

Connectionless services support the needs of protocols such as Switched Multimegabit Data Service (SMDS), Connectionless Broadband Data Services (CBDS), and Connectionless Network Services (CLNS), all examples of connectionless networks that would use ATM networks. They provide many features offered in LAN protocols on a Wide Area Network (WAN) scope. They offer speed and less processing by network nodes (by eliminating error correction in the network).

The Data Exchange Interface (DXI) is part of this service. Defined by Bellcore, the DXI places the Segmentation and Reassembly (SAR) sublayer of the AAL in a separate device, such as a Data Service Unit (DSU) or Channel Service Unit (CSU). The purpose was to make ATM more cost effective for data applications.

This is achieved by purchasing a software upgrade for routers, converting them to SMDS or CLNS. The router then sends Protocol Data Units (PDUs) in a special DXI frame to a DSU/CSU equipped with ATM AAL software. The PDU is then placed into an AAL frame, passed to the ATM layer in the DSU/CSU, and then placed into ATM cells for transmission over the network. The DSU/CSU has many input ports, so the cost can be shared by many departments instead of each department having to pay for its own ATM solution.

We have talked already about LES. It allows two different LANs to be connected as if they were one large LAN, even if geographically separated. This is an ideal solution for large corporate networks that want to maintain one large network without worrying about other protocols in the middle. Typically, a company may have Ethernet in their offices, Frame Relay interconnecting their offices, TCP/IP running on top of the Frame Relay providing e-mail over the Internet, and ISDN for voice applications. Using ATM they can reduce the number of interim protocols used for interconnectivity.

ATM video and audio services support the requirements of broadcast-quality video. We discussed some of the issues already when we talked about the video and audio applications. This service will meet the requirements of "raw" video, compressed video, and audio tracks which accompany the video.

9.1.4.7. ATM Bearer Services and Classes of Service ATM provides a connection-oriented cell transfer service between a source and a destination (end to end). Sequenced delivery is guaranteed. ATM also supports quality of service and throughput by negotiating these parameters at connection establishment. Switched Virtual Connections (SVCs) rely on the services of the control plane, adaptation layer, and signaling protocol.

There are four classes of service defined. They are:

- Class A—Circuit emulation, CBR video
- Class B—VBR audio and video
- Class C—Connection-oriented data transfer
- Class D—Connectionless data transfer

These class types provide a means for delivering the services defined. This is different from the AAL types, which provide a service. The AAL identifies the type of class to be supported and other services needed to deliver the data.

9.2. ATM Network Access

There are two interfaces defined in ATM standards, UNI and NNI. The UNI is the best defined and has been the focus of the ATM Forum for the last few years. This is the interface between the public network and the private network (or ATM user).

The NNI is still evolving, and is the interface between public networks. The ATM header is different between these two interfaces, and the services provided differ as well. Figure 9.1 shows the various ATM interfaces and their relationships.

Both the UNI and the NNI are defined as public or private. Public interfaces are provided by service providers and require more complexi-

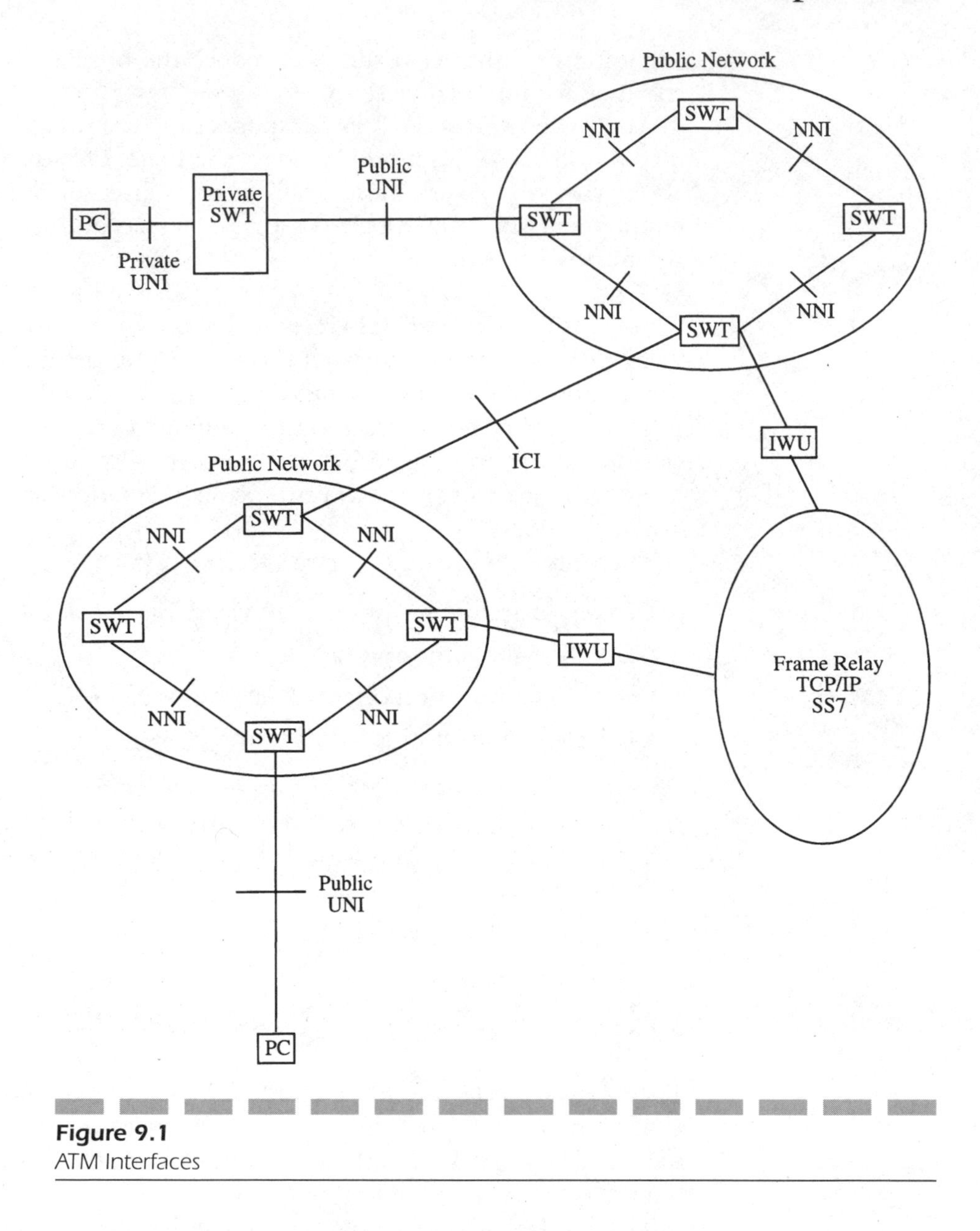

Figure 9.1
ATM Interfaces

ty than private interfaces. A private interface exists within a closed network, such as within a large corporation. While at some point access to a public interface may exist, the private interfaces link only to internal nodes (within the private network).

9.2.1. User-to-Network Interface

As mentioned above, UNI is the interface between the ATM user and the ATM public network. Since ATM is expected to support a variety of different types of applications, it should also be assumed that ATM must support a variety of different network types at the subscriber premise. The AAL was developed to interwork different network types with the ATM network and maintain transparency to the user.

The beauty of this concept is that subscribers can continue using the LAN technologies they already have in place, while enjoying the benefits of ATM access to the public network. All of their telecommunications networks, data, voice, and video, can be merged at the public network and transferred to the public network over a single ATM facility. The alternative is to lease different types of facilities to support each unique application. This would mean using a channelized service such as T-1 for voice applications, possibly a leased line or Frame Relay service for data, and a leased line or ISDN for the video network. The cost of this approach is somewhat prohibitive, not only for the subscriber but also for the service provider who must maintain several pairs of copper to the subscriber premise to support all of these services.

ATM is advantageous then to both the subscriber and the service provider, providing facility consolidation as well as additional bandwidth to support a variety of broadband applications. The key here is reduced cost and enhanced services as much as it is more bandwidth.

The physical interface for the UNI is currently defined as 44.736-, 100-, and two different 155.52-Mbps interfaces. The 155.52-Mbps interface uses Synchronous Optical Network Synchronous Transport Signal 3 (SONET STS-3c). The NNI interfaces range from 155.52 Mbps up to 600 Mbps (in concert with SONET).

9.2.2. Network-to-Network Interface

The NNI is used within the public network. This is where the various service providers interwork their networks with one another. NNI provides a different set of services than the UNI does simply because the requirements are very different. The UNI does not require the same types of network management and reliability that the public network does.

The adaptation layer is not supported over the NNI, mostly because it is not needed. The NNI is simply interconnecting ATM networks to one another and does not require interworking with unlike networks. There are some exceptions, but for brevity we will not go into details.

There are applications within the public network that will use the adaptation layer, such as SS7. The SS7 network must be able to connect to the ATM network but will remain a separate network for now. Work is under way to develop interfaces for SS7 signaling points, supporting CES and AAL for the transport of signaling messages over the ATM network. There are many cost advantages to this, but the principal reason is twofold.

As the ATM network migrates into the public network, replacing many of the interoffice facilities, the circuits used for SS7 will also be replaced. At some point, SS7 links will need to be placed on ATM facilities because the channelized facilities will be gone. The SS7 signaling points will remain an integral part of the network, especially the Signal Transfer Point (STP) and the Service Control Point (SCP) because of the services they provide to the public telephone network.

Another factor is the amount of traffic in the SS7 network. In today's network, SS7 is able to support the capacity. However, new applications are being deployed for SS7, and the traffic mix is changing rapidly. There are more and more database applications requiring SS7 services, which means SS7 traffic is increasing. This increase will soon begin taxing the 64-kbps links that are used today. ATM allows the telephone companies to integrate their interoffice facilities to carry all traffic, including SS7 signaling traffic.

9.3. ATM Overview

ATM uses a fixed-length cell of 53 bytes. It is capable of carrying any type of data. Even data with protocol headers from other networks can be carried in an ATM cell (or segmented into several ATM cells).

The concept is that small cells introduce less delay. Devices waiting to transmit do not have to wait as long if cells are short. Small cells can be processed more quickly than large cells, reducing the processing at network nodes. The small cell size was a compromise between the telephone and cable industries and the data industry. While large cell sizes are better for data, small cells work better for voice and video.

There are various "planes" supported in ATM. Think of each plane as a level of communications. The user plane transmits user data from one endpoint to another. This plane is also responsible for multiplexing among different connections (using the Virtual Path Identifier/Virtual Channel Identifier, or VPI/VCI) and for cell rate decoupling, cell discrimination, payload type discrimination, and traffic shaping. The user plane performs selective cell prioritization using the CLP field in the ATM cell header.

9.3.1. ATM Planes

The control plane establishes virtual connections and handles signaling traffic. The management plane is divided into two parts: plane management and layer management. The management plane provides alarm surveillance, connectivity verification, and verification of VPI/VCI. Operations, Administration and Management (OAM) cells are used to exchange alarm and connectivity information between nodes. If an invalid VPI/VCI is received, the cell is discarded and layer management is informed.

9.3.1.1. OAM Messages The OAM cell uses a different format than other cells. There are two types of cells supported, F4 and F5. Figure 9.2 shows both formats.

There are two types of indications given through OAM alarming: Alarm Indication Signal (AIS) and Far End Receive Failure (FERF). Both types provide a failure type and failure location in the "function-specific" field of the OAM cell.

The AIS message is used to alert downstream nodes of an alarm condition. Either an VPC/VCC failure or physical layer failure can cause AIS message generation. When a switch at the public UNI receives an AIS message, it returns an FERF message to alert the downstream nodes of the failure downstream. A virtual path AIS and virtual path FERF are always carried on VCI = 4. A virtual channel AIS and virtual channel FERF are always carried over cells with PT = 101.

9.3.1.2. OAM Connectivity Verification OAM payload consists of loopback indication, a correlation identifier, loopback location identifier, and source identifier. Loopback indication indicates that the cell should either be looped back or discarded. If looped back, the value is decremented (to prevent the originator from interpreting it as a loopback from another node and sending it back).

Correlation ID allows the originating node to keep track of OAM

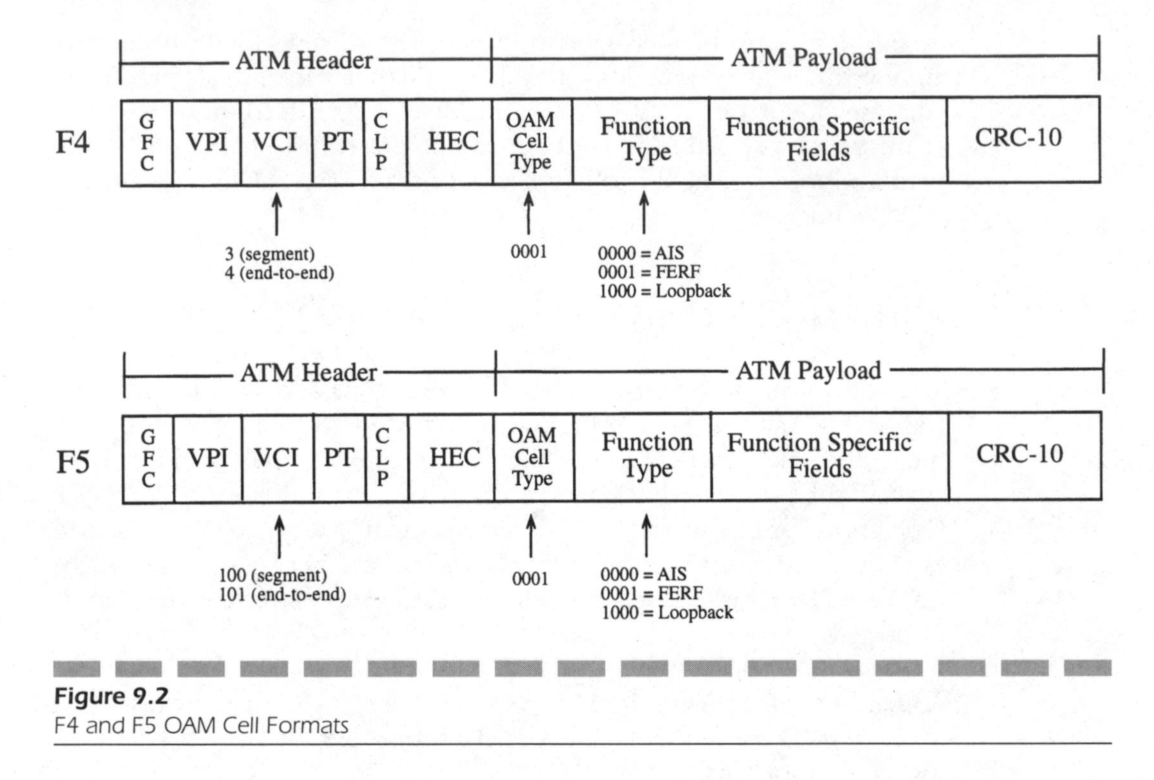

Figure 9.2
F4 and F5 OAM Cell Formats

responses when OAM cells are sent over the same virtual connection. This value is used only by the originator of an OAM cell and is ignored by other nodes.

Loopback location ID is a 96-bit field identifying where the loopback should occur within a virtual connection. A value of all 1s indicates that the loopback should occur at the remote endpoint. Source ID allows the originator to identify the loopback instruction as one it has sent. No values are defined, and the standard allows any kind of identification.

Segment loopbacks take place between the subscriber equipment and public ATM switch on either end of the UNI. In other words, the segment is defined as the link between these two points. The loopback message never travels beyond these points.

End-to-end loopbacks can be generated by any node in the connection but can only be discarded or looped back by endpoints (defined by the VPC or VCC). These messages are identified as:

- Payload type = 101 for VCC (end to end)
- VCI = 4 for VPC (end to end)

- Payload type = 100 for VCC (segment)
- VCI = 3 for VPC (segment)

Loopbacks are used to ensure that a logical connection can be established. In analog facilities, a connection must be established over the facility before any connection requests can be made with the remote end. This is when analog facilities are tested for continuity. In digital facilities, there is no way of testing a connection prior to connection establishment because there is no physical connection, only virtual connections. Signaling is often sent over different channels. Loopbacks over digital facilities are sometimes called continuity checks.

9.3.1.3. Interim Local Management Interface Interim Local Management Interface (ILMI) provides status, configuration, and control information about links and physical layer parameters at the UNI. It also provides address registration across the UNI. ILMI is a protocol used between two adjacent UNI management entities supporting bidirectional communications between the nodes.

Each ATM device is associated with a UNI management entity which resides within software in each device. The management entity interacts with software residing in network management stations. These stations typically exist in locations that are remote from the ATM nodes and are capable of communicating with the ATM switches throughout the network.

9.3.2. ATM Layers

There are many viewpoints as to how ATM fits within the OSI Model. If you examine the functions provided by ATM, the distinction is less clear. Perhaps Walter J. Goralski in his book *Introduction to ATM Networking* sums it up best: "The exact relationship of the ATM layers to the OSI layers is undefined." In other words, there is no alignment of ATM layers and functions to OSI layers and functions. This is because at the time OSI was developed, communications were very much different than they are today.

The OSI Model was developed for protocols used over unreliable facilities. Copper cable at fairly low speeds (in comparison with today's networks) was the only option available, and data transmission was very unreliable. This is why protocols such as X.25 have so much error detection/correction. The network nodes had to perform most of this function to get the data to where it belonged. If left to the endpoints, the

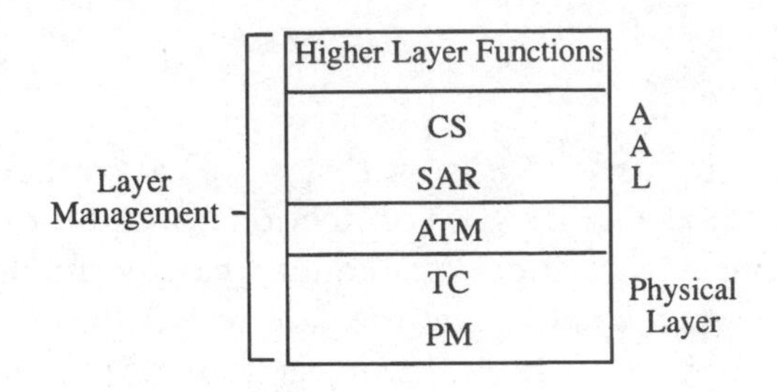

Figure 9.3
ATM Layers

networks would be clogged with retransmissions end to end, rather than node to node.

ATM does not suffer from these problems because today the transmission medium is highly reliable. Error detection/correction is not necessary within the network and is best left to the endpoints. This allows the network nodes to transmit data quickly with very little processing. The endpoints are left with the chore of checking the user data for errors and managing retransmissions.

ATM is divided into layers (see Fig. 9.3). The physical layer is divided into two parts. The ATM physical medium sublayer is responsible for the transmission of data over the physical medium, regardless of the type of medium used. ATM was originally designed to operate over fiber optics but because of the slow deployment of fiber, was later modified to operate over copper and coaxial facilities as well.

The physical medium sublayer is responsible for receiving and transmitting bit streams in a continuous method. This is important to channelized services, which rely on constant bit streams to maintain synchronization. When the bit stream stops, channelized equipment interprets the condition as an error and releases the virtual connection. Bit synchronization is also maintained by this sublayer.

The transmission convergence sublayer is responsible for the transmission and reception of frames over a framed transmission facility, such as T-3. ATM cells are packed into these frames and unpacked at the remote end. This sublayer also performs error detection/correction but only on the ATM header. This prevents the cells from being sent to the wrong destination.

Cell rate decoupling is used when a continuous data stream is required at the physical layer, as in SONET and channelized facilities such as DS1. Cell rate decoupling sends special "idle" cells over the framed facility and discards any idle cells it receives. Idle cells are necessary to maintain a connection in channelized facilities because the channel bank equipment must always see a constant bit rate transmission, or it disconnects the channel. When nothing is being sent over a channelized facility, idle flags are transmitted (this is also used to maintain clock synchronization between two endpoints). Idle cells are not recognized by the ATM layer.

The functions of the transmission Convergence Sublayer (CS) differ depending on the medium being used. For instance, if SONET is the medium, the physical layer requires a different set of functions than a DS-3 medium would require. This sublayer provides whatever services are needed by each type of medium.

There are some specific functions required for DS3 and 100-Mbps interfaces. The ATM physical layer provides a convergence protocol (Physical Layer Convergence Protocol, PLCP), which maps ATM cells onto a DS3. The interface supports 44.736 Mbps. ATM cells are mapped into a DS3 PLCP data unit, which is then mapped into the DS3 payload. The DS3 PLCP is not aligned to the DS3 framing bits.

The 100-Mbps access was intended for private UNIs. Private UNIs are not as complex as public UNIs, which must provide higher reliability and complex monitoring. The specification is based on the FDDI physical layer.

I mentioned earlier that two 155-Mbps interfaces were defined. One is for the public UNI, while the other is for the private UNI. The difference lies in the distances supported by each interface. The 155-Mbps private UNI interface can be used over fiber optics or twisted-pair copper. The public UNI requires fiber optics using single-mode fiber.

The ATM layer is responsible for multiplexing cells over the interface. ATM must read the VPI/VCI of incoming cells, determine which link cells are to be transmitted over, and place new VPI/VCI values into the header. At endpoints, the ATM layer generates and interprets cell headers (endpoints do not route cells).

The ATM layer supports the following connection types:

- Point-to-point Virtual Channel Connection (VCC)
- Point-to-multipoint VCC
- Point-to-point Virtual Path Connection (VPC)
- Point-to-multipoint VPC

A VCC is a single connection between two endpoints. A VPC is a bundle (or group) of VCCs carried transparently between two endpoints.

The AAL is used mostly by endpoints. It is divided into two sublayers: SAR and the Convergence Sublayer (CS). The SAR reconstructs data that has been segmented into different cells (reassembly). It is also responsible for segmenting data that cannot fit within a 48-byte payload of an ATM cell (segmentation).

The CS determines the class of service to be used for a transmission. This will depend on the bit rate (constant or variable bit rate), the type of data, and the type of service to be provided (connection-oriented or connectionless). The quality of service parameters necessary for the transmission are determined by the class of service assigned.

9.3.3. ATM Header and Payload

The ATM cell is 53 bytes in length. The first 5 bytes are used for the header. The payload portion of the cell is 48 bytes. Keep in mind as we discuss the protocol that the payload is also used for the AAL overhead and any other overhead from upper layers.

There are two formats for the ATM header. One header is used for the UNI, while the other is used for the NNI. The difference lies in the Generic Flow Control (GFC) parameter found in the UNI header. The GFC is not supported in the public network, nor was it intended to be. We will define the GFC below.

The header is placed in front of the payload (it arrives first). There is no trailer used in ATM. Figure 9.4 shows the two-header formats. In the UNI header, the GFC can be used to throttle traffic from a specific destination. The GFC values are not currently defined, but the intent is that the GFC could be used to provide flow control from the user to the network (and vice versa). This parameter is not used out in the public network and is overwritten by network nodes.

Two modes are defined for GFC, controlled and uncontrolled. Controlled GFC allows subscriber equipment to control flow of ATM cells; however, the values for this have not been defined. They are of local significance only, which means they are related to a link connection at a switch and used to communicate with the remote end of a link. Uncon-

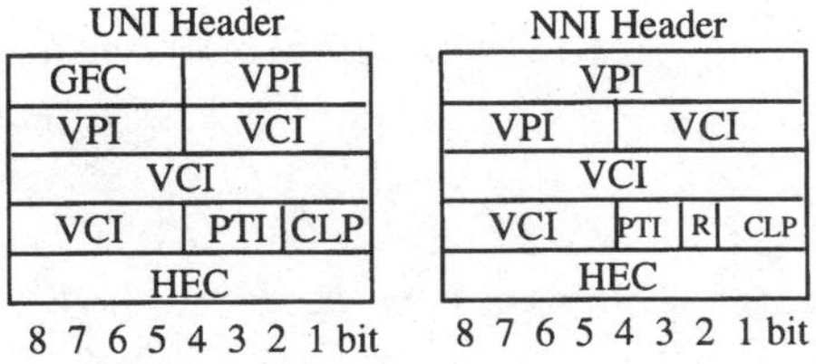

Figure 9.4 ATM Headers (UNI and NNI)

UNI Header

GFC	VPI	
VPI	VCI	
VCI		
VCI	PTI	CLP
HEC		

8 7 6 5 4 3 2 1 bit

NNI Header

VPI			
VPI	VCI		
VCI			
VCI	PTI	R	CLP
HEC			

8 7 6 5 4 3 2 1 bit

trolled GFC simply means that GFC is not supported, and the parameter is set to all zeroes.

The VPI is used to identify a group of virtual channels with the same endpoint. This is the form of addressing supported in ATM. Rather than identifying millions of unique nodes, ATM addressing identifies a connection. A virtual channel is used for a communication link. Each virtual channel is identified by the VCI.

The meta signaling channel (discussed later) is a dedicated channel used to establish virtual channel connections between two endpoints. The virtual paths are predetermined at the time of installation (either through administration or by the service provider). Virtual paths can be negotiated by the user or the network using meta signaling. This means the establishment or release of virtual paths can be controlled by the user or the network.

Work is ongoing in this area. There are a lot of different ways in which virtual paths and channels can be established. Read the UNI specification published by the ATM Forum for the latest specifications.

Following the VPI and the VCI is the Payload Type Indicator (PTI). This parameter indicates the type of data found in the payload portion of the cell. Remember that the payload is not always data. There could be signaling information, network management messages, and other forms of data. These are identified by the PTI.

The PTI is followed by the Cell Loss Priority (CLP) parameter. This is used to prioritize cells. In the event of congestion or some other trouble, a node can discard cells that have a CLP value of 1 (considered low priority). If the CLP value is 0, the cell has a high priority and should only be discarded if it cannot be delivered.

The last byte in the header is the Header Error Control (HEC) parameter, which is used for error checking and cell delineation. Only the header is checked for errors. The HEC works like other error checking methods, where an algorithm is run on the header and the value placed in the HEC. ATM is capable of fixing single bit errors but not multiple bit errors.

An ATM node places itself in error correction mode during normal operation. If a single bit error is detected in the header, the data in the header is small enough that the error correction algorithm can determine which bit is in error and correct it. If a multibit error is detected, the node places itself in error detection mode. Errors are not corrected while in this mode. The node remains in error detection mode as long as cells are received in error. When cells are received without error, the node places itself back into error correction mode.

9.3.4. Routing in ATM—VCI/VPI

In circuit-switched facilities, time slots are assigned to a transmission (or to a device). When a device wishes to transmit data, it must wait for its assigned time slot. The assignment of time slots can be dynamic, as is the case for ISDN. This means that a multiplexer assigns a time slot when a device requests a connection over the network.

In ATM, a specified number of cells are made available during a time period. A device can take any available cell to transmit data (or multiple cells, as is typically the case). The ATM multiplexer takes the data received from the device, adds the 5-byte header, and transmits a cell. In some cases, the data may have to first be processed by the AAL, which then adds a header and trailer and passes the frame to the SAR, which then segments the data into several data units. The data units are then sent to ATM, where the header is added.

ATM addressing consists of two identifiers, which identify the virtual path and the virtual connection. Figure 9.5 shows the relationship

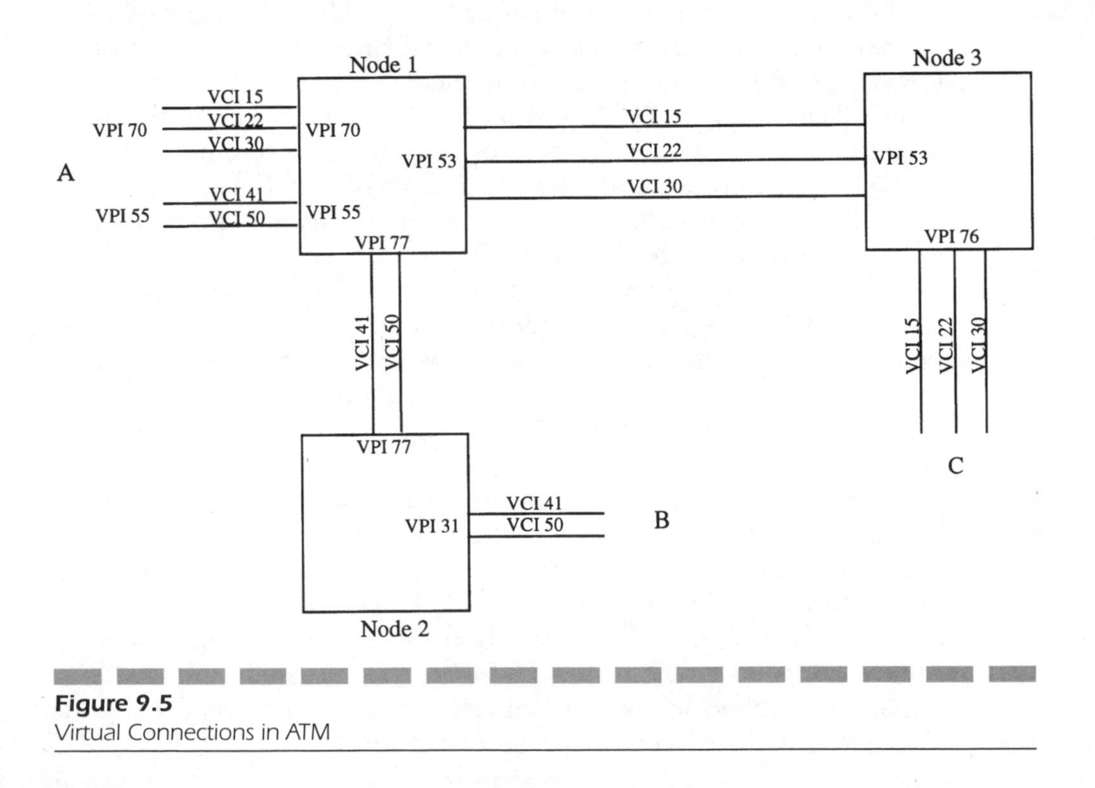

Figure 9.5
Virtual Connections in ATM

between virtual channels and virtual paths. A virtual path consists of multiple virtual channels to the same endpoint.

The virtual channels are static to a destination. The virtual paths change at each node. The virtual paths are prearranged either through system administration or by a signaling protocol. This is analogous to a trunk group assignment in voice switches, where the trunk group consists of multiple voice circuits going to the same endpoint, each with its own unique identifier. A connection from end-user VCI to end-user VCI is called a Virtual Channel Connection (VCC). A connection from end VPI to end VPI is called a Virtual Path Connection (VPC).

The UNI supports a maximum of 256 VPIs at any one node for each physical link. Remember that each VPI is an ATM link connection, with multiple virtual channels. A maximum of 64,000 virtual channels are supported for each virtual path. The NNI supports a maximum of 4096 VPIs because the NNI ATM header does not have the GFC parameter. Instead, these 4 bits are used to expand the VPI.

9.3.5. ATM Signaling

ATM meta-signaling is used for dynamic connections. These are made on demand and are released when transmission is complete. Both point-to-point and point-to-multipoint configurations can be supported. The alternative to dynamic connections is permanent connections, which are established at the time of installation. A permanent connection is just that; it remains connected all of the time, unless there is a failure. This is analogous to permanent virtual circuits.

Q2931 defines the signaling messages used to establish, maintain, and release connections at the UNI. The public network will not use Q2931; it uses SS7 instead. The SS7 protocol supports Q2931 through the Broadband ISUP protocol (BISUP). Q2931 was derived from the ISDN signaling protocol, Q931.

ATM will require much more complexity than existing signaling. For example, if a caller initiates a voice call, a signaling message is generated to establish a connection for the voice. If the caller then activates a camera for videoconferencing, another connection must be established for just the video segment of the call. Both connections must be correlated and synchronized.

Broadcast signaling virtual channels support connection establishment for applications where the same data must be sent to multiple destinations. There are two types, general and selective. General allows signaling to be sent to all endpoints at the user interface (not all endpoints within the network). Selective allows the network to broadcast signaling to endpoints which meet a particular service profile. Both general and selective are unidirectional, sent from the network to endpoints at a user interface.

This means that ATM could place significant demand on the existing SS7 network used within the NNI. The BISUP protocol has been defined to support these ATM services and is still evolving. This is also the reason for expanding the capacity of the SS7 links beyond the present 64 kbps (in addition to requirements being placed on SS7 to support more database applications).

Many ATM books may declare that SS7 is not required anymore or that it is unclear what role SS7 will play in the ATM world. However, the RBOCs are already busy planning expansions to their SS7 networks because they envision increased demand on SS7 services. The SS7 network provides much more than just connection establishment through the public network. It also supports database access for telephone switches providing Intelligent Network solutions, cellular applications, and now local number portability.

SS7 may remain a mystery to many in the data world mainly because it has always been an obscure telephone company technology. It is best understood by those directly involved with telephone networks (and in many cases not well understood even there). One thing is certain, SS7 is not going away, and ATM signaling will not replace SS7. ATM signaling is designed to meet the signaling requirements at the UNI, but it does not provide the services required of the NNI.

9.3.5.1. ATM Addressing Addressing endpoints in an ATM network is part of the UNI signaling. While ATM switches route based on connections (identified by the VPI/VCI), once the cell arrives at the destination network, there has to be an address to get it to the proper node.

Endpoint addresses are carried in the payload of an ATM cell as part of the signaling message. There are three different address formats presently defined for use at the private UNI and an additional format for the public UNI:

- Data Country Code (DCC) (private)
- International Code Designator (ICD) (private)
- E.164 ATM private format (private)
- E.164 ATM public format (public)

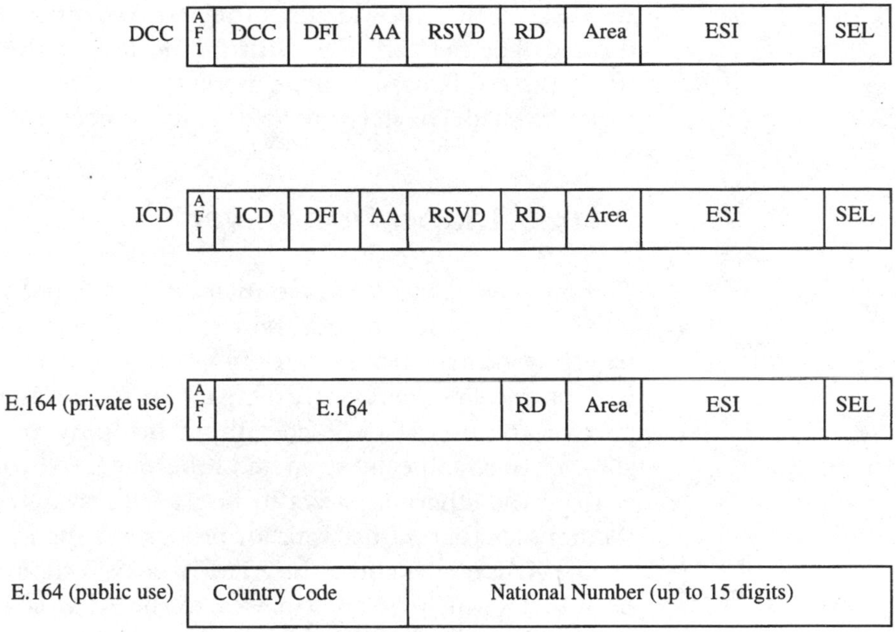

Figure 9.6
ATM Address Formats

DCC	AFI	DCC	DFI	AA	RSVD	RD	Area	ESI	SEL

ICD	AFI	ICD	DFI	AA	RSVD	RD	Area	ESI	SEL

E.164 (private use)	AFI	E.164	RD	Area	ESI	SEL

E.164 (public use)	Country Code	National Number (up to 15 digits)

Figure 9.6 shows the four address formats used in ATM. The Authority and Format Identifier (AFI) defines the authority responsible for address registration and the format used. The authority can be an ATM equipment manufacturer, service provider, telephone company, or administrator of a private network. The DCC is a 2-byte field that identifies the country in which the address is registered.

The Domain Specific Part Identifier (DFI) is a 1-byte field that specifies the structure of the rest of the fields. The Administrative Authority (AA) is a 3-byte field that identifies the authority that is responsible for the rest of the address. The Routing Domain (RD) field is a 2-byte field that specifies a unique routing domain. The area field is a 2-byte field used to identify an area within a routing domain.

The End System Identifier (ESI) is a 6-byte field that identifies an end system within an area. The selector field (SEL) is a 1-byte field used by the end system to select an endpoint within an end system. The ICD is a 2-byte field that identifies an international organization. Codes are maintained by the British Standards Institute. The E.164 field is an 8-byte field that uses the same addressing as defined for ISDN; it is used to identify ISDN numbers.

When a cell is transmitted, the subscriber equipment provides the ESI and SEL values, which identify the end system and endpoint originating the cell. The network then fills in the rest of the address information when the cell is passed to the network over the UNI. The addresses are registered with the network for future connections.

9.3.6. Adaptation Layer

The purpose of the AAL is to interwork ATM networks with other non-ATM networks. It provides the necessary information to ensure proper handling of data from a variety of different networks. Keep in mind that protocol headers remain intact in some cases, treated by AAL as part of the user data. For example, bridging an Ethernet LAN with another geographically separated Ethernet LAN would require the header from the Ethernet packet to be passed through the ATM network to the remote Ethernet network for proper routing in that network.

The AAL resides above the ATM layer. No cells are formed at this layer, only PDUs which are then passed to the ATM layer to be inserted into ATM cells. AAL is capable of viewing user data and is responsible for the segmentation and reassembly of user data.

There are two parts to the AAL. The convergence sublayer is the upper portion of the AAL. User data is first passed to the convergence sublayer, where it is encapsulated into a convergence PDU and passed to the lower part of AAL. The lower part of AAL is the SAR sublayer. The SAR is responsible for segmenting the user data, as well as any overhead added by the convergence sublayer, and passing the segments to the ATM layer for encapsulation into ATM cells.

The convergence sublayer is divided into two parts (see Fig. 9.7). There is the Service Specific Convergence Sublayer (SSCS) and the Common Part Convergence Sublayer (CPCS). User data is first presented to SSCS, where it is encapsulated into an SSCS PDU. SSCS provides clock recovery

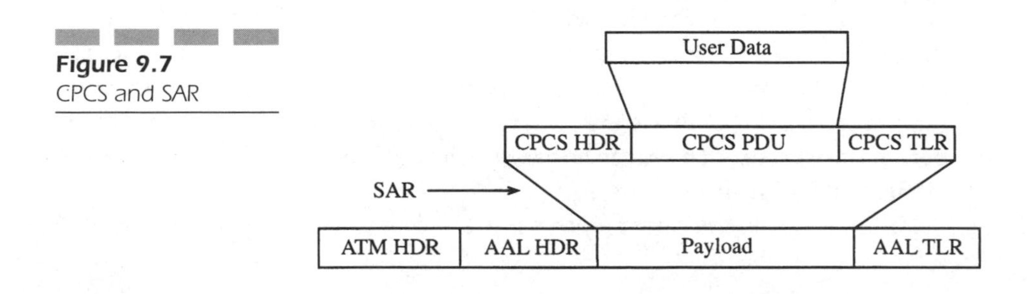

Figure 9.7
CPCS and SAR

	Class A	Class B	Class C	Class Y	N/A	Class X	Class D
	CBR	VBR					
	Connection-Oriented						Connectionless
	Timing Preserved		Variable Delay Acceptable				
Higher Layers	Any	Any	Frame Relay TCP/IP	Any	Q.2931	N/A	SIP-3, Others
Use	Circuit Emulation	VBR Voice, Video	Connection Oriented Data	Available Bit Rate	Signaling	Cell Relay	Connectionless Data
	AAL 1	AAL 2	AAL 5			AAL 0	AAL 3/4
Payload	47 bytes	45-47 bytes	48 bytes			N/A	44 bytes

Figure 9.8
AAL Services

and message identification. The CPCS provides message identification and detection of sequence errors. These are still evolving standards.

Figure 9.8 shows the services provided by the AAL. Earlier we defined the classes of service provided by ATM. A class is a means of providing a service, while the AAL type defines the service to be provided.

Layer 2 of the AAL provides Cyclic Redundancy Checking (CRC) and length identification of user data, buffer allocation size, and sequence numbering. Layer 3 provides buffer allocation size and sequence numbering. Layer 4 provides sequence numbering, segmentation, and reassembly.

There is additional overhead associated with the AAL. This overhead depletes the amount of available payload in the ATM cell because the overhead becomes part of the payload itself. In other words, AAL overhead has nothing to do with the ATM header. It is added to the user data in the form of a PDU, passed to the SAR sublayer for segmentation, and then passed down to the ATM layer for transmission as a series of cells.

AAL-1 is used for synchronous bit streams. Designed for voice and data transmitted over channelized facilities, such as T-1, AAL-1 delivers timing information from the source to the destination. AAL-1 provides segmentation and reassembly, management of cell delay variation, management of lost and misinserted cells, clock frequency recovery at the destination, and bit error recovery.

AAL-2 has been the most difficult to develop. The challenge has been defining a way by which AAL-2 can recover clock frequency at the destination when there has been a long idle period at the source. In synchronous networks, constant transmission of bits is necessary to maintain clock synchronization. When there are extended idle periods, clock synchronization must be reestablished.

This is not an issue with AAL-1 because there is a constant stream of bits. In AAL-2, video compression at the source eliminates a lot of the bit stream, resulting in a bursty-type traffic pattern. MPEG-2 (a video compression scheme developed by the ISO Motion Picture Experts Group) has a 10:1 compression rate (worst case). If there are no changes in the video scene, MPEG-2 can deliver ratios as high as 50:1, which means long idle periods. AAL-2 is still under development with both the ITU-T and the ATM Forum.

AAL-3/4 was at one time two separate services. They were combined because the only difference was the presence of one field in the protocol. AAL-4 used a 10-bit field to deliver connection-oriented services, and AAL-3 did not use this field. This was because developers developed two distinct sets of code for both services, when one set of code could perform both functions.

AAL-3/4 provides connection-oriented and connectionless services for data transfer. This service was intended for applications that are not sensitive to delay variations. It is thought that the majority of applications using ATM will be data applications (especially in the early implementation of ATM). This may not be the case, as the focus of ATM shifts further away from desktop or LAN deployment to backbone applications such as the telephone network.

AAL-5 is sometimes referred to as the Simple and Efficient Adaptation Layer (SEAL). This layer does not provide sequencing or error checking (of user data). Instead, it relies on the upper layers to provide these services. AAL-5 supports connection-oriented, variable bit rate, timing-insensitive data. It is intended for use in point-to-point ATM configurations.

ATM is not a perfect solution for data, nor is it a perfect solution for voice. There is still a lot of work to be completed before the standards are finished. Once standards work has been stabilized, and vendors begin selling their wares in the main marketplace (rather than in trial networks), the real work begins.

Telephone companies run painstaking interoperability tests on all equipment they place in their networks. No equipment is placed in the public network and simply turned on. Tests must prove that the equipment will not introduce any errors into the network and will not cause any outages in it. All of this testing takes time, and solving problems discovered during testing also takes time.

What all of this amounts to is a long duration between protocol development and deployment. It may be many years before ATM replaces the existing public network infrastructure and becomes as common as channelized facilities are today. Vendors are already busy developing their products, testing them in trial networks, and laying the groundwork for nationwide as well as international deployment.

9.4. Chapter Test

1. What are the two types of broadcast signaling virtual channels?
2. Telephone services over ATM require what services?
   a. Connectionless
   b. Connection-oriented
   c. Variable Bit Rate (VBR)
   d. None of the above
3. Class B supports:
   a. Circuit emulation, CBR video
   b. VBR audio and video
   c. Connection-oriented data transfer
   d. Connectionless data transfer
4. What is interworking with non-ATM networks supported by?
5. The SAR is part of what?
6. The signaling standard for ATM is defined in the ITU-T publication:
   a. Q.931
   b. Q.921
   c. Q.2931
   d. Q.2921
7. SS7 signaling for BISDN services is supported by:
   a. SCCP
   b. BISUP
   c. TCAP
   d. MTP
8. Common Part Convergence Sublayer (CPCS) is part of:
   a. The ATM layer
   b. The SAR
   c. The AAL

d. The AFLCIO

9. Interworking with channelized facilities is supported by:
   a. Circuit Emulation Services (CES)
   b. LAN Emulation Services (LES)
   c. Convergence Sublayer (CS)
   d. None of the above
10. Frame Relay over ATM requires:
   a. AAL-1
   b. AAL-3/4
   c. AAL-2
   d. AAL-5
11. The DXI places the _____ into a separate device, such as a data service unit?
12. LES supports IEEE 802.3 and IEEE _____ standards.
13. What two values of the endpoint address does the subscriber equipment assign?
14. Signaling (such as SS7) will rely on AAL-_____.
15. Cell relay services rely on which AAL?
   a. AAL-1
   b. AAL-5
   c. AAL-3/4
   d. None of the above
16. Connectionless and connection-oriented data transfer is provided by AAL-_____.
17. What cells are used to exchange alarm and connectivity information between two nodes?
18. The insertion of "idle" cells is a function of:
   a. Cell rate decoupling
   b. Channel service unit
   c. AAL-0
   d. The ATM layer
19. The CLP field is used for congestion control.
   a. True
   b.. False
20. The ATM Forum is responsible for defining ATM standards.
   a. True
   b. False

APPENDIX A

COMMUNICATIONS EVOLUTION

A.1. History of Computing

1614 John Napier develops algorithms.

1617 John Napier introduces "Napier's Bones."

1642 The Pascaline is invented by Blaise Pascal.

1804 Jacquard's Loom is invented by Joseph Marie Jacquard.

1822 The Difference Engine is invented by Charles Babbage.

1834 Analytical Engine is invented by Charles Babbage.

1847 Boolean Algebra is introduced by George Boole.

1853 A working Difference Engine is built by George Scheutz.

1890 The Punch Card Tabulator is invented by Herman Hollerith.

1906 The vacuum tube is invented by Lee DeForest.

1917 Frequency Division Multiplexing (FDM) is invented by Bell Laboratories.

1928 Eighty column punch cards are introduced by IBM.

1930 The Differential Analyzer is invented by Vannevar Bush of MIT.

1940 The Complex Number Calculator is invented by George Stibitz of Bell Labs.

1942 The Differential Analyzer is modified.

1942 The first electronic digital computer is invented by John Atansoff and Clifford Berry.

1943 The Colossus computer is invented.

1944 Harvard Mark I is invented by Howard Aiken and IBM.

1946 The ENIAC is invented by John W. Mauchly and J. Presper Eckert.

1947 The transistor is invented by William Shockley, Walter Brattain, and John Bardeen of Bell Labs.

1948	The Selective Sequence Electronic Calculator (SSEC) is invented by IBM.
1948	The term *Cybernetics* is first coined by Norbert Weiner.
1949	Magnetic tape storage and the BINAC are invented by Eckert and Mauchly.
1949	The Manchester Mark I is invented.
1950	The ERA 101 is invented by Engineering Research Associates.
1951	Lyon's Electronic Office (LEO) is invented by Lyon's Moving and Storage.
1951	The UNIVAC is introduced by Sperry Rand.
1951	The junction transistor is invented by William Shockley.
1952	The integrated circuit is invented by G.W.A. Dummer.
1953	Magnetic core memory is invented.
1954	Silicon-based junction transistor is invented by Texas Instruments.
1956	The first operating system is invented by Bob Patrick (General Motors) and Owen Mock (North American Aviation).
1957	FORTRAN is invented by John Backus of IBM.
1958	The integrated circuit is improved and introduced by Jack St. Clair Kilby of Texas Instruments.
1958	The 7000 Series Mainframe is introduced by IBM.
1960	The CDC 1604 is introduced by Seymour Cray of Control Data Corporation.
1960	The Dataphone modem is introduced by Bell System.
1961	The Unimate is introduced by Unimation.
1962	MOS transistors are invented by RCA.
1963	The ASCII standard is published by ANSI.
1964	The System 360 is introduced by IBM.
1964	BASIC is introduced by Thomas Kurtz and John Kemeny.
1965	The PDP-8 is introduced by DEC.
1967	The first RAM chip is introduced by Fairchild Semiconductor.
1968	CMOS is introduced.
1968	The Mouse is invented by Douglas Engelbart.

1969	The RS232 standard is published by Bell Labs and the EIA.
1969	UNIX is introduced by Kenneth Thompson and Dennis Ritchie of Bell Labs.
1969	The ARPANET is deployed by the Department of Defense.
1970	The first ATM machine is installed.
1970	Light Sensitive Chips (CCD) are invented by Willard Boyle and George Smith of Bell Labs.
1970	PASCAL is introduced by Niklaus Wirth.
1971	The microprocessor is introduced by Ted Hoff of Intel.
1971	The 8-in floppy disk drive is invented by Alan Shugart and IBM.
1972	The HP35 Electronic Slide Rule is introduced by Hewlett Packard.
1972	The first PONG game is invented by Nolan Bushnell of Atari.
1972	The C programming language is introduced by Dennis Ritchie of Bell Labs.
1972	The Intel 8008 is introduced by Intel.
1973	The Winchester Disk Drive is introduced.
1973	Universal Product Code (UPC) is introduced.
1973	The Ethernet LAN is introduced by Robert Metcalfe of Xerox
1974	The Intel 8080 CPU is introduced.
1974	The Z80 CPU is introduced by Zilog.
1975	The Altair 8800 computer is introduced by MITS.
1975	Microsoft is formed by Bill Gates and Paul Allen.
1975	The laser printer is introduced by IBM.
1976	The X.25 standard is introduced by the ITU.
1976	The Cray 1 is introduced.
1977	Apple II is introduced by Steve Wozniak and Steve Jobs.
1978	5.24-in floppy drive is invented by Radio Shack and Apple with Shugart Industries.
1979	The Motorola 68000 CPU is introduced.
1980	The Seagate Hard Drive is introduced by Shugart and Seagate.

1980	Optical WORM storage is introduced by Phillips.
1981	The Osborne 1 is introduced by Adam Osborne.
1981	The IBM PC is introduced.
1981	MS-DOS is introduced by Microsoft.
1984	The Macintosh computer is introduced by Apple.
1984	3.5-in floppy drives are introduced by Sony.
1985	The CD-ROM is invented.
1985	The Intel 80386 is introduced.
1991	The Intel 80486 is introduced.
1992	Apple and IBM join forces.
1993	The Information Super Highway is born.

A.2. History of Telephony

1851	First telegraph cable is laid under the English Channel.
1856	Western Union is founded.
1860	Philip Reis transmits sound electrically.
1875	Alexander Graham Bell invents the receiver.
1876	Alexander Graham Bell files a patent for his telephone.
1876	Elisha Gray files a patent for his telephone hours after Bell did.
1876	Bell makes famous "Watson, come here" transmission.
1877	Bell receives his second patent for an improved telephone device.
1877	The first permanent telephone wire is installed.
1877	Bell Telephone Company is founded.
1877	Bell marries Mabel Hubbard on July 11.
1877	American Speaking Telephone Company is founded.
1877	Thomas Edison invents the carbon transmitter.
1878	The first switchboard installed in New Haven, Connecticut.
1878	First operators were hired (all males).
1878	The first woman operator was hired.

1879	Bell resigns the board of Bell Telephone Company.
1879	Telephone numbers are used for the first time.
1880	Bell resigns from the Bell Telephone Company as Chief Engineer.
1881	Western Electric Company is purchased by Bell Telephone Company.
1882	New York is serviced by five telephone companies.
1885	Indiana begins deregulation of the telephone industry.
1891	Almon B. Strowger invents the step-by-step telephone switch.
1896	Almon B. Strowger invents the rotary dial.
1900	American Telephone and Telegraph (AT&T) is founded.
1907	Theodore N. Vail is named President of AT&T.
1913	The Kingsbury Commitment was signed.
1913	The crossbar switch was invented.
1915	The first wireless telephone call was placed.
1915	The first transcontinental telephone call was placed.
1920	AT&T entered into radio broadcasting.
1921	The first step-by-step switch was installed in a central office by Bell Telephone.
1922	Alexander Graham Bell dies.
1923	A short-distance telephony submarine cable is laid between Los Angeles and Catalina Island.
1925	Bell Laboratories is formed.
1925	AT&T sells all international plants to ITT.
1926	Bell Telephone introduces the first synchronous-sound motion picture.
1927	Time of Day service begins.
1928	Major studios sign a licensing agreement with Western Electric for sound.
1930	Bell Telephone purchases a working crossbar switch from Europe.
1934	Watson, Bell's assistant, dies.
1934	The Communications Act of 1934 signed, forming the FCC.
1936	The first coaxial cable is put into service.

1938	The first crossbar switch is installed in Brooklyn, New York.
1948	The first No. 5 crossbar switch is installed.
1949	The Government Anti-Trust Suit is filed against Bell Telephone.
1950	Microwave is used for the first time in telephony.
1951	Direct Distance Dialing (DDD) is offered commercially.
1956	Transatlantic-oceanic cable is laid.
1956	The AT&T Consent Decree is signed.
1960	There is a field trial of electronic switch in Morris, Illinois.
1962	Telstar, the first international communications satellite is launched.
1965	Earth satellite is available for extending telephone service around the world.
1965	The first ESS office is in service in Trenton, New Jersey.
1968	The Carterfone Decision is signed.
1969	The MCI Decision is signed.
1970	Picturephone service is offered to the public.
1974	A government antitrust suit is filed against Bell Telephone.
1984	Divestiture of the Bell System.
1995	FCC begins auctions for PCS frequency blocks and markets.
1996	AT&T splits into three different companies.
1996	A new telecommunications bill is passed, allowing local and long distance carriers to cross into each others territories.
1996	The Consent Decree is made obsolete by the Telecommunications Bill of 1996.

BIBLIOGRAPHY

Anschutz, Thomas A., "A Historical Perspective of CSTA," *IEEE Communications Magazine,* April 1996: 48–54.

Arellano, Michael, "Broadbands Present Sense," *tele.com Magazine,* April 1996: 34.

Augarten, Stan, *Bit by Bit.* New York: Ticknor & Fields (1984).

Bellcore, *Telecommunications Transmission Engineering: Volume One,* Bellcore (1990).

———, *Telecommunications Transmission Engineering: Volume Two,* Bellcore (1990).

———, *Telecommunications Transmission Engineering: Volume Three,* Bellcore (1990).

Black, Ulysses, *TCP/IP and Related Protocols,* 2d ed. New York: McGraw-Hill (1995).

Brooks, John, *Telephone: The First Hundred Years.* New York: Harper and Row (1976).

Comer, Douglas E., and David L. Stevens, *Internetworking with TCP/IP: Volume III,* 2d ed. Englewood Cliffs, N.J.: Prentice Hall (1994).

Cronin, Paul, "An Introduction to TSAPI and Network Telephony," *IEEE Communications Magazine,* April 1996: 48–54.

Dziatkiewicz, Mark, "Wireless Showdown," *America's Network,* February 1, 1995: 24–27.

Feit, Sidnie, *SNMP: A Guide to Network Management.* New York: McGraw-Hill (1995).

Fist, Stewart, "Will GSM and D-AMPS Give Way to the CDMA Push," *Australian Communications,* July 1993: 83–88.

Freeman, Roger L., *Practical Data Communications.* New York: John Wiley and Sons (1995).

Goralski, Walter J., *Introduction to ATM Networking.* New York: McGraw-Hill (1995).

Hou, Yiwei Thomas, Leandros Tassiulas, and H. Jonathon Chao, "Overview of Implementing ATM-Based Enterprise Local Area Network for Desktop Multimedia Computing," *IEEE Comunications Magazine,* April 1996: 70–76.

Ireland, Tom, "The Standards War in Digital Cellular," *Telecommunications Magazine,* March 1996: 46.

Kessler, Gary C., *ISDN,* 2d ed. New York: McGraw-Hill (1993).

King, Rachel, "ADSL, Take Two," *tele.com Magazine,* April 1996: 20–22.

LaPorta, Thomas F., Malathi Veereraghavan, Philip A. Treventi, and Ramachandran Ramjee, "Distributed Call Processing for Personal Communications Services," *IEEE Communications Magazine.* June 1995: 66–75.

Lee, William C. Y., *Mobile Cellular Telecommunications.* New York: McGraw-Hill (1995).

Macario, Raymond C. V., *Cellular Radio: Principles and Design.* London: MacMillan Press (1993).

Myers, Jason, "Wireless Nation," *Telephony Magazine,* March 4, 1996: 13–16.

Nixon, Toby, "Design Considerations for Computer-Telephony Application Programming Interfaces and Related Components," *IEEE Communications Magazine,* April 1996: 44.

O'Shea, Dan, "The Devices of Extension," *Telephony Magazine,* March 4, 1996: 20–21.

Pandya, Raj, "Mobility and Intelligent Networks," *IEEE Communications,* June 1995: 44–52.

Pollini, Gregory P., Kathleen S. Meier-Hellstern, and David J. Goodman, "Signaling Traffic Volume Generated by Mobile and Personal Communications," *IEEE Communications Magazine,* June 1995: 60–65.

Saadawi, Tarek N., Mostafa H. Ammar, and Ahmed El Hakeem, *Fundamentals of Telecommunication Networks.* New York: John Wiley & Sons (1994).

Saarela, Kimmo K., *ADSL,* White Paper. March 1996.

Sapronov, Walter, *Telecommunications and the Law.* Rockville, Md.: Computer Science Press (1988).

Schneiderman, Ron, "TDMA or CDMA? Is It Becoming a Tougher Call?" *Wireless Systems Design,* June 1996: 12–20.

Slekys, Dr. Arunas G., *High Capacity Cellular for Wireless Telephony,* Hughes Network Systems, Digital Cellular Networks Business Unit. October 1993.

Stallings, William, *ISDN and Broadband ISDN,* 2d ed. New York: MacMillan Publishing (1992).

Vittore, Vince, "ATM Could Realize Promise as Technology, Not Service," *America's Network,* April 15, 1995: 19–20.

Winch, Robert G., *Telecommunications Transmission Systems.* New York: McGraw-Hill (1993).

Wirth, Patricia E., "Teletraffic Implications of Database Architectures in Mobile and Personal Communications," *IEEE Communications Magazine,* June 1995: 54–59.

INDEX

10BASE2, 106
10BASE5, 106
2B1Q, 261

A

access tandems (AT), 58
ACD (Automatic Call Distributor), 80—81
Adaptive Differential PCM (ADPCM), 27
Add-Drop Multiplexer (ADM), 337
Address Resolution Protocol (ARP), 170—171
address signals, 200
ADM (Add-Drop Multiplexer), 337
ADPCM (Adaptive Differential PCM), 27
ADSL (Asymmetric Digital Subscriber Loop), 352
Advanced Research Projects Agency (ARPA), 143
alerting signals, 200
American National Standards Institute (ANSI), 16
American Speaking Telephone Company, 8
American Standard Code for Information Interchange (ASCII), 22—25
American Telephone & Telegraph (AT&T):
  divestiture, 11, 55—56
  history, 10
  reorganization, 12
Analytical Engine, 3
ANSI (American National Standards Institute), 16
antitrust suit, 10
APS (Automatic Protection Switching), 349
ARP (Address Resolution Protocol), 170—171
ARPA (Advanced Research Projects Agency), 143
ARPANET, 6, 143
ASAI, 87
ASCII (American Standard Code for Information Interchange), 22—25
Asymmetric Digital Subscriber Loop (ADSL), 352
Asynchronous Time Division Multiplexing (ATDM), 70
Asynchronous Transfer Mode (ATM), 88—89
  addressing, 382—384
  described, 358—362
  interfaces, 369—372
  overview, 372—379
  routing, 380—381
  services, 362—369
  signaling, 381—382
AT (access tandems), 58
ATM (*see* Asynchronous Transfer Mode)
ATM Forum, 20—21
AT&T (*see* American Telephone & Telegraph)
automated attendant, 80
Automatic Call Distributor (ACD), 80—81
automatic call routing, 78
Automatic Protection Switching (APS), 349
autonomous systems (TCP/IP), 144

B

Backward Explicit Congestion Notification (BECN), 295
Base Station Controller (BSC), 315
Base Station Subsystem (BSS), 315
Base Station Transceiver (BTS), 315
basic rate interface (BRI), 127—128, 261

BCD (Binary Coded Decimal), 25
BECN (Backward Explicit Congestion Notification), 295
Bell, Alexander Graham, 8, 10
Bell Communications Research (Bellcore), 16—17, 56—57
Bell System (*see* Bell Telephone Company)
Bell Telephone Company:
  divestiture, 11, 52, 55, 60
  history 8, 9, 10
  networks, 52—59
  organization, 52—60
BGP (Border Gateway Protocol), 169
Binary Coded Decimal (BCD), 25
BISDN (*see* Broadband Integrated Services Digital Network)
bit robbing (T-1), 74
Boolean Algebra, 3
Border Gateway Protocol (BGP), 169
BRI (basic rate interface), 127—128, 261
bridge, 102—103
Broadband Integrated Services Digital Network (BISDN), 90
  architecture, 290
  described, 286—290
broadcast mode, 98
BSC (Base Station Controller), 315
BSS (Base Station Subsystem), 315
BTS (Base Station Transceiver), 315
bus topology, 99
byte interleaving, 348

C

call centers, 84
call progress signaling, 199
Canadian Standards Association (CSA), 20
Carlson, Stromberg, 9
Carterfone Decision, 11
CBR (constant bit rate), 286—287
CDMA (Coded Division Multiple Access), 302, 309, 322—323
CDPD (Cellular Digital Packet Data), 326
CD-ROM, 7
cell relay, 47—48
Cellular Digital Packet Data (CDPD), 326
central processing unit (CPU), 94
Centrex, 81—82
channel bank, 72, 73, 74
CIR (committed information rate), 293
circuit switching, 45—46
class 1 telephone office, 55
class 2 telephone office, 55
class 3 telephone office, 55
class 4 telephone office, 55
class 5 telephone office, 55
client/server, 41, 123
clock synchronization, 73
cluster controller, 95
Coded Division Multiple Access (CDMA), 302, 309, 322—323
collision detection (Ethernet), 106
combined linkset (SS7), 213
committed information rate (CIR), 293
common channel signaling, 201—202
Communications Act of 1934, 10
communications protocol, 95
companding, 27
computer telephony, 82—87
concentrator, 104
connection control, 33
connectionless protocol, 35
connection-oriented protocol, 35—36
Consent Decree, 11
constant bit rate (CBR), 286—287
control signals, 199
control switching points, 55
controllers, 39—40
Cordless Telephony 2 (CT2), 309—310
CPU (central processing unit), 94
CSA (Canadian Standards Association), 20
CT2 (*see* Cordless Telephony 2)

D

data communications:
  history, 2—7

datagram, 145
Data Link Control Identifier (DLCI), 267
data link layer, 29
(*See also* protocols)
Data User Part (DUP), 236
DARPA (Defense Advanced Research Projects Agency), 143
DCA (Defense Communications Agency), 19
de facto standard (*see* standards)
Defense Advanced Research Projects Agency (DARPA), 143
Defense Communications Agency (DCA), 19
digital cross connect (DSX), 72
Digital Hierarchy, 71—75
digital signal 0 (DS0), 71
digital signal 1 (DS1), 71
digital signal 3 (DS3), 72
Digital Speech Interpolation (DSI), 28
Digital Subscriber Loop (DSL), 351
Difference Engine, 3
digital voice, 25—28
divestiture (*see* Bell Telephone Company)
DLCI (Data Link Control Identifier), 267
Domain Name System (DNS), 162—164
DS0 (digital signal 0), 71
DS1 (digital signal 1), 71
DS3 (digital signal 3), 72
DSI (Digital Speech Interpolation), 28
DSL (Digital Subscriber Loop), 351
DSS1, 255
dual ring topology, 100—101
DUP (Data User Part), 236

E

EBCDIC (Extended Binary Coded Decimal Information Code), 22—25
ECSA (Exchange Carriers Standards Association), 17
EFS (Extended Superframe), 75
EGP (Exterior Gateway Protocol), 169
Electronics Industries Association (EIA), 18
electronic mail (e-mail), 135
encapsulation, 33
end telephone office, 55
Equipment Serial Number (ESN), 315
error detection and correction, 35
ESN (Equipment Serial Number), 315
Ethernet, 6, 99, 105—112
ETSI (European Telecommunications Standards Institute), 16
European Digital hierarchy, 72
European Telecommunications Standards Institute (ETSI), 16
Exchange Carriers Standards Association (ECSA), 17
Extended Binary Coded Decimal Information Code (EBCDIC), 22—25
Extended Superframe (EFS), 75
Exterior Gateway Protocol (EGP), 169

F

FCC (Federal Communications Commission), 18
FDDI (Fiber Distributed Data Interface), 117—123
FDM (Frequency Division Multiplexing), 68—69
FECN (Forward Explicit Congestion Notification), 295
Federal Communications Commission (FCC), 18
Federal Telecommunications Standards Committee (FTSC), 19
fiber optics, 68
Fiber-to-the-Curb (FTTC), 351
Fiber-to-the-Loop (FTTL), 352
Fiber Distributed Data Interface (FDDI), 117—123
File Transfer Protocol (FTP), 135, 186
Fill-in Signal Unit (FISU), 219—221
firewalls, 130
first telephone switch, 9

first transcontinental telephone call, 9
first transoceanic cable, 11
FISU (Fill-in Signal Unit), 219—221
floppy disk drive, 6, 7
flow control, 34
Forward Explicit Congestion Notification (FECN), 295
FRAD (Frame Relay Access Device), 294
frame, 33
Frame Relay, 127, 292—296
Frame Relay Access Device (FRAD), 294
Frequency Division Multiplexing (FDM), 68—69
frequency reuse, 305
FTP (File Transfer Protocol), 135, 186
FTSC (Federal Telecommunications Standards Committee), 19
FTTC (Fiber-to-the-Curb), 351
FTTL (Fiber-to-the-Loop), 352

G

Gateway-to-Gateway Protocol (GGP), 169
Global System for Mobile Communications (GSM), 303, 309, 311—312, 324—325
global title digits, 217
Global Title Translation (GTT), 217
Grey, Elisha, 8
GSM (*see* Global System for Mobile Communications)
GTT (Global Title Translation), 217

H

HDSL (High-bit-rate Digital Subscriber Line), 354
HDT (Host Digital Terminal), 351
HFC (Hybrid Fiber Coax), 351
High-bit-rate Digital Subscriber Line (HDSL), 354
HLR (Home Location Register), 313
Hollerith Cards, 3—4
Home Location Register (HLR), 313
Host Digital Terminal (HDT), 351
HTML (Hypertext Markup Language), 131
HTTP (Hypertext Transport Protocol), 190—191
hunt group, 79
Hybrid Fiber Coax (HFC), 351
Hypertext Markup Language (HTML), 131
Hypertext Transport Protocol (HTTP), 190—191

I

IAB (Internet Advisory Board), 148
ICMP (Internet Control Message Protocol), 145, 150, 174—175
IEEE (Institute of Electrical and Electronics Engineers), 18
IESC (Internet Engineering Steering Committee), 148
IETF (Internet Engineering Task Force), 133
IGP (Interior Gateway Protocol), 169
IN (Intelligent Networks), 203—205
inband signaling, 200
independent telephone companies, 53—54
Information Highway, history, 7
Information Infrastructure Task Force (IITF), 63
Institute of Electrical and Electronics Engineers (IEEE), 18
integrated circuit, 5
Integrated Services Digital Network (ISDN), 90, 127—128
  described, 248—249
  functions, 264
  protocols, 265—286
  services and standards, 249—254
Intelligent Networks (IN), 203—205

Interim Standard - 41 (IS-41), 310
Interim Standard - 54 (IS-54), 321
Interim Standard - 136 (IS-136), 321
Interior Gateway Protocol (IGP), 169
interexchange carriers (IXCs), 58
International Organization for Standards (ISO), 15—16
International Telecommunications Union (ITU), 14—15
Internet, 129
Internet Advisory Board (IAB), 148
Internet Control Message Protocol (ICMP), 145, 150, 174—175
Internet Engineering Steering Committee (IESC), 148
Internet Engineering Task Force (IETF), 133
Internet Protocol (IP), 149—175
  addressing, 155—162
  routing, 164—167
  routing protocols, 168—173
  services, 173—174
Internet Protocol next generation (IPng), 150
Internet Research Steering Committee (IRSC), 149
Internet Research Task Force (IRTF), 149
Internet Service Provider (ISP), 133
InterNIC, 149
Intranet, 131
IP (*see* Internet Protocol)
IPng (Internet Protocol next generation), 150
IRSC (Internet Research Steering Committee), 149
IRTF (Internet Research Task Force), 149
IS-41 (Interim Standard - 41), 310
IS-54 (Interim Standard - 54), 321
IS-136 (Interim Standard - 136), 321
ISDN (*see* Integrated Services Digital Network)
ISDN Forum, 261
ISDN User Part (ISUP), 236—243
ISO (International Organization for Standards), 15—16
ISP (Internet Service Provider), 133
ISUP (ISDN User Part), 236—243
ITU (International Telecommunications Union), 14—15

L

LAN (*see* Local Area Network)
LAPB (Link Access Protocol - B), 125
LAPD (Link Access Protocol - D), 266—275
LATAs (Local Access and Transport Areas), 52, 58—59
LCP (Link Control Protocol), 191
LEC (local exchange carrier), 59
Link Access Protocol - B, 125
Link Access Protocol - D, 266—275
Link Control Protocol (LCP), 191
linksets (SS7), 213
Link Status Signal Unit (LSSU), 221
LLC (logical link layer), 110
LNP (local number portability), 61, 206
Local Access and Transport Areas (LATAs), 52, 58—59
Local Area Network (LAN), 41—43, 97—124
  addressing, 42
  routing, 42—43
  services, 42
local exchange carrier (LEC), 59
local loop, 89
local number portability (LNP), 61, 206
logarithms, 2
logical address, 30
logical connection, 36
logical link layer (LLC), 110
LSSU (Link Status Signal Unit), 221

M

MAC (media access control), 107—109
machine address, 30
mainframe computer, 94

Management Information Base (MIB), 193
MAU (media access unit), 104
MCI Decision, 11
media access control (MAC), 107—109
media access unit (MAU), 104
Message Signal Unit (MSU), 221—222
Message Transfer Part (MTP), 222—227
Metropolitan Statistical Areas (MSAs), 312—313
Metropolitan Trading Areas (MTAs), 313
MIB (Management Information Base), 193
microprocessor, 6
MIME (Multipurpose Internet Mail Extensions), 135
MIN (*see* Mobile Identification Number)
Mobile Identification Number (MIN), 314
Mobile Switching Center (MSC), 313
mouse, 5—6
MSA (Metropolitan Statistical Areas), 312—313
MSC (*see* Mobile Switching Center)
MSU (Message Signal Unit), 221—222
MTA (Metropolitan Trading Areas), 313
MTP (Message Transfer Part), 222—227
multiplexing, 46, 68—71
Multipurpose Internet Mail Extensions (MIME), 135

N

name resolver (TCP/IP), 163
National Bureau of Standards (NBS), 19
National Information Infrastructure (NII), 61
National Science Foundation (NSF), 132
NBS (National Bureau of Standards), 19
NCIH (North Carolina Information Highway), 65—66
network access layer, 29
(*See also* protocols)
network computers, 41
network interface card (NIC), 104
Network News Transport Protocol (NNTP), 137, 189—190
Network Operating Systems (NOSs), 42
Network Personal Computers (NPCs), 97—98
Network Reliability Council (NRC), 18—19, 63
Network Termination Point 1 (NT1), 264
Network Termination Point 2 (NT2), 264
Network Time Protocol (NTP), 166
newsgroups, 136—137
NIC (network interface card), 104
NII (National Information Infrastructure), 61
NMT (Nordic Mobile Telephone), 311
NNTP (Network News Transport Protocol), 137, 189—190
Nordic Mobile Telephone (NMT), 311
North American ISDN Users Forum, 261
North Carolina Information Highway (NCIH), 65—66
NOSs (Network Operating Systems), 42
NPC (Network Personal Computers), 97—98
NRC (Network Reliability Council), 18—19, 63
NSF (National Science Foundation), 132
NT1 (Network Termination Point 1), 264
NT2 (Network Termination Point 2), 264
NTP (Network Time Protocol), 166
NyQuest Theorem, 27

O

ONU (Optical Network Unit), 351
Open Shortest Path First (OSPF), 172—173
Open System Interconnection (OSI):
  described, 36
  layers, 37—39
operating system, 5
Optical Network Unit (ONU), 351
ordered delivery, 34
OSI (Open System Interconnection), 36, 37—39
OSPF (Open Shortest Path First), 172—173
out-of-band signaling, 200

P

packet, 33
packet switching, 46—47
PAM (Pulse Amplitude Modulation), 26—27, 73
Pascaline, 2
PBXs (Private Branch Exchanges), 77—78
PCM (Pulse Coded Modulation), 27, 68
PCS (Personal Communications Services), 326—328
PDU (protocol data unit), 33
Personal Communications Services (PCS), 326—328
personal mobility, 307
physical layer, 29
  (*See also* protocols)
point-of-presence, 58
point-to-multipoint, 98
point-to-point, 98
primary rate interface (PRI), 128, 262
Private Branch Exchanges (PBXs), 77—78
process layer, 29
  (*See also* protocols)
Point-of-Presence Protocol (POP), 135
Point-to-Point Protocol (PPP), 134, 191
POP (Point-of-Presence Protocol), 135
POP (TCP/IP; *see* Post Office Protocol)
Post Office Protocol (POP), 188—189
PPP (Point-to-Point Protocol), 134, 191
protocols:
  addressing, 30
  defined, 28
  data link layer, 29
  network access layer, 29
  physical layer, 29
  process layer, 29
  services, 28—35
  tasks, 30—35
  transport layer, 29
protocol data unit (PDU), 33
Public Switched Telephone Network (PSTN), 91
Pulse Amplitude Modulation (PAM), 26—27, 73
Pulse Coded Modulation (PCM), 27, 68

Q

Quality of Service (QoS), 70
quasi-associated signaling, 237
quantizing scale, 27

R

RAD (Remote Antenna Driver), 315
radiotelephone, 304
Random Access Memory (RAM), 5
RARP (Reverse Address Resolution Protocol), 171
RASP (Remote Antenna Signal Processing), 315—316
RBOCs (Regional Bell Operating Companies), 57—58
reassembly, 32
recordable CD-ROM, 7
Regional Bell Operating Companies (RBOCs), 57—58
Reis, Philip, 8.
Remote Antenna Driver (RAD), 315

Remote Antenna Signal Processing (RASP), 315—316
repeaters, 102
Request for Comments (RFCs), 147—148
Reverse Address Resolution Protocol (RARP), 171
RFCs (Request for Comments), 147—148
ring topology, 99
RIP (Routing Information Protocol), 171—172
rotary dial, 9—10
route (SS7), 213
routers, 103—104
routeset (SS7), 214
Routing Information Protocol (RIP), 171—172
RSA (Rural Statistical Areas), 313
RTA (Rural Trading Areas), 313
Rural Statistical Areas (RSAs), 313
Rural Trading Areas (RTAs), 313

S

SAPI (Service Access Point Identifier), 267
SAPs (Service Access Points), 155
SCAI, 87
SCCP (Signaling Connection Control Part), 228—230
SCP (Service Control Point), 218—219
SDH (Synchronous Digital Hierarchy), 332
SDSL (Symmetric Digital Subscriber Line), 354
segmentation, 32
sequence numbering, 34
Serial Line Interface Protocol (SLIP), 191
Service Access Point Identifier (SAPI), 267
Service Access Points (SAPs), 155
Service Control Point (SCP), 218—219
service portability, 307
Service Profile Identifier (SPID), 260
Service Switching Point (SSP), 216
session, 34
SGML (Standard Generic Markup Language), 131
Signal Transfer Point (STP), 217—218
signaling, 199—201
Signaling Connection Control Part (SCCP), 228—230
Signaling System #7 (SS7), 198—243
  architecture, 211—219
  broadband, 210
  described, 198—206
  protocols, 219—243
  services, 206—210
signaling link management, 224
signaling route management, 225
signaling traffic management, 225—226
Signaling Unit Error Rate Monitor (SUERM), 220
Simple Mail Transfer Protocol (SMTP), 135, 187—188
Simple Network Management Protocol (SNMP), 192—194
SLIP (Serial Line Interface Protocol), 191
SMTP (Simple Mail Transfer Protocol), 135, 187—188
SNMP (Simple Network Management Protocol), 192—194
sockets, 30, 146, 156
SONET (*see* Synchronous Optical Network)
source routing (TCP/IP), 165
SPID (Service Profile Identifier), 260
SSP (Service Switching Point), 216
SS7 (*see* Signaling System #7)
standards:
  de facto standards, 12
  de jure standards 12
  organizations, 12—21, 14—21
  standards process, 13—14
  regulatory standards, 12
  voluntary standards, 13
Standard Generic Markup Language (SGML), 131
star topology, 98
Statistical ATDM, 70
Step-By-Step telephone switch, 9

STM (Synchronous Transfer Mode), 291
STP (Signal Transfer Point), 217—218
Strowger Switch (Almon B. Strowger), 9
subnet masking, 159—162
subnetworks (TCP/IP), 144
SUERM (Signaling Unit Error Rate Monitor), 220
supervisory signaling, 199
switched 56, 126
switching, 45—48
  circuit switching, 45—46
  packet switching, 46—47
  cell relay, 47—48
Symmetrical Digital Subscriber Line (SDSL), 354
Synchronous Digital Hierarchy (SDH), 332
Synchronous Optical Network (SONET), 72, 75—77
  described, 332—337
  protocol, 340—350
Synchronous Transfer Mode (STM), 291

T

T-1, 72—75, 126
TA (Terminal Adapter), 264
TACS (Total Access Communication Systems), 311
tandem office (TO), 58
tandem switch, 46, 55
TAPI (Telephony Application Programming Interface), 83
TCAP (Transaction Capabilities Application Part), 230—235
TCP (*see* Transport Control Protocol)
TCP/IP (Transport Control Protocol/Internet Protocol), 128—129, 142—194
TDM (Time Division Multiplexing), 69—71
TDMA (Time Division Multiple Access), 302, 320—322
TE1 (Terminal Equipment 1), 264
TE2 (Terminal Equipment 2), 264
TEI (Terminal Endpoint Identifier), 267
telecommunications:
  defined, 2
  history, 7—12
Telecommunications Act of 1996, 10, 52, 59, 60—61
Telephone User Part (TUP), 235—236
Telephony Application Programming Interface (TAPI), 83
telephony history, 7
TELNET, 136, 185
temporary local directory number (TLDN), 319
10BASE 2, 106
10BASE 5, 106
terminals, 94
Terminal Adapter (TA), 264
Terminal Endpoint Identifier (TEI), 267
Terminal Equipment 1 (TE1), 264
Terminal Equipment 2 (TE2), 264
terminal mobility, 307
terminator, 99
TFTP (Trivial File Transfer Protocol), 186—187
Time Division Multiplexing (TDM), 69—71
Time Division Multiple Access (TDMA), 302, 320—322
TLDN (temporary local directory number), 319
TO (tandem office), 58
token ring, 112—117
topology, 98
Total Access Communication Systems (TACS), 311
Transaction Capabilities Application Part (TCAP), 230—235
Transport Control Protocol (TCP), 175—183
  addressing, 180
  header, 176—179
  services, 180—183
Transport Control Protocol/Internet Protocol (TCP/IP), 128—129, 142—194

transistor, 4
transport layer, 29
(*See also* protocols)
Trivial File Transfer Protocol (TFTP), 186—187
TUP (Telephone User Part), 235—236
2BiQ, 261

U

UDP (User Datagram Protocol), 145, 183—184
Underwriters Laboratories (UL), 19
Uniform Resource Locator (URL), 190—191
UNIVAC, 4—56
URL (Universal Resource Locator), 190—191
User Datagram Protocol (UDP), 145, 183—184
User-to-user signaling, 278

V

vacuum tubes, 4
variable bit rate (VBR), 286—287
virtual tributaries (VTs), 346—348
Visitor Location Register (VLR), 314
voice and data integration, 81
voice announcers, 79
voice mail, 79
VTs (virtual tributaries), 346—348

W

WANs (*see* Wide Area Networks)
Western Electric, 8
Western Union, 7
Wide Area Networks (WANs), 43—45
defined, 43
routing, 44—45
services provided, 43—44
World Wide Web (WWW), 132, 137—138

X

X.25, 125
X.75, 125

About the Author

Travis Russell is a manager in the Product Management Group of the Network Switching Division at Tekelec in North Carolina. He has been a field engineer in the telecommunications business for more than fifteen years and lectures on basic telecommunications at area colleges and universities as well as industry seminars. He is the author of McGraw-Hill's *Signaling System #7* and the co-author of *CDPD: Cellular Digital Packet Data Standards and Technologies.*